Multistage
Fuzzy Control

Multistage Fuzzy Control

A model-based approach to
fuzzy control and decision making

Janusz Kacprzyk
Systems Research Institute,
Polish Academy of Sciences,
Warsaw, Poland

JOHN WILEY & SONS
Chichester • New York • Weinheim • Brisbane • Singapore • Toronto

Other Wiley Editorial Offices

John Wiley & Sons, Inc., 605 Third Avenue,
New York, NY 10158-0012, USA

VCH Verlagsgesellschaft mbH, Pappelallee 3,
D-69469 Weinheim, Germany

Jacaranda Wiley Ltd, 33 Park Road, Milton,
Queensland 4064, Australia

John Wiley & Sons (Canada) Ltd, 22 Worcester Road,
Rexdale, Ontario M9W 1L1, Canada

John Wiley & Sons (Asia) Pte Ltd, 2 Clementi Loop #02-01,
Jin Xing Distripark, Singapore 0512

British Library Cataloguing in Publication Data

A catalogue record for this book is available from the British Library

ISBN   0 471 96347 X

Produced from camera-ready copy supplied by the author
Printed and bound in Great Britain by Bookcraft (Bath) Ltd
This book is printed on acid-free paper responsibly manufactured from sustainable forestation,
for which at least two trees are planted for each one used for paper production.

Contents

Foreword

Applications of fuzzy logic to control have a long history. When I wrote my first paper on fuzzy sets in 1965, my expectation was that the theory of fuzzy sets would find its principal applications in the realm of fields such as economics, linguistics and psychology – fields in which human perceptions, emotions and cognition play pivotal roles. In these fields, the classes that one deals with are usually fuzzy and conventional methods of analysis are of questionable effectiveness.

My 1970 paper with Bellman, "Decision-Making in a Fuzzy Environment," represented an attempt to come to grips with the fact that, in most real-world settings, decisions are made in an environment in which the goals, constraints and dependencies are not sharply defined. At about that time, I began to realize that the theory of fuzzy sets could be applied to the control of mechanistic systems in which the objective is to maintain a set point or perform a specified task. My views on the applicability of fuzzy set theory to problems of this type were described in my 1972 paper, "A Rationale for Fuzzy Control," and in a joint paper with S. S. L. Chang, "On Fuzzy Mapping and Control."

The impetus for industrial applications of fuzzy control was provided by the seminal work of Mamdani and Assilian in 1975 in which they demonstrated the workability of the basic ideas underlying fuzzy control by applying them to the control of an experimental steam engine.

Today, close to sixty percent of the applications of fuzzy logic relate in one way or another to control. What are the reasons for this preponderance?

In the first place, applications to control involve an easy-to-use subset of fuzzy logic which centers on what might be called the *calculus of fuzzy rules*. In the calculus of fuzzy rules, the dependencies and commands are expressed in the form of fuzzy "if-then" rules in which the antecedents and consequents involve linguistic variables. The use of such rules makes it possible for the designer to express what a system is expected to do in a form that is close to human intuition. In effect, the point of departure in fuzzy control is a human solution which is articulated by the designer in the language of fuzzy rules. In this sense, fuzzy logic solutions are usually descriptive rather than prescriptive in nature.

In the second place, the use of fuzzy rules provides a way of exploiting the tolerance for imprecision to achieve tractability, robustness, and low solution cost. Exploitation of the tolerance for imprecision is one of the main reasons why the use of fuzzy control is rapidly growing in popularity, especially in the realm of consumer products.

This is the background against which the contents and contributions of "Multistage

Fuzzy Control" should be viewed. The main concern of Professor Kacprzyk's work is not fuzzy control in its mundane sense – a sense which defines the use of fuzzy control in consumer products. Rather, it is focused on the much more complex issues which arise when what is involved is a multistage decision process with specified fuzzy goals and fuzzy constraints. In this sense, "Multistage Fuzzy Control" is much closer in spirit to "Decision-Making in a Fuzzy Environment" than to the literature on fuzzy control which is centered on consumer products and industrial systems design.

Reflecting the wide-ranging advances in fuzzy logic since 1970, Professor Kacprzyk's work goes far beyond my joint paper with Bellman in its treatment of the basic issues underlying multistage decision analysis. Of particular importance is the author's treatment of fuzzy events, fuzzy probabilities and fuzzy quantifiers – a subject to which he has made important contributions. The point at issue is that in most real-world applications, the events and their probabilities are not known with sufficient precision to justify the use of conventional probability-based methods in which the events and their probabilities are assumed to be crisp. This is one of the serious shortcomings of decision analysis as it is practiced today.

As a simple illustration of this point, consider the following problem. A box contains approximately ten balls of various sizes of which several are large and a few are small. What is the probability that a ball drawn at random is neither large nor small?

Conventional probability-based methods cannot lead to an answer to this question because what is missing in probability theory is the semantics of natural languages. More specifically, in the example under consideration what is needed is a machinery for defining the meaning of *approximately ten, large, small, several large balls*, and *neither large nor small*. In the absence of this machinery, the fuzzy probability of drawing a ball at random which fits the fuzzy description *neither large nor small* cannot be computed.

The point of this example is that most real-world probabilities – whether subjective or not – are rooted in fuzzy perceptions of cardinalities of fuzzy sets. Viewed in this perspective, an understanding of how fuzzy probabilities and fuzzy utilities can be dealt with within fuzzy logic is an essential prerequisite to applying decision analysis to the solution of real-world problems.

The issue of fuzzy probabilities and the related issue of fuzzy quantifiers are just two of the many important issues which are addressed in "Multistage Fuzzy Control" in depth and with authority. The last chapter presents a panoply of examples of applications which reflect Professor Kacprzyk's extensive experience in applying decision analysis to the solution of real-world problems.

Viewed in its totality, "Multistage Fuzzy Control" is a major contribution to the literature which should be on the desk of anyone who has an interest in developing an understanding of how real-world problems in decision analysis can be formulated and solved through the use of fuzzy-logic-based methods. Professor Kacprzyk deserves our thanks and congratulations for authoring a text which is superlative in all respects.

Berkeley, California   Lotfi A. Zadeh
June, 1996

1

Introduction

Fuzzy control, or – as it is often termed – *fuzzy logic control*, has recently been without doubt one of the much talked about new technologies. It has been very important for a couple for reasons. First, it has shown that even for complicated processes, whose models are unknown (or their development or identification is too difficult or too costly), one can devise an automatic controller employing only an imprecise (e.g., verbal) knowledge of an experienced process operator. And this controller does work incredibly well!

Such a controller may be viewed to be a simple expert system, and – from this point of view – a fuzzy controller may be considered to be a convincing proof that the expert system technology does work.

Second, for Zadeh's (1965a) *fuzzy sets theory* itself, the advent of fuzzy (logic) control has been a turning point, too, in the sense that it has shown a real applicability of fuzzy sets, and at what a scale – in just a couple of years fuzzy-control-based products exemplified by still and video cameras, washing machines, refrigerators, vacuum cleaners, etc. have been widely available. Clearly, these products have been accompanied by those less visible to the general public and media as, for example, cranes, subway control systems, etc. This applicability and wide availability of fuzzy-control-based products have been in marked contrast to widely employed arguments of dubious applicability of fuzzy sets theory which have been raised and used against Zadeh and his followers.

Fuzzy control has its roots in two important events that occurred between 1965 and 1975. First, for a long time it has been clear that probability theory is not in a position to cover all possible aspects of uncertainty (or, better, imperfection of information) since it concerns only randomness-related uncertainty which is relevant in situations when all we can say about, say, the air temperature tomorrow is some probability distribution that may be obtained, for example, from statistical data. In this case we can imagine that we are concerned with (precise) events such as, for example, "temperature will be 23°C," "temperature will be 24°C," etc. and we can only say that, for example, the first will be with probability 0.5, the second with probability 0.3, etc.

It is easy to notice that the above probabilistic treatment of uncertainty does not cover other aspects of information imperfection in such a case as, for example, it does not make it possible to deal with imprecise events exemplified by "temperature will be *high*," "temperature will be *moderate*," etc. which are a clear consequence of the use of natural language that is the only fully natural means of expression and communication for humans, but is strange to the "machine" (including all formal means such as mathematics). Such an information imperfection reflects *vagueness* whose essence may be stated as that the transition from elements exhibiting such a vague property (e.g., *high* values of temperature) to elements not exhibiting it (e.g., *not-high* values of temperature) is not abrupt as in the case of crisp, non-vague concepts exemplified by "temperature *above the freezing point*," but gradual in the sense that some temperatures may be, for example, for sure *high*, some may be for sure *not high*, and for some these two facts can be ascertained only to some degree.

Such a gradual transition between the elements exhibiting some property (i.e. belonging to the set of such elements) and those which do not, may make questionable the use of the powerful conventional set theory, and – since it is the basis of the whole pure and applied mathematics – of all powerful and well developed classical modeling, optimization, etc. in all problems in which vague elements play a relevant role. Such problems are omnipresent in the analysis of all non-trivial real-world problems in which a human being is somehow involved, that is in virtually all problems people are facing in practice.

The above limitations of conventional mathematics, which might be termed its low "human consistency," had been recognized for quite a long time. The rationale behind multi-valued logic developed in the early 1920's by Lukasiewicz and Post might be viewed similarly, and Black's (1937, 1963, 1970) works concerned related topics.

However, it was Zadeh (1965a) who first proposed a simple and constructive theory of *fuzzy sets*. Basically, he replaced the characteristic function in the conventional set theory, which takes on the value 0 or 1 that stands for "element belongs to the set" and "element does not belong to the set," respectively, by the so-called *membership function* taking values in the whole unit interval, from 0 for "element does not belong to the set" to 1 for "element belongs to the set" through all intermediate values between 0 and 1 standing for the belongingness to some degree. Notice that this is virtually what is needed for the representation and manipulation of vague properties. Needless to say that fuzzy sets theory has just provided a convenient modeling apparatus for an effective and efficient dealing with vague concepts, relationships, etc., and by no means a "new set theory" in the strict mathematical sense. If such a point of view had been accepted, many later criticisms and even hostilities against fuzzy sets theory would obviously have been immaterial.

Being quite a considerable departure from the conventional precise and crisp modeling approach, fuzzy sets theory has faced opposition since the very beginning. However, Zadeh and his followers have been making steady progress in the theory and applications. Zadeh's papers have provided steady stimulation and new ideas. From the point of view of this book, the first milestone is Bellman and Zadeh's (1970) paper on a general, simple and efficient framework for decision making under fuzziness (under fuzzy constraints and fuzzy goals). This framework encompasses multistage decision making problems too, that is, *control*. Bellman and Zadeh have devised a dynamic programming scheme for solving such multistage control problems. This paper will be in fact our point of departure, and virtually all our considerations will dwell upon it.

Another milestone is Zadeh's (1972) paper in which he outlined a rationale for fuzzy control. Zadeh, a prominent systems and control theorist himself, has been fully aware of the relevance of "soft" and ill-defined elements in virtually all non-trivial practical control problems, and has clearly seen the role fuzzy sets may be able to play in this respect.

And, again, it was Zadeh (1973) who proposed a general fuzzy-set-based approach to the analysis of systems whose main elements were:

- *Linguistic variables* taking on values like *high*, *medium*, *low*, etc. which are equated with fuzzy sets,
- *Fuzzy conditional statements* expressing relations between linguistic variables by means of IF–THEN statements like "if the temperature is *high*, then the pressure is *medium*," and
- *Compositional rule of inference* which – if we know a value of a primary variable as, say, the temperature is *very high*, and a relation between a primary and a secondary variable as, say, "if the temperature is *high*, then the pressure is *medium*" (notice that the current value of the primary variable and its value in the fuzzy conditional statement need not be the same!) – makes it possible to find the induced value of the secondary variable as, say, the pressure is *high*.

These concepts constitute in fact all what has been needed for fuzzy control since they provide tools for the representation of expert knowledge (expressed in a natural way by means of natural language) and its manipulation. However, it was Mamdani (1974, 1976) and his collaborators (e.g., Mamdani and Assilian, 1975) who showed how these ideas could be combined into a working control system. They have been followed by numerous other contributors who have both introduced some conceptually new architectures (e.g., Takagi and Sugeno, 1985) or have just proposed improvements and refinements. These are to numerous to cite, and we can refer the reader to, say, the books of Driankov, Hellendoorn and Reinfrank (1993), or Yager and Filev (1994), or the edited volume by Kandel and Langholz (1991).

The essence of fuzzy (logic) control meant in the above sense is that we do not intend to build a model of the control process itself, as is customary in traditional control, because this may be too difficult or too costly, or even such a model may be unknown. We just acquire knowledge from, say, experienced process operators or other experts as to *how to control the process*. This knowledge need not be expressed precisely (e.g., using numbers) but may be given just in the form of linguistic statements exemplified by "if the pressure is *low*, then increase the temperature *slightly*." The linguistic terms (written in italics) are represented as fuzzy sets. Then, such statements, which stand for control laws for particular situations, are used to infer the control for a current situation. In Section 1.1 we will provide a brief overview of fundamentals, basic architectures, etc. of fuzzy (logic) control.

Looking at the very essence of fuzzy (logic) control sketched above, we can notice that this is clearly a *descriptive, non-model-based* approach since we explicitly *describe how to control the process*, and do not rely on a model of the process under control.

This descriptive character of fuzzy control may be, however, viewed as counter-intuitive, and somehow contradicting the long tradition of control. Namely, the essence of control may be generally meant as to *find* a *best* course of action assuming knowledge of the *behavior* of the system to be controlled, and some *goals* to be attained and *constraints* under which we are to operate. Therefore, the essence of the problem is not to describe how the system is to be controlled but to determine, or *prescribe*, how to control the system.

The above mentioned *prescriptive approach* to fuzzy control is even earlier than the descriptive one formerly mentioned as it appeared as early as in the late 1960's and early 1970's (Chang, 1969a, b; Bellman and Zadeh, 1970). It had initially attracted much attention [cf. Esogbue and Bellman (1981, 1984), Kacprzyk (1977–1982b), etc.], and even Kacprzyk's (1983b) book discussed it in detail. Then the attention shifted to the descriptive approach to fuzzy control.

It seems, however, that a revival of the prescriptive approach to fuzzy control should happen in the near future. First, even if the traditional fuzzy (logic) control is really very effective and has found so many applications, it seems that its potentials from the conceptual point of view are somehow exhausted. There are evidently still open questions but they seem to be related to technical issues such as, for example, which form of implication to use, etc. Even such apparently new approaches as the so-called neuro-fuzzy control do not provide conceptually new ideas related to the very essence of the approach. Fuzzy (logic) control becomes more and more an engineering field than scientific, which is natural.

It is our opinion that fuzzy control needs a *conceptually new* approach, going beyond its traditional descriptive paradigm. A *new generation* of fuzzy control should emerge, and we think that some sort of a prescriptive approach will be one of its relevant parts.

This book may be viewed as a step toward such a new generation of fuzzy control by explicitly incorporating tools which will make it possible to determine *how a system should best be controlled*, i.e. to *prescribe* control. In Section 1.2 we will discuss in more detail such a prescriptive approach to be discussed in this book.

Finally, in Section 1.3 we will summarize the contents of the consecutive chapters.

1.1   FUZZY LOGIC CONTROL – A DESCRIPTIVE APPROACH

Fuzzy (logic) control was proposed by Mamdani and his collaborators [cf. Mamdani (1974, 1976), Mamdani and Assilian (1975)] using Zadeh's (1973) fuzzy-sets-based approach with linguistic variables, fuzzy conditional statements, and the compositional rule of inference.

The rationale for fuzzy control may be summarized as:

- the model of the process to be controlled is not available as, e.g., the process is complicated or the identification of its model is too costly or time consuming,
- in spite of the above, the process is (reasonably) well controlled by experienced operators, or there are experts (even in the form of the operator's manuals!) who know how to control it, and
- if asked for, these experts express their knowledge in a verbal form (remember that natural language is the only fuzzy natural human means of communication!), and not in a precise (e.g., numerical) form.

Probably the best way to start explaining the essence of fuzzy control is to use its pictorial representation shown in Figure 1.

So, there is a human expert who does not care about the model of the system under control either because he or she does not know it or has no chance or intention to

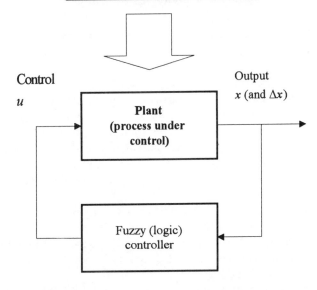

Figure 1 A pictorial representation of the descriptive approach to fuzzy control

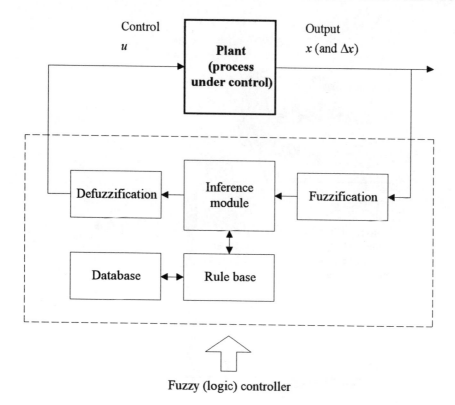

Figure 2   Main elements of the fuzzy (logic) control system

build it – the system under control is therefore a "black box" to him or her. Moreover, he or she cannot precisely articulate – in terms of a performance function – what the goal or purpose of control is; for instance, he or she does not know if the sum or squared deviations from some predefined trajectory over a planning horizon should be minimized. However, the expert feels what is really needed in the control of the process considered, and he or she knows from experience how to control this process to obtain really "good" results. Therefore, if asked for, the expert can *describe* how to control the process.

This suggests that the above can be mimicked by a simple expert system shown in Figure 2 which is what is meant by a *fuzzy (logic) control* system.

We will now briefly describe the particular elements of a fuzzy control system taking into account the flow of information and computations in the development and running phases.

The *knowledge base* consists of a rule base and a database which incorporate knowledge about how to control the process. The rule base is a collection of IF–THEN rules that relate the state of the control process to the control needed; for simplicity, throughout this book we will equate the state with an output which is clearly "visible" to the expert.

As already mentioned, for a human expert the most natural way for expressing the

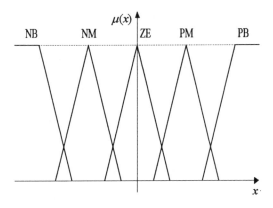

Figure 3 Five standard linguistic values

above rules is to use (quasi)natural language terms as, e.g., *high, medium, small*, etc.

So, first, we need to define a linguistic *universe of discourse* which consists of such linguistic terms. It is well known that a human being can relevantly distinguish at most 5–9 different verbal descriptions (Miller's magic number 7 ± 2!).

Usually, we introduce some standard linguistic values ranging from *positive big* to *negative big*, that is, in the case of 5 standard values:

- **NB**: *negative big,*
- **NM**: *negative medium,*
- **ZE**: *zero,*
- **PM**: *positive medium,* and
- **PB**: *positive big*

and similarly in the case of 7 or 9 standard values (cf. Driankov, Hellendoorn and Reinfrank, 1993).

All these standard linguistic values are equated with fuzzy sets. Since we wish to use the above standard terms for describing various entities as, e.g., temperature, pressure, change of speed, etc. they should be defined in some standard universe of discourse as, e.g., in the interval $[-1, 1]$. Thus, the (real) universes of discourse of all the entities should be first normalized into this interval.

Then, usually, the above standard values are defined as trapezoid or triangular fuzzy sets (numbers) in [-1, 1] as shown in Figure 3 for 5 standard values.

Note that, first, the number of standard values is to be chosen while developing a fuzzy controller, and depends on the situation and available knowledge. In general, there should be *sufficiently many* standard values so that they cover all possible values. On the other hand, their membership functions should be "overlapping" (cf. Figure 3) since – as we will see later – otherwise we may have problems with inference.

Now, we are ready to use experts' testimonies to construct the rule base. Suppose therefore that the experts watch the state of the process which consists of the value of a primary variable and its change, denoted by x and Δx, respectively, and employ another variable, say a secondary variable, denoted by Δu which means the change of control, to control the process; these may be exemplified by pressure and its change, and the change of temperature (e.g., controlled by gas inflow).

The experts are therefore asked to articulate verbally their knowledge as to relations between Δu, and x and Δx.

Basically, two "architectures" of fuzzy control are used. In the so-called *Mamdani architecture* (Mamdani, 1974), the knowledge acquired from the experts might be, for example:

$$\text{IF } x = \text{NS AND } \Delta x = \text{NS THEN } \Delta u = \text{PB}$$
$$\text{IF } x = \text{NS AND } \Delta x = \text{NM THEN } \Delta u = \text{PB}$$
$$\ldots$$
$$\text{IF } x = \text{NS AND } \Delta x = \text{PB THEN } \Delta u = \text{NM}$$
$$\ldots$$
$$\text{IF } x = \text{NM AND } \Delta x = \text{NS THEN } \Delta u = \text{PM}$$
$$\ldots$$
$$\text{IF } x = \text{ZE AND } \Delta x = \text{ZE THEN } \Delta u = \text{ZE}$$
$$\ldots$$
$$\text{IF } x = \text{PM AND } \Delta x = \text{ZE THEN } \Delta u = \text{ZE}$$
$$\ldots$$
$$\text{IF } x = \text{PB AND } \Delta x = \text{NS THEN } \Delta u = \text{NM}$$
$$\ldots$$
$$\text{IF } x = \text{PB AND } \Delta x = \text{PB THEN } \Delta u = \text{NB}$$

In the so-called *Takagi–Sugeno architecture* (Takagi and Sugeno, 1985), the rules are assumed to be in the form:

$$\text{IF } x = \text{NS AND } \Delta x = \text{NS THEN } \Delta u = h_{1,1}(x, \Delta x)$$
$$\text{IF } x = \text{NS AND } \Delta x = \text{NM THEN } \Delta u = h_{1,2}(x, \Delta x)$$
$$\ldots$$
$$\text{IF } x = \text{NS AND } \Delta x = \text{PB THEN } \Delta u = h_{1,5}(x, \Delta x)$$
$$\ldots$$
$$\text{IF } x = \text{NM AND } \Delta x = \text{NS THEN } \Delta u = h_{2,1}(x, \Delta x)$$
$$\ldots$$
$$\text{IF } x = \text{ZE AND } \Delta x = \text{ZE THEN } \Delta u = h_{3,1}(x, \Delta x)$$
$$\ldots$$
$$\text{IF } x = \text{PM AND } \Delta x = \text{ZE THEN } \Delta u = h_{4,3}(x, \Delta x)$$
$$\ldots$$
$$\text{IF } x = \text{PB AND } \Delta x = \text{NS THEN } \Delta u = h_{5,1}(x, \Delta)$$
$$\ldots$$
$$\text{IF } x = \text{PB AND } \Delta x = \text{PB THEN } \Delta u = h_{5,5}(x, \Delta x)$$

Therefore, in the Takagi–Sugeno architecture, for linguistically specified premises (x and Δx) the (real-valued) consequence Δu is not specified linguistically but as a function of the (real) values of x and Δx.

In this section we will basically deal with the Mamdani architecture since it should better suit the purpose of this short introduction, but a similar argument can proceed for the Takagi–Sugeno architecture.

The control rules mentioned above are to be derived, and the most commonly used technique is their direct acquisition from experienced process operators, experts, or even some operator's manuals. One can also use other techniques such as a neural network representation of the rule set which is first trained on some examples of (successful) control runs by experienced operators. This issue, which is very important in itself, will not be considered here, and details can be found in, say, Driankov, Hellendoorn and Reinfrank (1993).

The rule base needs a *database* which contains parameters to be used. First of all, it stores the membership functions with which all linguistic terms used are equated. Moreover, it contains scaling factors, etc. to be mentioned later.

The rule base (together with the database) is then used for *inference*. Basically, the problem is as follows. Suppose that we have a set of IF–THEN rules as given previously, and some fuzzy values of the primary variables x and Δx – which need not be exactly equal to any of the values of x and Δx specified in the rules! We need to infer the fuzzy value of the secondary variable Δu induced by the current x and Δx and the rule base. Details will be given in Section 2.1.5.

Now, following in a sense the fuzzy controller's operations, specific values of x and Δx are observed or measured. These values are usually crisp (non-fuzzy) as they concern mainly some values of technological variables (e.g., pressure, water level, ...). Such crisp values cannot be directly employed since our knowledge as to how to control the process, i.e. the control rules, concerns fuzzy (linguistic) entities. So, they need to be *fuzzified*.

As a result of inference we obtain therefore a fuzzy value of Δu which should then be *defuzzified*, i.e. we need to find its corresponding crisp (non-fuzzy) values, since the value of the (change of) control to be applied to the control process is to be usually crisp (e.g., the change of gas inflow). This defuzzification, which is a very important aspect in the design of fuzzy controllers, may be performed in various ways exemplified by the so-called method of the center-of-area/gravity presented in Section 2.1.9 which is by far the most widely used. For other, more conventional methods, see for instance Driankov, Hellendoorn and Reinfrank (1993), while for some less conventional methods we may refer the reader to Yager and Filev (1994).

This basic scheme of fuzzy control has proven to be very effective and efficient in many real-world applications ranging from everyday products to specialized equipment. Needless to say that though the latter may be more relevant and sophisticated, the former are certainly more visible to the general public and media, and have contributed a lot to the present exposure and recognition of fuzzy control.

Looking at the essence of the fuzzy (logic) control paradigm sketched above one can see that, from our point of view, its main strength is that the lack of a mathematical model of the process – no matter if caused by its non-existence or unavailability or just too high a cost of its derivation – is not an obstacle provided an adequate human experience or expertise on how to control the process is available. Fuzzy (logic) control is therefore a *non-model-based* approach.

Many other advantages of fuzzy (logic) control are also reported, exemplified by a shorter product development so that a new product may reach the market faster. This is clearly extremely relevant in practice.

However, one may see at first glance one serious disadvantage which is extremely relevant for us, and in fact which may be considered to be one of those that have triggered our approach. That is to say, the rule base reflects some good control strategy known by the operators for some (possibly unknown) unspecified performance criterion as, for example, the lowest fuel consumption. However, first, though the operator knows that this is a (very) good strategy, he or she has no idea whether it would be possible to attain a better result. Second, if requirements as to the control process change as, for example, into the fastest possible attainment of some output level, this would need the development of a new rule base. The same applies for all further, even slight, changes of control requirements. This all is because the traditional fuzzy (logic) control does not *explicitly* account for a *performance function*.

This concludes our brief introduction to fuzzy (logic) control whose purpose was to provide the reader with a basic knowledge on this attractive and much discussed topic. Moreover, thanks to this section the reader will hopefully be able to better understand the rationale behind and the advantages of the approach presented and advocated in this book.

1.2 MULTISTAGE FUZZY CONTROL – A PRESCRIPTIVE APPROACH

The traditional fuzzy (logic) control the essentials of which were sketched in the previous section has already reached maturity in the sense of entering industrial practice. Even if there are still unresolved problems, this technology is becoming more and more a part of engineering than science. Though, from the scientific point of view, as is always the case with every product's life cycle, it seems to have reached its limits, an attempt should be made to at least trigger the development of some *new generation* of fuzzy control.

Some steps in this direction have already been undertaken by introducing the so-called neuro-fuzzy control, hierarchical fuzzy control, etc. However, from the point of view of our discussion, these attempts do not constitute a new generation as they do not go beyond the traditional descriptive character of fuzzy (logic) control.

Notice, first, that this descriptive character may be viewed as to be somehow contradicting the traditional control paradigm. Namely, in control one normally assumes that:

- the model (dynamics) of the system under control is known,
- requirements as to the control process, exemplified by some constraints
  to be satisfied and goals to be attained, are known – i.e. an explicit
  performance function is given, and
- a (best) control is to be found by an algorithm,

so that this is clearly a *prescriptive, model-based* approach since one does not describe how the (known!) process is controlled but prescribes how to do this best.

The essence of this prescriptive approach to fuzzy control is shown in Figure 4.

It is the purpose of this book to present an alternative, prescriptive, model-based approach to fuzzy control which might be a relevant ingredient of a next generation of fuzzy control.

It may be advantageous to look at the essence of our approach from the following perspective:

Where does fuzziness enter in the control problem formulation?

In the traditional, fuzzy logic control approach fuzziness is clearly in the *descriptive*

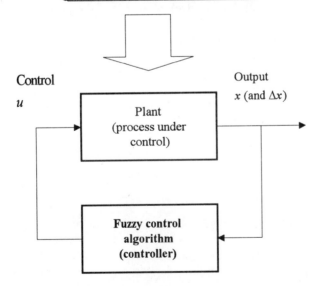

Figure 4   A pictorial representation of the prescriptive, model-based approach
to fuzzy control

control rules.

On the other hand, in the approach advocated and presented in this book, fuzziness occurs in the following elements:

- first, fuzziness occurs in the definition of the *environment* within which the process is to proceed, i.e. in the definition of *fuzzy constraints* on the controls to be applied and in *fuzzy goals* on the states (outputs) to be attained; notice that an aggregation of fuzzy constraints and fuzzy goals constitute an *explicit performance function* (which is to be minimized or maximized),
- second, fuzziness may also occur in the definition of the model (dynamics or state transitions) of the control systems when, first, the controls and states themselves may be only imprecisely known in the sense that their values may only be known to take on linguistic (fuzzy) values, and second, the system's temporal evolution is not precisely known and may be described in linguistic (fuzzy) terms only.

However, fuzziness does not occur in the algorithmic determination of (best) controls which is "crisp" in the sense that conventional nonfuzzy optimization-type algorithms are employed for the solution.

A very illustrative example of a fuzzily known model of the system under control may be the one in which only a verbal (fuzzy) description of the system's dynamics, i.e. its state transitions, is available.

These state transitions mean that from a known fuzzy state at control stage t, X_t, and under a known fuzzy control at control stage t, U_t, the system proceeds to a known fuzzy state at control stage $t + 1$, X_{t+1}.

One can well imagine here that such transitions may be described by IF–THEN rules (which essentially play the role of an imprecisely specified function relating the current, at control stage t, fuzzy state X_t and current fuzzy control U_t to the next fuzzy state at control stage $t + 1$, X_{t+1}.

Such IF–THEN rules may be similar to those used in the two fuzzy (logic) control architectures mentioned in Section 1.1, that is:

- by the "Mamdani-type" state transitions

$$\text{IF } X_t = \text{NS AND } U_t = \text{NS THEN } X_{t+1} = \text{PB}$$
$$\text{IF } X_t = \text{NS AND } U_t = \text{NM THEN } X_{t+1} = \text{PB}$$

. . .

$$\text{IF } X_t = \text{NS AND } U_t = \text{PB THEN } X_{t+1} = \text{NM}$$

. . .

$$\text{IF } X_t = \text{NM AND } U_t = \text{NS THEN } X_{t+1} = \text{PM}$$

. . .

$$\text{IF } X_t = \text{ZE AND } U_t = \text{ZE THEN } X_{t+1} = \text{ZE}$$

. . .

$$\text{IF } X_t = \text{PM AND } U_t = \text{ZE THEN } X_{t+1} = \text{ZE}$$

. . .

$$\text{IF } X_t = \text{PB AND } U_t = \text{NS THEN } X_{t+1} = \text{NM}$$

. . .

$$\text{IF } X_t = \text{PB AND } U_t = \text{PB THEN } X_{t+1} = \text{NB}$$

i.e. fuzzy values of the states and control only are involved, and
- by the "Takagi–Sugeno-type" state transitions

$$\text{IF } X_t = \text{NS AND } U_t = \text{NS THEN } x_{t+1} = f_{1,1}(x_t, u_t)$$
$$\text{IF } X_t = \text{NS AND } U_t = \text{NM THEN } x_{t+1} = f_{1,2}(x_t, u_t)$$

. . .

$$\text{IF } X_t = \text{NS AND } U_t = \text{PB THEN } x_{t+1}u = f_{1,5}(x_t, u_t)$$

. . .

$$\text{IF } X_t = \text{NM AND } U_t = \text{NS THEN } x_{t+1} = f_{2,1}(x_t, u_t)$$

. . .

$$\text{IF } X_t = \text{ZE AND } U_t = \text{ZE THEN } x_{t+1} = f_{3,1}(x_t, u_t)$$

. . .

$$\text{IF } X_t = \text{PM AND } U_t = \text{ZE THEN } x_{t+1} = f_{4,3}(x_t, u_t)$$

. . .

$$\text{IF } X_t = \text{PB AND } U_t = \text{NS THEN } x_{t+1} = f_{5,1}(x_t, u_t)$$

. . .

$$\text{IF } X_t = \text{PB AND } U_t = \text{PB THEN } x_{t+1} = f_{5,5}(x_t, u_t)$$

i.e. both fuzzy and nonfuzzy values are involved;

and the meaning of the above two forms of state transitions will become clear in Section 2.4.3.

One may therefore readily see that the descriptive and prescriptive approaches to fuzzy control have something in common, i.e. usually they both rely on human expertise. However, in the former it enters into the definition of control strategies, and in the latter it enters into the specification of behavior of the system under control and the definition of control requirements (performance criteria).

The new prescriptive approach to fuzzy control outlined above needs some general framework within which to operate just as the descriptive approach has needed, say, tools of fuzzy sets and approximate reasoning.

Now, in addition to fuzzy sets, fuzzy logic and approximate reasoning – which are needed for the specification of the system's dynamics and performance function – it

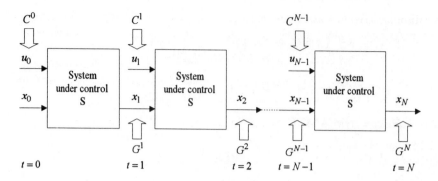

Figure 5   A general framework of the prescriptive approach to fuzzy control adopted in the book

clearly needs some decision making (maybe even optimization) related tools since a best (optimal) control is to be found.

Luckily enough such a general framework is available, and stems from Bellman and Zadeh's (1970) works on decision making in a fuzzy environment [Chang's (1969a, b) works on a similar topic should also be cited], and more specifically on fuzzy dynamic programming. Note that these works are earlier than that by Mamdani (1974) which introduced the idea of fuzzy (logic) control.

Bellman and Zadeh's (1970) seminal paper had then been followed by numerous contributions exemplified by Esogbue and Ramesh (1970), Kacprzyk (1977–1983a), Fung and Fu (1973, 1975, 1977), etc. These developments were presented in detail in Kacprzyk's (1983b) book in which a systematic classification of problem classes and basic solution techniques had been proposed and discussed.

This book, while preserving a basic general philosophy and structure of Kacprzyk's (1983b) former book, goes much further by examining new theoretical developments and tools (e.g., neural networks and genetic algorithms), and also relevant applications, both the traditional ones and the new ones that have recently been introduced.

Basically, the general control related framework adopted in this book may be depicted as in Figure 5. We start from an initial state at control stage (time) $t = 0$, x_0, apply a control at $t = 0$, u_0, attain a state at time $t = 1$, x_1, apply u_1, ... Finally, being at control stage $t = N - 1$ in state x_{N-1} we apply control u_{N-1} and attain the final state x_N.

The dynamics of the system under control, S, is assumed known and given by state transitions from state x_t to x_{t+1} under control u_t, the consecutive controls applied u_t are subjected to fuzzy constraints C^t, and on the states attained x_{t+1} fuzzy goals G^{t+1} are imposed, $t = 0, 1, \ldots, N - 1$.

The performance (goodness) of the control process is gauged by some (aggregation) measure of how well, at all the consecutive control stages, the fuzzy constraints on controls and fuzzy goals on states are satisfied. And an optimal sequence of controls at the consecutive control stages, $u_0^*, \ldots, u_{N-1}^*$, is sought (to be determined by an algorithm).

Since dynamics is explicitly involved, and for practical reasons it makes sense to account for discrete time instance (control stages) only, we perform our control analysis in terms of multistage control, hence the approach presented is termed a *multistage fuzzy control*, tacitly assuming that it is a *prescriptive* approach.

One may see that the above general scheme of a prescriptive approach to multistage fuzzy control may be viewed to give rise to the following general *problem classes*, for the following types of the termination time and system under control:

- type of *termination time*:
  1. fixed and specified in advance,
  2. explicitly specified (as the moment of entering for the first time a termination set of states),
  3. fuzzy, and
  4. infinite;
- type of *system under control*:
  1. deterministic,
  2. stochastic, and
  3. fuzzy.

This seems to be the most appropriate classification (cf. Kacprzyk, 1983b), and will be adopted in this book too.

1.3   A BRIEF SURVEY OF THE CONTENTS

In the consecutive chapters we will consider all the above mentioned problem classes and related basic issues. The book will be in general self-contained in the sense that, first, the reader will only need some basic knowledge of mathematics exemplified by set theory, analysis, calculus or mathematical logic. Second, we will assume that the reader has been exposed to fundamentals of, say, optimization and mathematical programming – in particular dynamic programming and branch-and-bound, control theory, probability theory, etc. Therefore, we will not discuss these aspects here referring the interested reader to a vast array of books (e.g., on operations research) which are available from all major scientific publishers. On the other hand, we will explain all the more advanced issues arising.

In Chapter 2, in the first part, we will discuss basic elements of fuzzy sets theory. This part is quite informative and extensive and provides the reader with comprehensive, yet readable, survey of all that will be needed to follow the arguments, not only in this book but in many other books on fuzzy sets and their applications. Our discussion will be widely illustrated by examples.

We will consider the concept of a fuzzy set, basic operations on fuzzy sets, the concept of a fuzzy relation and its related issues exemplified by various compositions. We will deal with the linguistic approach in which variables are assumed to take on (quasi)natural language linguistic values equated semantically with fuzzy sets, relationships between the primary and secondary linguistic variables are represented by fuzzy conditional statements (IF–THEN rules), and the compositional rule of inference is used to determine the value of a secondary variable induced by the value of a primary variable and a fuzzy conditional statement.

We will discuss the so-called extension principle which makes possible the extension of conventional (nonfuzzy) relations, operations, etc. to their counterparts operating on fuzzy quantities. We will consider the concept of a fuzzy number and fuzzy arithmetic. We will discuss many more related issues as, for example, probabilities of fuzzy events, defuzzification, etc.

In the second part of Chapter 2 we will discuss dynamic systems under control, mainly from the point of view of modeling their dynamics, i.e. state transitions. We will start with the basic case of a deterministic system under control described by a deterministic state transition equation.

Then we will proceed to the case of a stochastic system, assumed to be a Markov chain, whose state transitions are described by a conditional probability. Finally, we will discuss the most interesting case of a fuzzy system under control governed by a fuzzy state transition equation. We will also show how dynamics of a fuzzy system under control may be described linguistically by IF–THEN rules or represented by a neural network. We will sketch extremely relevant issues related to the identification of fuzzy models.

In Chapter 3 we will present the essence of Bellman and Zadeh's (1970) approach to decision making in a fuzzy environment (called also decision making under fuzziness) which is a general, simple and efficient framework that has been used as a point of departure in virtually all fuzzy approaches to decision making, optimization, control, etc. including multistage control dealt with in this book.

We will consider how fuzzy constraints and fuzzy goals imposed on the controls and states (outputs) at the consecutive control stages can play the role of an explicit performance function. In this context we will show the crucial issue of their proper aggregation which results in the so-called fuzzy decision being such an explicit performance function.

In Chapter 4 we will proceed to the analysis of the basic case, i.e. with a fixed and specified termination time. We will consider first a deterministic system under control, and formulate respective multistage control problems. We will show how to solve them both by using the traditionally employed techniques of (fuzzy) dynamic programming and branch-and-bound, and by using the new tools based on neural networks and genetic algorithms.

We will proceed to the case of a stochastic system, and present two basic problem formulations due to Bellman and Zadeh (1970), and Kacprzyk and Staniewski (1980a). We will arrive at their solution by using dynamic programming and an iterative implicit enumeration algorithm, respectively. Finally, we will proceed to the case of a fuzzy system under control. First, we will show how the basic problem formulation needs to be redefined, and then show the application of dynamic programming, branch-and-bound, and a genetic algorithm.

In Chapter 5 we will consider multistage fuzzy control problems with an implicitly specified termination time given as that control stage at which a terminating control occurs (when a termination set of states is first reached). We will show the solution by using the following approaches: iterative, graph-theoretic, and branch-and-bound.

In Chapter 6 we will discuss an interesting case of a fuzzy termination time exemplified by *as soon as possible*, and *around five control stages*. We will consider as before the cases of deterministic, stochastic, and fuzzy systems under control, and will discuss the solution by using dynamic programming, branch-and-bound, and a genetic algorithm.

In Chapter 7 we will deal with a difficult, yet relevant and interesting, problem of multistage fuzzy control with an infinite termination time. Such a class of problems is particularly suited for long-lasting and low-varying control processes. Here again, we will consider the cases of deterministic, stochastic, and fuzzy systems under control. Basically, we will show how in each case a policy iteration algorithm can be devised which, by a subsequent improvement of control policies, makes it possible to find an optimal stationary policy in a finite number of steps.

In Chapter 8, which concludes the book, we will present some applications. Though all our previous considerations have been illustrated by examples, they have been by necessity very small, and intended to show only how a particular method works. In this chapter, on the other hand, we will show some real-world (though clearly simplified) applications which will make it possible to follow, first of all, how an adequate multistage fuzzy control model is developed for a particular situation. Its solution is basically by the application of one previously described technique. Moreover, we will introduce some more application-specific concepts and tools which may be of interest and use to many readers.

The applications presented in this chapter will include:

- socioeconomic regional development,
- flood control,
- research and development (R&D) planning,
- scheduling of unit commitment in a power system,
- anesthesia administration during surgery,
- resource allocation,
- inventory control, and
- a review of some other applications in scheduling of power generator maintenance, design of a fuzzy controller, designing a distillation column in chemical engineering, determination of shortest paths for the transportation of hazardous waste, scheduling of autonomous guided vehicles in flexible manufacturing systems, and optimizing spare parts inventory in power stations.

Finally, an extensive up-to-date bibliography and an index are provided.

1.4 ACKNOWLEDGEMENTS

This book is a result of research supported, first of all, by the State Committee for Scientific Research (KBN) under grant 3 P403 009 05, and by the Systems Research Institute, Polish Academy of Sciences, and this support is greatly appreciated.

The author is indebted to many people, in particular to all his coauthors and close collaborators.

Roslyn Meredith of John Wiley & Sons, Chichester, UK has done a wonderful job in making this project come true. She has combined a high level of professionalism with an ability to create a synergistic collaboration of the author and the editor.

Ultimately, the author is fully responsible for all errors and omissions in this book.

2

Basic Elements of Fuzzy Sets and Fuzzy Systems

The purpose of this chapter is, first, to briefly expose a novice reader to basic elements of fuzzy sets theory which will be basic formal tools in the analysis of the control problems considered. Second, even for a more advanced reader a brief reminder of basic concepts, in their particular form of presentation adopted in this book, may be useful too. For all readers this chapter will provide the terminology and notation to be used throughout. Moreover, and we wish to emphasize this strongly, our exposition will only be as formal as necessary, of more intuitive a character since the problems discussed in this book may be of interest and relevance to quite a wide audience: applied mathematicians, control engineers, economists, systems analysts, decision analysts, operations research specialists, etc., of a different degree of familiarity with mathematics.

For the readers requiring or interested in a more thorough exposition of fuzzy sets and related concepts, we can recommend more specialized books which are offered now by virtually all major publishers such as, for example, Dubois and Prade (1980), Klir and Folger (1988), Klir and Yuan (1995), Lowen (1996), Zimmermann (1996), etc.

Our discussion will proceed, on the other hand, in the "pure" fuzzy setting, and we will not discuss possibility theory (which is related to fuzzy sets theory); the reader interested in possibility theory is referred to Dubois and Prade (1988).

As to the scope of this chapter, we will provide – on the one hand – an exposition of all concepts and topics related to fuzzy sets which will be needed throughout, and which are of a more general character, and not too specific to a particular application (in such a case they will be introduced later, when needed). On the other hand, we will discuss many elements (concepts, operations, etc.) of fuzzy sets theory that will not be employed in this book. This is motivated by at least two reasons. First, even if they are not used now, they may find their place in future developments of the models to be dealt with. Second, this chapter may serve as a source of information on fundamentals of fuzzy sets theory for a much wider audience than that interested in multistage fuzzy control (or decision making).

2.1   BASIC ELEMENTS OF FUZZY SETS THEORY

Fuzzy sets theory, introduced by Zadeh in 1965 (Zadeh, 1965a), is – as we have already indicated in Chapter 1 – a simple yet very powerful, and effective and efficient means to represent and handle imprecise information (of vagueness type) exemplified by *"tall* buildings," *"large* numbers," etc. It will also constitute a major formal tool in our discussion.

2.1.1   The idea of a fuzzy set

First, note that the purpose of a (conventional) set in mathematics is to characterize some concept. For instance, the concept of "integer numbers which are greater than or equal to 3 and less than or equal to 10" may be uniquely represented by the following set $\{x \in I : 3 \le x \le 10\} = \{3, 4, 5, 6, 7, 8, 9, 10\}$ where I is the set of integers.

To begin, we need to specify a *universe of discourse* (also called a universe, universal set, referential, reference set, etc.) that contains all those elements which are relevant for the particular concept to be represented. For instance, in the above example, the universe of discourse is clearly the set of all integer numbers.

A conventional set, say A, may be equated with its *characteristic function*

$$\varphi_A : X \longrightarrow \{0, 1\} \qquad (2.1)$$

which associates with each element x of a universe of discourse $X = \{x\}$ (which is in our example $X = I$, the set of integers) a number $\varphi(x) \in \{0, 1\}$ such that $\varphi_A(x) = 0$ means that such an $x \in X$ does not belong to the set A, and $\varphi_A(x) = 1$ means that such an x belongs to the set A.

Thus, for the former example of "integer numbers which are greater than or equal to 3 and less than or equal to 10" the set $A = \{3, 4, 5, 6, 7, 8, 9, 10\}$ may be represented by its characteristic function

$$\varphi_A(x) = \begin{cases} 1 & \text{for } x \in \{3, 4, 5, 6, 7, 8, 9, 10\} \\ 0 & \text{otherwise} \end{cases}$$

which may be depicted as in Figure 6.

Note that in a conventional set there is a clear-cut differentiation between elements belonging to the set and not belonging, i.e. the transition from the belongingness to nonbelongingness is abrupt, and this is proper for the example considered above.

However, it is easy to see that a serious difficulty arises when we try to formalize (by means of a set) vague concepts such as "integer numbers which are *more or less* equal to 6" – the source of imprecision (vagueness) is here clearly the term *more or less*. Evidently, the (conventional) set as sketched above cannot be used to adequately characterize such an imprecise concept as an abrupt and clear-cut differentiation between the elements belonging and not belonging to the set is artificial here, and no clear-cut borderline cannot be imposed.

This has led Zadeh (1965a) to the idea of a *fuzzy set* which is a class of objects (elements of a universe of discourse) with unsharp boundaries, i.e. in which the transition from the belongingness to nonbelongingness is not abrupt; thus, elements of a fuzzy set may belong to it to partial degrees, from the full belongingness to the full nonbelongingness through all intermediate values.

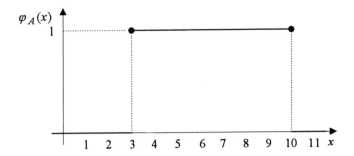

Figure 6 Characteristic function of the (conventional) set "integer numbers which are *greater than or equal* to 3 and *less than or equal* to 10"

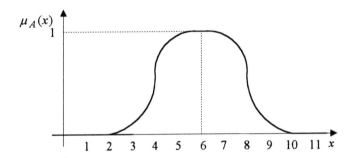

Figure 7 Membership function of a fuzzy set "integer numbers which are *more or less* 6"

One should again start with a universe of discourse containing all elements relevant for the (imprecise) concept to be formalized. Then, the above line of reasoning suggests that Zadeh replaces the characteristic function $\varphi : X \longrightarrow \{0,1\}$ by a *membership function* defined as

$$\mu_A : X \longrightarrow [0,1] \tag{2.2}$$

such that $\mu_A(x) \in [0,1]$ expresses the degree to which an element $x \in X$ belongs to the fuzzy set A: from $\mu_A(x) = 0$ for the full non-belongingness to $\mu_A(x) = 1$ for the full belongingness, through all intermediate $(0 < \mu_A(x) < 1)$ values. So, the membership function takes on its values in the unit interval, $[0,1]$, instead of in a set $\{0,1\}$.

Now, if we consider our example of "integer numbers which are *more or less* 6," then $x = 6$ certainly belongs to this set so that $\mu_A(6) = 1$, the numbers 5 and 7 belong to this set "almost surely" so that $\mu_A(5)$ and $\mu_A(7)$ are very close to 1, and the more a number differs from 6, the less its $\mu_A(.)$. Finally, the numbers below 1 and above 10 do not belong to this set, so that their $\mu_A(.) = 0$. This may be given as in Figure 7, though we should bear in mind that although in our example the membership function is evidently defined for the integer numbers (x's) only, it is depicted in a continuous form for clarity.

In practice the membership function of a fuzzy set is usually assumed to be piecewise linear as shown in Figure 8 for the same fuzzy set as in Figure 7, i.e. "integer numbers

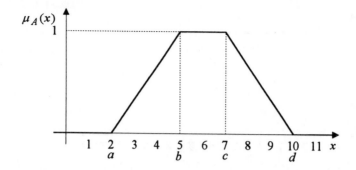

Figure 8   Piecewise linear membership function of a fuzzy set "integer
numbers which are *more or less* 6"

which are *more or less* 6". In such a case for the specification of a membership function
we need four numbers only: a, b, c, and d; in Figure 8 $a = 2$, $b = 5$, $c = 7$, and $d = 10$.
Such a piecewise linear representation will be used in principle throughout this book.

Note that in a fuzzy set the transition from the belongingness to non-belongingness
is gradual rather than abrupt, and hence it may provide us with what we have been
looking for to characterize imprecise (vague) concepts. However, we should bear in
mind that the form of a membership function is *subjective* as opposed to an "objective"
form of a characteristic function. However, this may be viewed to be quite natural,
as concepts fuzzy sets are meant to characterize are subjective since, for example,
"integer numbers *more or less* 6" depend on the individual, and is "in the eye of
the beholder." Clearly, precise concepts as, e.g., "integer numbers which are greater
than or equal to 3 and less than or equal to 10" are objective, and do not depend
on the individual. Unfortunately, this inherent, and clearly natural subjectivity of the
membership function of a fuzzy set may lead to some problems in our formal models
that would "prefer" objective elements alone. This will be considered later.

We will now define formally a fuzzy set in a form that will be suitable for our
analysis. Note that the definition of a membership function is considered by many to
be equivalent to the definition of a fuzzy set. This view will also be often, but not
always, adopted here.

A *fuzzy set* A in a universe of discourse $X = \{x\}$, written A in X, is defined as a
set of pairs

$$A = \{(\mu_A(x), x)\} \tag{2.3}$$

where $\mu_A : X \longrightarrow [0,1]$ is the *membership function* of A and $\mu_A(x) \in [0,1]$ is the
grade of membership (or *membership grade*) of an element $x \in X$ in a fuzzy set A.
So, a fuzzy set is a set of pairs consisting of the particular elements of the universe of
discourse and their degrees of membership in this particular fuzzy set.

Needless to say, our definition of a fuzzy set (2.3) is clearly equivalent to the
definition of the membership function (2.2) because a function may be represented by
a set of pairs "argument–value of the function for this argument." For our purposes,
however, the definition (2.3) is more "set-theoretic-like" which will often be more
convenient.

So, in the following we will equate fuzzy sets with their membership functions saying, e.g., "a fuzzy set $\mu_A(x)$," and also very often we will equate fuzzy sets with their labels saying, e.g., "a fuzzy set *large numbers*" with the understanding that the label *large numbers* is equivalent to the fuzzy set mentioned, written "$A = $ *large numbers*." However, we will use $\mu_A(x)$ for the membership function of a fuzzy set A in X, and not just put $A(x)$, to be consistent with most contributions that are relevant to our discussion.

For practical reasons, we will assume throughout this book that all the sets (universes of discourse) are finite as, e.g., $X = \{x_1, \ldots, x_n\}$. In such a case the pair $\{(\mu_A(x), x)\}$ will be denoted by "$\mu_A(x)/x$" which is called a *fuzzy singleton*. Then, the fuzzy set A in X will be written as

$$A = \{(\mu_A(x), x)\} = \{\mu_A(x)/x\} =$$
$$= \mu_A(x_1)/x_1 + \cdots + \mu_A(x_n)/x_n = \sum_{i=1}^{n} \mu_A(x_i)/x_i \qquad (2.4)$$

where "$+$" and "$\sum$" are meant in the set-theoretic sense. By convention, the pairs "$\mu_A(x)/x$" with $\mu_A(x) = 0$ are omitted here and throughout this book.

A conventional (nonfuzzy) set may obviously be written in our notation, e.g., in the case of the set "integer numbers greater than or equal to 3 and less than or equal to 10" as

$$A = 1/3 + 1/4 + 1/5 + 1/6 + 1/7 + 1/8 + 1/9 + 1/10$$

Throughout this book we will use the same notation for both the nonfuzzy and fuzzy sets and their related properties, notions, etc. This should not lead to confusion; otherwise, we will clearly state if a nonfuzzy or fuzzy set, notion, etc. is meant.

The family of all fuzzy sets defined in X is denoted by $\mathcal{A}$; evidently, it also includeso the empty fuzzy set to be defined by (2.5), page 24, i.e. $A = \emptyset$ such that $\mu_A(x) = 0$, for each $x \in X$, and the whole universe of discourse X which may be written in our convention as $X = 1/x_1 + \cdots + 1/x_n$.

It should be noted that the presented definition of a fuzzy set by the membership function of the type $\mu_A : X \longrightarrow [0, 1]$ is the simplest and most straightforward, and serves very well the purpose of allowing for a gradual transition from belongingness to nonbelongingness. However, the same role may be played by a generalized definition of a membership function of the type $\mu_A : X \longrightarrow L$ where L is some (partially) ordered set such as a lattice. This was introduced by Goguen (1967) as an *L-fuzzy set*. We will not, however, consider such a case in this book, and will refer the interested reader to Goguen's (1967) source paper or to one of books discussing this concept in more detail (e.g., Dubois and Prade, 1980).

The concept of a fuzzy set as defined above has been the point of departure for the *theory of fuzzy sets* (or *fuzzy sets theory*) which will be briefly sketched below. We will again follow a more intuitive and less formal presentation, which is better suited to this book.

2.1.2 *Basic definition and properties related to fuzzy sets*

In this section we will briefly list basic definitions and properties related to fuzzy sets illustrating them by simple examples.

A fuzzy set A is said to be *empty*, written $A = \emptyset$, if and only if

$$\mu_A(x) = 0, \qquad \text{for each } x \in X \tag{2.5}$$

and, clearly, since we omit the pairs "$0/x$," an empty fuzzy set is really void in the notation (2.4) as there are no singletons in the right-hand side.

Two fuzzy sets A and B defined in the same universe of discourse X are said to be *equal*, written $A = B$, if and only if

$$\mu_A(x) = \mu_B(x), \qquad \text{for each } x \in X \tag{2.6}$$

Example 2.1 Suppose that $X = \{1, 2, 3\}$ and

$$A = 0.1/1 + 0.5/2 + 1/3$$
$$B = 0.2/1 + 0.5/2 + 1/3$$
$$C = 0.1/1 + 0.5/2 + 1/3$$

then $A = C$ but $A \neq C$ and $B \neq C$. $\square$

It is easy to see that this classic definition of the equality of two fuzzy sets (2.6) is rigid and clear-cut, contradicting in a sense our intuitive feeling that the equality of fuzzy sets should be "softer" (to some degree), and not abrupt.

Namely, two fuzzy sets A and B defined in X are said to be *equal to a degree* $e(A, B) \in [0, 1]$, written $A =_e B$. The degree of equality $e(A, B)$ may be defined in many ways, and the following four definitions (Bandler and Kohout, 1980) are basic.

First, to simplify further notation, let us denote:

Case 1: $A = B$ in the sense of (2.6);

Case 2: $A \neq B$ in the sense of (2.6) and $T = \{x \in X : \mu_A(x) \neq \mu_B(x)\}$;

Case 3: $A \neq B$ in the sense of (2.6) and there exists an $x \in X$ such that

$$\mu_A(x) = 0 \text{ and } \mu_B(x) \neq 0 \quad \text{or} \quad \mu_A(x) \neq 0 \text{ and } \mu_B(x) = 0$$

Case 4: $A \neq B$ in the sense of (2.6) and there exists an $x \in X$ such that

$$\mu_A(x) = 0 \text{ and } \mu_B(x) = 1 \quad \text{or} \quad \mu_A(x) = 1 \text{ and } \mu_B(x) = 0$$

Now, the following *degrees of equality* of two fuzzy sets, A and B, may be defined:

$$e_1(A, B) = \begin{cases} 1 & \text{for case 1} \\ \bigwedge_{x \in T} [\mu_A(x) \wedge \mu_B(x)] & \text{for case 2} \\ 0 & \text{for case 3} \end{cases} \tag{2.7}$$

$$e_2(A, B) = \begin{cases} 1 & \text{for case 1} \\ \bigwedge_{x \in T} [\mu_A(x)/\mu_B(x) - \mu_B(x)/\mu_A(x)] & \text{for case 2} \\ 0 & \text{for case 3} \end{cases} \tag{2.8}$$

$$e_3(A, B) = \begin{cases} 1 & \text{for case 1} \\ 1 - \max_{x \in X} |\mu_A(x) - \mu_B(x)| & \text{for case 2} \\ 0 & \text{for case 4} \end{cases} \tag{2.9}$$

$$e_4(A, B) =$$

$$= \begin{cases} 1 & \text{for case 1} \\ \max_{x \in X}\{[(1 - \mu_A(x)] \wedge [\mu_A(x) \vee (1 - \mu_B(x))]\} & \text{for case 2} \\ 0 & \text{for case 4} \end{cases} \quad (2.10)$$

Now we will proceed to the second basic concept of the containment between two fuzzy sets.

A fuzzy set A defined in X is said to be *contained in* or, alternatively, is said to be a *subset of* a fuzzy set B in X, written $A \subseteq B$, if and only if

$$\mu_A(x) \leq \mu_B(x), \qquad \text{for each } x \in X \qquad (2.11)$$

Example 2.2 Suppose that $X = \{1, 2, 3\}$ and

$$A = 0.1/1 + 0.5/2 + 1/4$$
$$B = 0.1/1 + 0.4/2 + 0.9/3$$
$$C = 0.1/1 + 0.6/2 + 1/3$$

then only $B \subseteq A$. □

The above traditional definition of containment is clearly rigid and clear-cut, and hence there have been proposed many other definitions in which a *degree of containment*, $c(A, B) \in [0, 1]$, has been employed. Once again, Bandler and Kohout's (1980) definitions can be mentioned here, and these basically follow the line of reasoning analogous to that behind the degree of equality, i.e. (2.7)–(2.10), but we will not give more details as the degree of containment will not be used in this book.

Let us proceed now to some further relevant notions.

A fuzzy set A defined in X is said to be *normal* if and only if

$$\max_{x \in X} \mu_A(x) = 1 \qquad (2.12)$$

i.e. when the membership function takes on the value of 1 for at least one argument. Otherwise, the fuzzy set is said to be *subnormal*.

Example 2.3 If $X = \{1, 2, 3\}$, $A = 0.1/1 + 0.5/2 + 1/3$ and $B = 0.1/1 + 0.6/2 + 0.9/3$, then A is normal and B is subnormal. □

It may be clearly desirable to work with normal fuzzy sets since they may provide for some sort of "context-free" comparability. However, in many instances we obtain in the course of an algorithm or procedure subnormal fuzzy sets. They are then often normalized although, unfortunately, this may be misleading in some cases. This important issue will be mentioned later in the book when a need for normalization will occur.

We have now some important concepts of nonfuzzy sets associated with a fuzzy set. The *support* of a fuzzy set A in X, written supp$A$, is the following (nonfuzzy) set:

$$\text{supp}A = \{x \in X : \mu_A(x) > 0\} \qquad (2.13)$$

and, evidently, $\emptyset \subseteq \text{supp}A \subseteq X$.

Example 2.4 If $X = \{1, 2, \ldots, 7\}$ and $A = 0.1/3 + 0.5/4 + 0.8/5 + 1/6$, then $\operatorname{supp} A = \{3, 4, 5, 6\} \subset \{1, 2, \ldots, 7\}$.   □

The *α-cut*, or *α-level set*, of a fuzzy set A in X, written A_α, is defined as the following (nonfuzzy) set:

$$A_\alpha = \{x \in X : \mu_A(x) \geq \alpha\}, \qquad \text{for each } \alpha \in (0, 1] \qquad (2.14)$$

and if "$\geq$" in (2.14) is replaced by "$>$," then we have the *strong α-cut*, or *strong α-level set*, of a fuzzy set A in X. In principle, we will use the α-cuts given by (2.14) if not otherwise specified.

Example 2.5 If $X = \{1, 2, 3, 4\}$ and $A = 0.1/1 + 0.5/2 + 0.8/3 + 1/4$, then we obtain the following α-cuts:

$$A_{0.1} = \{1, 2, 3, 4\} \quad A_{0.5} = \{2, 3, 4\} \quad A_{0.8} = \{3, 4\} \quad A_1 = \{4\}$$

□

The α-cuts play an extremely relevant role in both formal analyses and applications as they make it possible to uniquely replace a fuzzy set by a sequence of nonfuzzy sets. We will widely use them in the following, and the interested reader is referred for details and properties to any book on fuzzy sets theory as, e.g., Dubois and Prade (1980), Klir and Folger (1988) or Klir and Yuan (1995).

Among more important properties of the α-cuts, one can mention

$$\alpha_1 \leq \alpha_2 \iff A_{\alpha_1} \subseteq A_{\alpha_2}, \qquad \text{for each } \alpha_1, \alpha_2 \in (0, 1] \qquad (2.15)$$

The following theorem, called the *representation theorem* (cf. Negoita and Ralescu, 1975), is very relevant both in theoretical analyses and applications.

Theorem 2.1 *Each fuzzy set A in X can be represented as*

$$A = \sum_{\alpha \in (0,1]} \alpha A_\alpha \qquad (2.16)$$

where A_α is an α-cut of A defined as (2.15), "$\sum$" is in the set-theoretic sense, and αA_α denotes the fuzzy set whose degrees of membership are

$$\mu_{\alpha A_\alpha}(x) = \begin{cases} \alpha & \text{for } x \in A_\alpha \\ 0 & \text{otherwise} \end{cases} \qquad (2.17)$$

The expression (2.16) is also called the *resolution identity*.

Example 2.6 Let $X = \{1, 2, \ldots, 10\}$, and $A = 0.1/2 + 0.3/3 + 0.6/4 + 0.8/5 + 1/6 + 0.7/7 + 0.4/8 + 0.2/9$.
Then:

$$A = \sum_{\alpha \in (0,1]} \alpha A_\alpha =$$

$$
\begin{aligned}
=\ & 0.1(1/2 + 1/3 + 1/4 + 1/5 + 1/6 + 1/7 + 1/8 + 1/9) + \\
& + 0.3(1/3 + 1/4 + 1/5 + 1/6 + 1/7 + 1/8 + 1/9) + \\
& + 0.6(1/4 + 1/5 + 1/6 + 1/7) + 0.7(1/5 + 1/6 + 1/7) + \\
& + 0.8(1/5 + 1/6) + 1(1/6) = \\
=\ & 0.1/2 + 0.3/3 + 0.6/4 + 0.8/5 + 1/6 + 0.7/7 + 0.4/8 + 0.2/9
\end{aligned}
$$

$\square$

An important issue, both in theory and application, is to be able to define the *cardinality* of a fuzzy set, i.e. to define how many elements it contains. Unfortunately, this is a difficult problem, and the definitions proposed have been criticized. Two definitions, presumably the most widely used, will be discussed below.

A *nonfuzzy cardinality* of a fuzzy set $A = \mu_A(x_1)/x_1 + \cdots + \mu_A(x_n)/x_n$, the so-called *sigma-count*, denoted $\sum\mathrm{Count}(A)$, is defined as (Zadeh, 1978a, 1983a)

$$
\sum\mathrm{Count}(A) = \sum_{i=1}^{n} \mu_A(x_i) \tag{2.18}
$$

Example 2.7 If $A = 1/x_1 + 0.8/x_2 + 0.6/x_3 + 0.2/x_4 + 0/x_5$, then

$$
\sum\mathrm{Count}(A) = 1 + 0.8 + 0.6 + 0.2 = 2.6
$$

$\square$

The $\sum\mathrm{Count}$ is very simple, and hence is widely used, also in this book (cf. Sections 2.3 and 4.1.5). However, an immediate objection may be that the set is fuzzy but its cardinality is not. A solution in this respect, a "fuzzy cardinality," was proposed by Zadeh (1983a), and is shown below.

Let A be a fuzzy set defined in X, and A_α, for each $\alpha \in (0, 1]$, its α-cuts defined by (2.15). First, Zadeh (1983a) introduces the $FG\mathrm{Count}(A)$ as the fuzzy integer defined as

$$
FG\mathrm{Count}(A) = \{1/0\} \sum_{\alpha \in (0,1]} \alpha/\mathrm{card}(A_\alpha) \tag{2.19}
$$

where "$\sum$" is in the set-theoretic sense as in (2.4), $\mathrm{card}(A_\alpha)$ is the usual number of elements in A_α, and "$1/0$" means the integer number 0.

Equivalently, if A is defined in X such that $\mathrm{card}(X) = n$, then for each nonnegative integer $i = 0, 1, \ldots, n$, we denote

$$
FG\mathrm{Count}(A)_i = \sum_{\alpha \in (0,1]} \{\alpha : \mathrm{card}(A_\alpha) \geq i\} \tag{2.20}
$$

Semantically, $FG\mathrm{Count}(A)_i$ is the truth of the proposition "A contains at least i elements."

Next, Zadeh (1983a) introduces the $FL\mathrm{Count}(A)$ which is defined as the truth of the proposition "A contains at most i elements," i.e.

$$
FL\mathrm{Count}(A) = \neg[FG\mathrm{Count}(A)] - 1 \tag{2.21}
$$

where $\neg[.]$ is the complement to be defined by (2.28), page 29, and "$-$" is the subtraction in the sense of fuzzy numbers (2.72), page 42.

Alternatively, similarly as in (2.20), if A is defined in $X = \{1, 2, \ldots, n\}$, then we can denote

$$FL\text{Count}(A)_i = \sup_{\alpha \in (0,1]} \{\alpha : \text{card}(A_\alpha) \geq n - i\} \qquad (2.22)$$

Note that

$$FG\text{Count}(A)_i = 1 - FL\text{Count}(A)_{i+1}, \qquad \text{for } i = 1, 2, \ldots, n \qquad (2.23)$$

Finally, Zadeh (1983a) introduces the $FE\text{Count}(A)$ as

$$FE\text{Count}(A) = FG\text{Count}(A) \cap FL\text{Count}(A) \qquad (2.24)$$

or, similarly as above,

$$FE\text{Count}(A)_i = FG\text{Count}(A)_i \wedge FL\text{Count}(A)_i, \qquad i = 1, 2, \ldots, n \qquad (2.25)$$

where "$\cap$" and "$\wedge$" denote the intersection of two fuzy sets and the minimum operations, respectively, to be defined by (2.31), page 30.

Example 2.8 For the same fuzzy set as in Example 2.7, i.e. $A = 1/x_1 + 0.8/x_2 + 0.6/x_3 + 0.2/x_4 + 0/x_5$, we obtain

$$FG\text{Count}(A) = 1/0 + 1/1 + 0.8/2 + 0.6/3 + 0.2/4 + 0/5$$
$$FL\text{Count}(A) = 0/0 + 0.2/1 + 0.4/2 + 0.6/3 + 0.2/4 + 0/5$$
$$FE\text{Count}(A) = 0/0 + 0.2/1 + 0.4/2 + 0.6/3 + 0.2/4 + 0/5$$

$\square$

The above classical definitions of cardinality of a fuzzy set are widely employed, in particular the $\sum$Count. For some criticism, and a new definition, we refer the reader to Ralescu (1995), and Wygralak's (1996) book.

An important issue, which will be widely used throughout this book (in virtually all problems related to fuzzy systems under control), is a *distance* between two fuzzy sets. In practice, normalized distances are clearly more interesting. In the literature, and in this book too, the following two basic definitions are used.

Suppose that we have two fuzzy sets, A and B, both defined in $X = \{x_1, \ldots, x_n\}$. Then, we have the following two basic (normalized) distances between A and B:

- the *normalized linear* (Hamming) *distance* defined as

$$l(A, B) = \frac{1}{n} \sum_{i=1}^{n} \mid \mu_A(x_i) - \mu_B(x_i) \mid \qquad (2.26)$$

- the *normalized quadratic* (Euclidean) *distance* defined as

$$q(A, B) = \sqrt{\frac{1}{n} \sum_{i=1}^{n} [\mu_A(x_i) - \mu_B(x_i)]^2} \qquad (2.27)$$

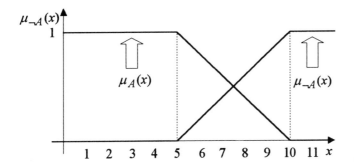

Figure 9   The complement of a fuzzy set

Example 2.9 If $X = \{1, 2, \ldots, 7\}$, $A = 0.7/1 + 0.2/2 + 0.6/4 + 0.5/5 + 1/6$ and $B = 0.2/1 + 0.6/4 + 0.8/5 + 1/7$, then:

$$l(A, B) = 0.37 \qquad q(A, B) = 0.49$$

$\square$

The next important issue to be discussed below is that of basic set-theoretic and algebraic operations on fuzzy sets.

2.1.3 Basic operations on fuzzy sets

As in the conventional (nonfuzzy) set theory, the basic operations in fuzzy set theory are also the complement, intersection and union. We will define them now briefly, in terms of the respective membership functions, mentioning their relations with the negation and the two connectives "or" and "and" acting on the labels that are represented by their respective fuzzy sets.

The *complement* of a fuzzy set A in X, written $\neg A$, is defined as

$$\mu_{\neg A}(x) = 1 - \mu_A(x), \qquad \text{for each } x \in X \tag{2.28}$$

and the complement corresponds to the negation "not."

Example 2.10 If $X = \{1, 2, 3\}$ and $A = 0.1/1 + 0.7/2 + 1/3$, then $\neg A = 0.9/1 + 0.3/2$.
$\square$

The idea of the complement can be visualized as in Figure 9.

In many applications, however, it may be more appropriate to define the complement in a different way. Suppose that the set X is a (finite) real interval $[0, K]$, then the *complement* of a fuzzy set A in X, written $\overline{A}$, may be defined as

$$\mu_{\overline{A}}(x) = \mu_A(K - x), \qquad \text{for each } x \in [0, K] \tag{2.29}$$

and, clearly, this definition may also be used if A is defined as a fuzzy set in the set of integers $\{0, 1, \ldots, K - 1, K\}$ yielding

$$\mu_{\neg A}(x) = \mu_A(K - x), \qquad \text{for each } x \in \{0, 1, \ldots, K - 1, K\} \tag{2.30}$$

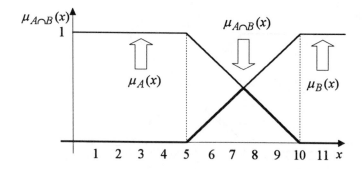

Figure 10   The intersection of two fuzzy sets

Example 2.11 Suppose that A in $\{0, 1, \ldots, 6\}$ is given as $A = 0.1/1 + 0.3/2 + 0.5/3 + 0.7/4 + 0.8/5 + 1/6$.

Then, by (2.29), we obtain $\overline{A} = 1/0 + 0.8/5 + 0.7/4 + 0.5/3 + 0.3/4 + 0.1/5$.   □

It is easy to see that the definition of a complement (corresponding to "not") given by (2.29) is quite intuitively appealing. In the following we will, however, use in principle the "traditional" definition of the complement given as (2.28) with the understanding that the non-standard definition (2.29) may readily be employed, if needed.

The *intersection* of two fuzzy sets A and B in X, written $A \cap B$, is defined as

$$\mu_{A \cap B}(x) = \mu_A(x) \wedge \mu_B(x), \qquad \text{for each } x \in X \tag{2.31}$$

where "$\wedge$" is the minimum operation, i.e. $a \wedge b = \min(a, b)$.

The intersection of two fuzzy sets corresponds to the connective "and."

Example 2.12 If $X = \{1, 2, 3, 4\}$, and $A = 0.2/1 + 0.5/2 + 0.8/3 + 1/4$ and $B = 1/1 + 0.8/2 + 0.5/3 + 0.2/4$, then we obtain by using (2.31) $A \cap B = 0.2/1 + 0.5/2 + 0.5/3 + 0.2/4$.   □

The intersection can be portrayed as in Figure 10 where $\mu_{A \cap B}(x)$ is indicated by a bold arrowhead.

The definition of intersection given above is the original Zadeh's (1965) one which is also by far the most widely used. This definition will also be used in principle throughout this book, though we will mention if and how other definitions of the intersection (to be discussed later) may be employed.

The *union* of two fuzzy sets A and B in X, written $A + B$, is defined as

$$\mu_{A+B}(x) = \mu_A(x) \vee \mu_B(x), \qquad \text{for each } x \in X \tag{2.32}$$

where "$\vee$" is the maximum operation, i.e. $a \vee b = \max(a, b)$.

The union of two fuzzy sets corresponds to the connective "or."

Example 2.13 If $X = \{1, 2, 3, 4\}$, and $A = 0.2/1 + 0.5/2 + 0.8/3 + 1/4$ and $B = 1/1 + 0.8/2 + 0.5/3 + 0.2/4$, then $A + B = 1/1 + 0.8/2 + 0.8/3 + 1/4$.   □

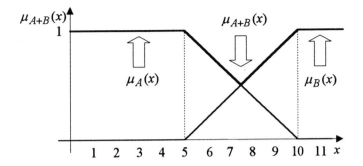

Figure 11   The union of two fuzzy sets

The essence of the union can be portrayed as in Figure 11 in which $\mu_{A+B}(x)$ is shown by a bold arrowhead.

The definition of the union given above is the original Zadeh's (1965) one which is also by far the most widely used, also throughout this book, though we will mention if and how other definitions of the union (to be discussed later) may be employed.

It is clear that though the intersection and union of fuzzy sets defined as above are quite well founded, intuitively appealing, and have proved useful in many applications, it might be useful to have a more general definition of these operations which would encompass the classic definitions but go beyond them. Promising, widely accepted options here are the so-called t-norms and s-norms (t-conorms).

A *t-norm* is defined as

$$t : [0, 1] \times [0, 1] \longrightarrow [0, 1] \tag{2.33}$$

such that, for each $a, b, c \in [0, 1]$:

1. it has 1 as the unit element, i.e.

$$t(a, 1) = a$$

2. it is monotone, i.e.

$$a \leq b \Longrightarrow t(a, c) \leq t(b, c)$$

3. it is commutative, i.e.

$$t(a, b) = t(b, a)$$

4. it is associative, i.e.

$$t[a, t(b, c)] = t[t(a, b), c]$$

Evidently, a t-norm is monotone non-decreasing in both arguments, and $t(a, 0) = 0$.

We will use interchangeably the notations $t(a, b)$ and atb; moreover, we will denote $a_1 t a_2 t \ldots t a_{k-1} t a_k = t(a_1, \ldots, t(a_{k-1}, a_k) \ldots) = t_{i=1}^{k} a_i$.

Some more relevant examples of t-norms are:

- the minimum

$$t(a, b) = a \wedge b = \min(a, b) \tag{2.34}$$

which is the most widely used, also in our context,

- the algebraic product

$$t(a,b) = a \cdot b \qquad (2.35)$$

- the so-called Lukasiewicz t-norm

$$t(a,b) = \max(0, a + b - 1) \qquad (2.36)$$

and note that we have written above both $t(a,b)$ and atb.

An s-norm (or a t-conorm) is defined as

$$s : [0,1,] \times [0,1] \longrightarrow [0,1] \qquad (2.37)$$

such that, for each $a, b, c \in [0,1]$:

1. it has 0 as the unit element, i.e.

$$s(a,0) = a$$

2. it is monotone, i.e.

$$a \leq b \Longrightarrow s(a,c) \leq s(b,c)$$

3. it is commutative, i.e.

$$s(a,b) = s(b,a)$$

4. it is associative, i.e.

$$s[a, s(b,c)] = s[s(a,b), c]$$

We will use interchangeably the notations $s(a,b)$ and asb; moreover, we will denote $a_1 s a_2 s \ldots s a_{k-1} s a_k = s(a_1, \ldots, s(a_{k-1}, a_k) \ldots) = s_{i=1}^k a_i$.

Some more relevant examples of s-norms are:

- the maximum

$$s(a,b) = a \vee b = \max(a,b) \qquad (2.38)$$

which is the most widely used, also in our context,

- the probabilistic product

$$s(a,b) = a + b - ab \qquad (2.39)$$

- the so-called Lukasiewicz s-norm

$$s(a,b) = \min(a + b, 1) \qquad (2.40)$$

Note that a t-norms is $dual$ to an s-norms in that

$$s(a,b) = 1 - t(1 - a, 1 - b) \qquad (2.41)$$

The above examples of t-norms and s-norms are simple enough and provide much freedom in the choice of an "adequate" operation. However, in many cases it may be expedient to have some parameterized definitions which – by an appropriate choice of a parameter – would allow an even broader characterization of operations representing the intersection and union. We will give three of them below.

The *parameterized Weber family* of t-norms and s-norms (Weber, 1983) is defined as, for $\lambda \in (-1, \infty)$:

- for t-norms

$$t_\lambda^W(a,b) = \max\left(\frac{a+b-1+\lambda ab}{1+\lambda}, 0\right) \tag{2.42}$$

- for s-norms

$$s_\lambda^W(a,b) = \min\left(a+b-\frac{\lambda ab}{1+\lambda}, 0\right) \tag{2.43}$$

Note that for $\lambda = 1$ we have the Lukasiewicz t-norm and s-norm, i.e. $t(a,b) = a+b-1$ and $s(a,b) = \min(a+b,a)$, for $\lambda \longrightarrow \infty$ we have $t^W(a,b) = ab$ and $s^W(a,b) = a+b-ab$, and for $\lambda \longrightarrow -1$, we have the so-called *drastic product*

$$t_1 = \begin{cases} a & \text{if } b = 1 \\ b & \text{if } a = 1 \\ 0 & \text{otherwise} \end{cases} \tag{2.44}$$

and the so-called *drastic sum*

$$s_1 = \begin{cases} a & \text{if } b = 0 \\ b & \text{if } a = 0 \\ 1 & \text{otherwise} \end{cases} \tag{2.45}$$

The *Hamacher family* of t-norms and s-norms (Hamacher, 1978) is defined as, for $\gamma > 0$:

- for t-norms

$$t_\gamma^H(a,b) = \frac{ab}{\gamma + (1-\gamma)(a+b-ab)} \tag{2.46}$$

- for s-norms

$$s_\gamma^H(a,b) = \frac{a+b-ab-(1-\gamma)ab}{1-(1-\gamma)ab} \tag{2.47}$$

The *Yager family* of t-norms and s-norms (Yager, 1980) is defined as, for $p > 0$:

- for t-norms

$$t_p^Y(a,b) = 1 - \min\{[(1-a)^p + (1-b)^p]^{1/p}, 1\} \tag{2.48}$$

- for s-norms

$$s_p^Y(a,b) = \min\{(a^p + b^p)^{1/p}, 1\} \tag{2.49}$$

Note that for $p \longrightarrow \infty$ we obtain $\min(a,b)$ and $\max(a,b)$, and for $p \longrightarrow 0$, we obtain the drastic product (2.44) and drastic sum (2.45), respectively.

The t-norms and s-norms are very useful, and commonly employed operators in fuzzy sets theory, and will also be used in our further discussions though, basically, our discussions will proceed for the minimum and maximum with the understandings that these may be replaced by other suitable operations, notably t-norms.

As to some other operations on fuzzy sets that may be of use in this book, one should also mention the following.

The *product of a scalar $a \in R$ and a fuzzy set A in X*, written aA, is defined as

$$\mu_{aA}(x) = a\mu_A(x), \qquad \text{for each } x \in X \tag{2.50}$$

where, by necessity, $0 \leq a \leq 1/\mu_A(x)$, for each $x \in X$.

The k-th power of a fuzzy set A in X, written A^k, is defined as

$$\mu_{A^k}(x) = [\mu_A(x)]^k, \qquad \text{for each } x \in X \tag{2.51}$$

where $k \in R$ and, evidently, $0 \leq [\mu_A(x)]^k \leq 1$.

These are the basic operations on fuzzy sets which will be used in this book. Note that the very purpose of these operations is to provide some *aggregation*. In Section 2.3 we will discuss some other aggregation operations, with different origins and motivations.

An important issue is the *adequacy* of the operations on fuzzy sets. The original Zadeh's (1965a) definitions of the complement (2.28), intersection (2.31 and union (2.32), pages 29–30, have much intuitive appeal, and have been very widely used. However, the question whether they do reflect the real human perception of their essence, i.e. whether they really reflect the semantics of "not," "and" and "or," has been asked since the very beginning. Diverse approaches have been used to find and justify a particular definition. These approaches may be classified as:

- *intuitive*, exemplified by, e.g., the original Zadeh's (1965a, 1973, ...) works in which it is shown by a rational argument that the operations defined are proper,
- *axiomatic*, whose line of reasoning is to assume some set of plausible conditions to be fulfilled by the particular operations, and then to show, using some analytic tools, that definitions assumed are the only possible ones; this may be exemplified by Bellman and Giertz (1973),
- *experimental*, whose essence is to devise some psychological tests for a group of selected individuals, and then use the responses to find which operation is best justified; this may be exemplified by Zimmermann and Zysno (1980).

The issue of how adequate the definitions of operations on fuzzy sets are is extremely important, in particular for applications. Unfortunately, the results of many investigations (in particular in the third, experimental, direction) are not conclusive enough. We will not therefore deal here with this important issue, and refer the interested reader to a the literature exemplified by Zimmermann (1987, 1996). Some aspects will only be mentioned in Section 3.1 where we will look at this problem from the perspective of control problems considered in this book, and in Section 8.1 where the adequacy of operation will be somehow the result of the analyst's (or decision maker's) intention and attitude. In this section we have listed a bunch of various operations on fuzzy sets. The main reason has been to provide the interested reader with a comprehensive summary, or reference, of what has been done (and is being done) in this respect. It should be noted that the derivation of new, improved, definitions of operations is still one of main areas of research in fuzzy sets theory.

However, this abundance is often an obstacle to a more pragmatically inclined user to whom the real merit of an operation is its usefulness (adequacy) and not nice formal properties.

In this book, due to some optimization-related aspects involved, the choice of operations is limited to simpler ones as, e.g., minimum and product for the intersection or maximum for the union. These will give us, on the one hand, adequate means to reflect various perceptions of the very essence of the problem. On the other hand, they will be simple enough for optimization-type analyses.

Now we will proceed to the next important element of fuzzy sets theory which will be used here, i.e. fuzzy relations.

2.1.4 Fuzzy relations

The concept of a relation is crucial for virtually all areas of mathematics and its applications, and the same holds true for fuzzy relations in fuzzy sets theory and its applications.

A *fuzzy relation* R between two (nonfuzzy) sets $X = \{x\}$ and $Y = \{y\}$ is defined as a fuzzy set in the Cartesian product $X \times Y$, i.e.

$$R = \{(\mu_R(x,y),(x,y))\} =$$
$$= \{\mu_R(x,y)/(x,y)\}, \qquad \text{for each } (x,y) \in X \times Y \qquad (2.52)$$

where $\mu_R(x,y) : X \times Y \longrightarrow [0,1]$ is the membership function of the fuzzy relation R, and $\mu_R(x,y) \in [0,1]$ gives the degree to which the elements $x \in X$ and $y \in Y$ are in relation R to each other.

Since this basic type of a fuzzy relation is defined in the Cartesian product of two sets, X and Y, such a fuzzy relation is sometimes called a binary fuzzy relation. One can well imagine a fuzzy relation defined in the Cartesian product of k sets, i.e. in $X_1 \times \cdots \times X_k$, and this might be called a k-ary fuzzy relation. In this perspective, a fuzzy set is a unary fuzzy relation.

Example 2.14 If $X = \{\text{horse}, \text{donkey}\}$ and $Y = \{\text{mule}, \text{cow}\}$, then the fuzzy relation R labelled "similarity" may be exemplified by

$$R = \text{"similarity"} =$$
$$= \quad 0.8/(\text{horse, mule}) + 0.4/(\text{horse, cow}) +$$
$$+ \, 0.9/(\text{donkey, mule}) + 0.2/(\text{donkey, cow})$$

which is to be read as follows: the horse and the mule are similar (with respect to "our own" subjective aspects!) to degree 0.8, i.e. to a very high extent, the horse and the cow are similar to degree 0.4, i.e. to quite a low extent, etc. □

It may easily be seen that the concept of a fuzzy relation makes it possible to express a partial (imprecise) relationship between elements of some sets, as opposed to a precise and abrupt one in the case of a nonfuzzy relation in which any two elements can either be or not be related. In the fuzzy relation we have a strength of relation, from 1 for being fully in relation to 0 for not being in relation at all, through all intermediate values.

Note that a fuzzy relation R in $X \times Y$ for X and Y of sufficiently low dimensionality may be conveniently represented in matrix form. For instance, the fuzzy relation $R = \text{"similarity"}$ given in Example 2.14 may be equivalently shown in the following matrix form:

$R = \text{"similarity"} =$		$y = \text{mule}$	cow
	$x = \text{horse}$	0.8	0.4
	donkey	0.9	0.2

As we will see later, such a matrix form is very convenient for the representation and handling of fuzzy relations.

Since a fuzzy relation is a fuzzy set, all the definitions, properties, operations, etc. on fuzzy sets presented in Sections 2.1.1–2.1.3 hold as well. So, we will concentrate below on some concepts which are more specific for fuzzy relations, and are also relevant in the following discussion.

The *max–min composition* of two fuzzy relations R in $X \times Y$ and S in $Y \times Z$, written $R \circ_{max-min} S$ is defined as a fuzzy relation in $X \times Z$ such that

$$\mu_{R \circ_{max-min} S}(x, y) =$$
$$= \max_{y \in Y}[\mu_R(x, y) \wedge \mu_S(y, z)], \qquad \text{for each } x \in X, z \in Z \qquad (2.53)$$

and since this type of composition will be used throughout this book, if not otherwise specified, then it will be briefly denoted as $R \circ S$.

Example 2.15 If $X = \{1, 2\}$, $Y = \{1, 2, 3\}$ and $Z = \{1, 2, 3, 4\}$, and the fuzzy relations R and S are as below. Its resulting max–min composition, $R \circ S$, is then:

$$R \circ S =$$

$$=\quad \begin{array}{c|ccc} & y = 1 & 2 & 3 \\ \hline x = 1 & 0.3 & 0.8 & 1 \\ 2 & 0.9 & 0.7 & 0.4 \end{array} \quad \circ \quad \begin{array}{c|cccc} & z = 1 & 2 & 3 & 4 \\ \hline y = 1 & 0.7 & 0.6 & 0.4 & 0.1 \\ 2 & 0.4 & 1 & 0.7 & 0.2 \\ 3 & 0.5 & 0.9 & 0.6 & 0.8 \end{array} \quad =$$

$$=\quad \begin{array}{c|cccc} & z = 1 & 2 & 3 & 4 \\ \hline x = 1 & 0.5 & 0.9 & 0.7 & 0.8 \\ 2 & 0.7 & 0.7 & 0.7 & 0.4 \end{array}$$

For illustration, let us show the calculation of an element of the fuzzy relation $R \circ S$, say the one corresponding to $x = 2$ and $z = 3$:

$$\mu_{R \circ S}(2, 3) = \max_{y \in \{1,2,3\}}[\mu_R(2, y) \wedge \mu_S(y, 3)] =$$
$$= [\mu_R(2, 1) \wedge \mu_S(1, 2)] \vee [\mu_R(2, 2) \wedge \mu_S(2, 3)] \vee \vee [\mu_R(2, 3) \wedge \mu_S(3, 2)] =$$
$$= (0.9 \wedge 0.4) \vee (0.7 \wedge 0.7) \vee (0.4 \wedge 0.8) = 0.4 \vee 0.7 \vee 0.4 = 0.7$$

□

This max–min composition is the original Zadeh's definition (cf. Zadeh, 1973), and is certainly the most widely used. However, if we recognize that the two basic operations involved in the definition of their composition, i.e. "min" ($\wedge$) and "max" ($\vee$) are just specific examples of the t-norm and s-norm (t-conorm) discussed in Section 2.1.3, then one can well define a much more general type of composition, i.e. the following one.

The *s–t norm composition* of two fuzzy relations R in $X \times Y$ and S in $Y \times Z$, written $R \circ_{s-t} S$, is defined as a fuzzy relation in $X \times Z$ such that

$$\mu_{R \circ_{s-t} S}(x, z) = s_{y \in Y}[\mu_R(x, y) \, t \, \mu_S(y, z)], \qquad \text{for each } x \in X, z \in Z \qquad (2.54)$$

Among such s–t compositions the max–product composition is certainly one of the more relevant.

The *max–product composition* of two fuzzy relations R in $X \times Y$ and S in $Y \times Z$, written $R \circ_{\text{max–prod}} S$, is defined as a fuzzy relation in $X \times Z$ such that

$$\mu_{R \circ_{\text{max–prod}} S}(x, z) =$$
$$= \max_{y \in Y}[\mu_R(x, y) \cdot \mu_S(y, z)], \qquad \text{for each } x \in X, z \in Z \qquad (2.55)$$

For other types of composition we refer the interested reader to, e.g., Kaufmann (1975), or any other book on fuzzy sets theory.

Note that, as we have already mentioned, we will basically use the max–min composition, and this will be simply denoted by "$\circ$."

Finally, let us mention two concepts concerning the fuzzy sets that are related to fuzzy relations.

The *Cartesian product* of two fuzzy sets A in X and B in Y, written $A \times B$, is defined as a fuzzy set in $X \times Y$ such that

$$\mu_{A \times B}(x, y) = [\mu_A(x) \wedge \mu_B(y)], \qquad \text{for each } x \in X, y \in Y \qquad (2.56)$$

A fuzzy relation (set) R in $X \times Y \times \cdots \times Z$ is said to be *decomposable* if and only if it can be represented as

$$\mu_R(x, y, \ldots, z) = \mu_{R_x}(x) \wedge \mu_{R_y}(y) \wedge \ldots$$
$$\ldots \wedge \mu_{R_z}(z), \qquad \text{for each } x \in X, y \in Y, \ldots, z \in Z \qquad (2.57)$$

where $\mu_{R_x}(x), \mu_{R_y}(y), \ldots, \mu_{R_z}(z)$ are projections of the fuzzy relation $\mu_R(x, y)$ on X, Y, ..., Z, respectively, defined as

$$\mu_{R_x}(x) = \sup_{\{y, \ldots, z\} \in Y \times Z} \mu_R(x, y, \ldots, z), \qquad \text{for each } x \in X \qquad (2.58)$$

2.1.5 *Linguistic variable, fuzzy conditional statement, and compositional rule of inference*

In this subsection we will briefly present the essence of Zadeh's (1973) *linguistic approach* to the analysis of systems. This approach has actually triggered the whole area of fuzzy (logic) control, i.e. a descriptive approach to fuzzy control, as we have mentioned in Section 1.1, which have resulted in so many real-world applications in diverse areas. In this book, i.e. in prescriptive approaches to fuzzy control, we will also often use elements of this approach in our future analyses but as a supporting tool, not the main one.

Basically, the rationale behind Zadeh's (1973) linguistic approach to the analysis of complex systems and decision (and control) processes is that the basic element is a *linguistic variable* exemplified by "temperature" which takes on as their values not conventional numerical values as, e.g., 150°, but linguistic values such as "high," "low," etc. that are in turn equated semantically with some fuzzy sets. Notice that such linguistic values are common in human discourse as natural language is the only fully natural human means of communication. Clearly, one can then form more complex linguistic expressions such as "not very low and not very high," "more or less medium," etc. by using some connectives (e.g., and, or, ...), modifiers (e.g., more or less, very,

...), etc. More information about these modifiers, and other elements that may be used, will be given in Section 2.2 in which fuzzy logic will be outlined. These more complex terms are also equated semantically with some fuzzy sets which are derived by performing first a syntactic analysis of a term, and then apply corresponding semantic rules. For details we will refer the reader to, e.g., Kaufmann (1975), Zadeh (1973, 1978a) or practically any book on fuzzy sets theory.

To represent a relationship between linguistic variables, *fuzzy conditional statements* are employed. For instance, if we have two linguistic variables, a primary one L and a secondary one K, such that the value of L is a fuzzy set A in X, and the value of K is a fuzzy set B in Y, then a relationship between L and K, in terms of their values A and B, respectively, may be written as

$$\text{IF } L = A \text{ THEN } K = B \tag{2.59}$$

or, in abbreviated form,

$$\text{IF } A \text{ THEN } B \tag{2.60}$$

This fuzzy conditional statement is now assumed to be equivalent to

$$\text{IF } A \text{ THEN } B = A \times B \tag{2.61}$$

i.e. to the Cartesian product (2.56) of the two fuzzy sets A and B which is in turn a fuzzy relation in $X \times Y$.

Note that this is what Mamdani (1974) used in his controller, and which is often called "Mamdani's implication" (though it is not an implication!). It is clear that this definition is the simplest one, and we can devise more sophisticated ones. In particular, if we view the fuzzy conditional statement as related to a "fuzzy implication," we could define the fuzzy conditional statement "IF A THEN B" in other ways as, for example, by employing various definitions of implication (2.90)–(2.95), page 47.

It is easy to see that the fuzzy conditional statement (2.60) may be extended to account for multiple values of A and B obtaining, if we use (2.61),

$$\text{IF } A_1 \text{ THEN } B_1 \text{ ELSE } \ldots \text{ ELSE IF } A_n \text{ THEN } B_n =$$
$$= A_1 \times B_1 + \cdots + A_n \times B_n \tag{2.62}$$

where A_i's are fuzzy sets in X and B_i's are fuzzy sets in Y, $i = 1, \ldots, n$.

And again, similarly as for (2.61), using different definitions of implication as, e.g., (2.90)–(2.95), page 47, we can use different forms of (2.62).

In the above two basic types of fuzzy conditional statements, (2.60) and (2.62), we specify what happens if the primary variable takes on some value. In many cases it is, however, also relevant what happens otherwise (i.e. when it does not take that value). In such a case the fuzzy conditional statement (2.60) becomes

$$\text{IF } A \text{ THEN } B \text{ ELSE } C \tag{2.63}$$

and it is represented as

$$\text{IF } A \text{ THEN } B \text{ ELSE } C = A \times B + \neg A \times C \tag{2.64}$$

Evidently, one can generalize this basic definition (2.64) by employing different definitions of the implication as, e.g., (2.90)–(2.95), page 47, similarly as for (2.61) and (2.62).

We have therefore some tool to represent a relation between a primary and secondary variable that is represented by some fuzzy relation. An immediate question is then: if the primary takes on some fuzzy value, say A' in X, and we have a relation between the values of the primary and secondary variables "IF A THEN B," then what will be the implied (inferred) value of the secondary variable B'?

This may be represented by the inference scheme

$$A'$$
$$\text{IF } A \text{ THEN } B \qquad\qquad\qquad (2.65)$$
$$\overline{}$$
$$B' =?$$

and, what is of prime importance, the fuzzy values A' and A need not be the same.

The answer to the question in (2.65) is provided by the *compositional rule of inference* which states that if R in $X \times Y$ is a fuzzy relation representing a dependence between a primary and secondary variable, represented by a fuzzy conditional statement, and the primary variable takes on a fuzzy value A' in X, then the implied fuzzy value of the secondary variable B' in Y is given by the $(s - -t)$ composition (2.54), page 36, of A' and R, i.e.

$$B' = A' \circ R \qquad\qquad\qquad (2.66)$$

which, if we assume the most popular max–min composition (2.53), page 36, becomes

$$\mu_{B'}(y) = \max_{x \in X}[\mu_A(x) \wedge \mu_R(x,y)], \qquad \text{for each } y \in Y \qquad (2.67)$$

and note that here the fuzzy set A' is considered to be a unary fuzzy relation as mentioned in Section 2.1.4.

Evidently, similarly as in the case of all compositions of fuzzy relations dealt with in Section 2.1.4, we can replace the above max–min compositional rule of inference by another, as, e.g., the $s - t$-norm , max-product, etc. rules.

Example 2.16 Suppose that $X = \{x\} = \{1, 2, 3\}$ and $Y = \{y\} = \{1, 2, 3, 4\}$, and the fuzzy conditional statement representing the dependence between the primary and secondary linguistic variables L and K, respectively, is

$$\text{IF } L = \text{"low" THEN } K = \text{"high"} = (\text{"low"}) \circ (\text{"high"})$$

where

$$\text{"low"} = 1/1 + 0.7/2 + 0.3/3 \quad \text{"high"} = 0.2/1 + 0.5/2 + 0.8/3 + 1/4$$

which is equivalent to the following fuzzy relation:

$$R = (\text{"low"}) \circ (\text{"high"}) =$$

	$y = 1$	2	3	4
$x = 1$	0.2	0.5	0.8	1
2	0.2	0.5	0.7	0.7
3	0.2	0.3	0.3	0.3

If now $L =$ "medium" $= 0.5/1 + 1/2 + 0.5/3$, then the K induced is given by

$$
\begin{aligned}
K &= (\text{"medium"}) \circ R = \\
&= \max_{x \in \{1,2,3\}} [\mu_L(x) \wedge \mu_R(x,y)] = 0.2/1 + 0.5/2 + 0.7/3 + 0.7/4
\end{aligned}
$$

$\square$

The fuzzy conditional statements may be used to represent simple dependencies and relations between linguistic variables. For more complicated dependencies and relations, fuzzy algorithms may be used (cf. Zadeh, 1973) which will not be dealt with here.

2.1.6 The extension principle

Now we will briefly present the essence of Zadeh's classic *extension principle* (cf. Zadeh, 1973) which is one of the most important and powerful tools in fuzzy sets theory. The extension principle addresses the following issue which is fundamental from the point of view of both the theory and application of fuzzy sets:

> If there is some relationship (e.g., a function) between "conventional" (nonfuzzy) entities (e.g., variables taking on nonfuzzy values, then what is its equivalent relationship between fuzzy entities (e.g., variables taking on fuzzy values)?

The extension principle makes it therefore possible, for instance, to extend some known conventional models, algorithms, etc. involving nonfuzzy variables to the case of fuzzy variables. One may well imagine that in such a way one the huge apparatus of science and technology (models, algorithms, procedures, etc.) may be readily available for use with fuzzy variables. Unfortunately, though this might eventually be possible in principle, its practical implementation may be doubtful from the point of view of efficiency, as it is the case with any general tool. One may do much better by devising anew a specialized fuzzy model, algorithm or procedure, as we will see later in this book. In spite of this, however, the extension principle is an extremely relevant tool indeed.

To formally present the extension principle, let $A_1, \ldots, A_n$ be fuzzy sets in $X_1 = \{x_1\}, \ldots, X_n = \{x_n\}$, respectively, and

$$
f : X_1 \times \cdots \times X_n \longrightarrow Y \tag{2.68}
$$

be some (nonfuzzy) function such that $y = f(x_1, \ldots, x_n)$.

Then, according to the *extension principle*, the fuzzy set B in $Y = \{y\}$ induced by the fuzzy sets $A_1, \ldots, A_n$ via the function f is

$$
\mu_B(y) = \max_{(x_1,\ldots,x_n) \in X_1 \times \ldots \times X_n : y = f(x_1,\ldots,x_n)} \bigwedge_{i=1}^{n} \mu_{A_i}(x_i) \tag{2.69}
$$

Example 2.17 Suppose that: $X_1 = \{1, 2, 3\}$, $X_2 = \{1, 2, 3, 4\}$, and f represents the addition, i.e. $y = x_1 + x_2$ and

$$
A_1 = 0.1/1 + 0.6/2 + 1/3 \qquad A_2 = 0.6/1 + 1/2 + 0.5/3 + 0.1/4
$$

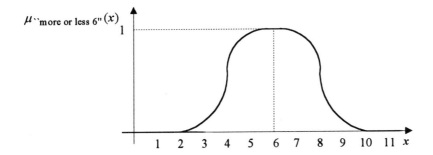

Figure 12   The membership function of a fuzzy number "more or less 6"

then
$$B = A_1 + A_2 = 0.1/2 + 0.6/3 + 0.6/4 + 1/5 + 0.5/6 + 0.1/7$$

where "+" is used here in both the arithmetic (the sum of real and fuzzy numbers – cf. Section 2.1.7) and set-theoretic senses which should not lead to confusion.   □

Equivalently, the extension principle (2.69) may be written in terms of the respective α-cuts defined by (2.14), page 26. Namely, suppose for simplicity that we only have $f : X \longrightarrow Y$, $X = \{x\}$ and $Y = \{y\}$, instead of (2.68), and A_α, for each $\alpha \in (0, 1]$, are α-cuts of the fuzzy set A in X. Then, the fuzzy set B in Y, induced by A via the extension principle is given as

$$B = f(A) = f\left(\sum_{\alpha \in (0,1]} \alpha \cdot A_\alpha \right) = \sum_\alpha \alpha f(A_\alpha) \tag{2.70}$$

and it is clear that the representation theorem (Theorem 2.16, page 26) has been employed.

Evidently, the extension principle may be used not only for the arithmetic operations, making it possible to handle fuzzy numbers as we will see in the next section, but for other operations as, e.g., the maximum or minimum, making it possible to, say, compare the fuzzy numbers (cf. Kaufmann and Gupta, 1985).

2.1.7   Fuzzy numbers

The same fundamental role as nonfuzzy (real, integer, ...) numbers play in conventional models, the fuzzy numbers play in fuzzy models.

A *fuzzy number* is defined as a fuzzy set in R, the real line. Usually, but not always, it is assumed to be a normal and convex fuzzy set. For example, the membership function of a fuzzy number "more or less 6" may be as shown in Figure 12, i.e. as a bell-shaped function. Notice that the membership function of this fuzzy number is the same as that in Figure 7, page 21.

For our purposes operations on fuzzy numbers are the most relevant. It is easy to see, just from the last example in the previous section (Example 2.17), that their definitions may readily be obtained by applying the extension principle (2.69) to their respective nonfuzzy equivalents (defined for the real numbers).

Suppose therefore that A and B are two fuzzy numbers in $R = \{x\}$ characterized by their membership functions $\mu_A(x)$ and $\mu_B(x)$, respectively. Then, the extension principle yields the following definitions of the four basic *arithmetic operations on fuzzy numbers*:

- the *addition*

$$\mu_{A+B}(z) = \max_{x+y=z} [\mu_A(x) \wedge \mu_B(y)], \qquad \text{for each } z \in R \qquad (2.71)$$

- the *subtraction*

$$\mu_{A-B}(z) = \max_{x-y=z} [\mu_A(x) \wedge \mu_B(y)], \qquad \text{for each } z \in R \qquad (2.72)$$

- the *multiplication*

$$\mu_{A \cdot B}(z) = \max_{x \cdot y=z} [\mu_A(x) \wedge \mu_B(y)], \qquad \text{for each } z \in R \qquad (2.73)$$

- the *division*

$$\mu_{A/B}(z) = \max_{x/y=z, y \neq 0} [\mu_A(x) \wedge \mu_B(y)], \qquad \text{for each } z \in R \qquad (2.74)$$

In some applications, the following one-argument operations on fuzzy numbers may also be of use:

- the *opposite* of a fuzzy number

$$\mu_{-A}(x) = \mu_A(-x), \qquad \text{for each } x \in R \qquad (2.75)$$

- the *inverse* of a fuzzy number

$$\mu_{A^{-1}}(x) = \mu_A(\frac{1}{x}), \qquad \text{for each } x \in R \setminus \{0\} \qquad (2.76)$$

In practice, however, such a general definition of fuzzy numbers and operations on them is seldom used. Normally, a further simplification is made, namely the fuzzy numbers are assumed to be *triangular* and, eventually, *trapezoid* fuzzy numbers whose membership functions are sketched in Figure 13.

Basically, we will use those simpler triangular and trapezoid fuzzy numbers (or, more generally, fuzzy sets) throughout this book (cf. Section 8.1).

For triangular and trapezoid fuzzy numbers, for the four basic operations, i.e. the addition, subtraction, multiplication and division, special formulas can be devised whose calculation is simpler that of (2.71)–(2.74). Moreover, a crucial problem of comparison of two fuzzy numbers may also be simplified. Since these issues are not directly relevant for our discussion, we will refer the reader to a rich literature exemplified by Dubois and Prade (1980), Kaufmann and Gupta (1985), Kacprzyk (1987a), Klir and Folger (1988), Klir and Yuan (1995), Zimmermann (1996) etc.

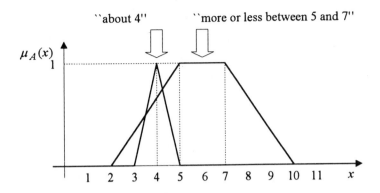

Figure 13 Triangular ("*about* 4") and trapezoid ("*more or less* between 5 and 7") fuzzy numbers

2.1.8 Fuzzy events and their probabilities

In this book we differentiate between fuzziness and randomness, which are two kinds of *information imperfection*. Fuzziness, on the one hand, is meant to concern entities and relations which are not crisply defined, i.e. which may adequately be represented by classes with a gradual transition between the elements belonging and not belonging to a class. Randomness, on the other hand, concerns situations in which the event is well defined but its occurrence is uncertain.

However, it is easy to imagine that in practice and in everyday discourse there is an abundance of situations in which there jointly occur fuzziness and randomness. For instance, when we ask about the probability of *cold weather* tomorrow or about the probability of a *high inflation* in the next year, we have an imprecise (fuzzy) event – cold weather and high inflation, respectively. To be able to formally deal with such problems, we need a concept of a *fuzzy event*, and a *probability of a fuzzy event*.

2.1.8.1 Zadeh's approach

The first approach in this respect is due to Zadeh (1968b). Its point of departure is the concept of a *fuzzy event* which is simply a fuzzy set A in $X = \{x\} = \{x_1, \ldots, x_n\}$ whose membership function is Borel measurable. We assume that the probabilities of the (nonfuzzy) elementary events $x_1, \ldots, x_n \in X$ are known and equal to $p(x_1), \ldots, p(x_n) \in [0, 1]$, respectively, with $p(x_1) + \cdots + p(x_n) = 1$.

As to some more important concepts related to fuzzy events, the following may be stated.

Two fuzzy events A and B in X are *independent* if and only if

$$p(AB) = p(A)p(B) \tag{2.77}$$

The *conditional probability* of a fuzzy event A in X with respect to a fuzzy event B in X is denoted $p(A \mid B)$ and defined as

$$p(A \mid B) = \frac{p(AB)}{p(B)}, \qquad p(B) > 0 \tag{2.78}$$

and if the fuzzy events A and A are independent, then

$$p(A \mid B) = p(A) \tag{2.79}$$

Observe that both of the above concepts are analogous to their nonfuzzy counterparts.

The (*nonfuzzy*) *probability of a fuzzy event* A in $X = \{x_1, \ldots, x_n\}$ is denoted $p(A)$ and defined as

$$p(A) = \sum_{i=1}^{n} \mu_A(x_i) p(x_i) \tag{2.80}$$

i.e. as the expected value of the membership function of A, $\mu_A(x)$.

Example 2.18 Suppose that $X = \{1, 2, \ldots, 5\}$, $p(x_1) = 0.1$, $p(x_2) = 0.1$, $p(x_3) = 0.1$, $p(x_4) = 0.3$, $p(x_5) = 0.4$, and $A = 0.1/2 + 0.5/3 + 0.7/4 + 0.9/5$.

Then

$$p(A) = 0.1 \times 0.1 + 0.1 \times 0.5 + 0.3 \times 0.7 + 0.4 \times 0.9 = 0.73$$

$\square$

See the following main properties of the above Zadeh's (1968b) (nonfuzzy) probability of a fuzzy event:

$$p(\emptyset) = 0$$

$$p(\neg A) = 1 - p(A)$$

$$p(A + B) = p(A) + p(B) - p(A \cap B)$$

$$p\left(\sum_{i=1}^{r} A_i\right) =$$

$$= \sum_{i=1}^{r} p(A_i) - \sum_{j=1}^{r} \sum_{k=1, k<j}^{r} p(A_j \cap A_k) +$$

$$+ \sum_{j=1}^{r} \sum_{k=1, k<j}^{r} \sum_{l=1, l<k}^{r} p(A_j \cap A_k \cap A_l) + \cdots$$

$$\cdots + (-1)^{r+1} p(A_1 \cap A_2 \cap \ldots \cap A_r)$$

so it does make sense to term the expression (2.80) a "probability."

Zadeh's (1968b) classic definition of a (nonfuzzy) probability of a fuzzy event is by far the most popular and most widely used, and will also be employed in principle, if not otherwise indicated, throughout this book.

Note that in Zadeh's (1968b) approach outlined above, the event is fuzzy but its probability is nonfuzzy, i.e. a real number from the unit interval. This may be viewed as counterintuitive, and hence some concepts of a fuzzy probability have been proposed later. We will present some of them below, notably those which will be employed in this book.

2.1.8.2   Yager's approach

In Yager's (1979) approach, the fuzzy event is again defined as a fuzzy set A in $X = \{x_1, \ldots, x_n\}$ whose membership function $\mu_A(x)$ is Borel measurable. Moreover, we again assume that $p(x_1), \ldots, p(x_n) \in [0, 1]$ are known, and $p(x_1) + \cdots + p(x_n) = 1$.

The *(fuzzy) probability of a fuzzy event A* in $X = \{x_1, \ldots, x_n\}$ is denoted $P(A)$ and defined as the following fuzzy set in $[0, 1]$:

$$P(A) = \sum_{\alpha \in (0,1]} \alpha / p(A_\alpha) \qquad (2.81)$$

or in terms of the membership function

$$\mu_{P(A)}[p(A_\alpha)] = \alpha, \qquad \text{for each } \alpha \in (0, 1] \qquad (2.82)$$

where A_α is the α-cut of A given by (2.14), page 26; (2.81) is evidently a result of application of the extension principle (2.69), page 40, and it also represents the use of the representation theorem (Theorem 2.16, page 26).

Example 2.19 Suppose that: $X = \{1, 2, 3, 4\}$, $p(1) = 0.1$, $p(2) = 0.3$, $p(3) = 0.5$ and $p(4) = 0.1$, and $A = 0.2/1 + 0.5/2 + 0.8/3 + 1/4$.

Therefore, $\alpha \in \{0.3, 0.5, 0.8, 1\}$ and

$$A_1 = \{4\} \quad A_{0.8} = \{3, 4\} \quad A_{0.5} = \{2, 3, 4\} \quad A_{0.3} = \{1, 2, 3, 4\}$$

Hence

$$p(A_1) = 0.1 \quad p(A_{0.8}) = 0.6 \quad p(A_{0.5}) = 0.9 \quad p(A_{0.3}) = 1$$

$\square$

Yager's definition is somewhat controversial as it does not satisfy all the conditions that Zadeh's nonfuzzy probability does. We will not, however, deal with this issue, and will refer the reader to the literature (e.g., Klir and Folger, 1988 or Klir and Yuan, 1995). In this book we will use elements of Yager's definition in Section 4.2.1.2.

2.1.9   Defuzzification of fuzzy sets

In many applications, we employ some fuzzy-sets-based technique and obtain a fuzzy result. However, in virtually all real world systems it is a crisp (non-fuzzy) result that should be applied (implemented). A notable example here is fuzzy (logic) control outlined in Section 1.1.

Suppose therefore that we have a fuzzy set A defined in $X = \{x_1, x_2, \ldots, x_n\}$, written as $A = \mu_A(x_1)/x_1 + \mu_A(x_2)/x_2 + \ldots + \mu_A(x_n)/x_n$. We need to find a crisp (nonfuzzy) number $a \in [x_1, x_n]$ which "best" represents A. Notice that we assume here, as in the whole book, that the fuzzy set A is defined in a finite universe of discourse, but its corresponding defuzzified number a need not be in general any of the finite values of the set X, but should be between the lowest and highest elements of X (evidently, this requires some ordering of x_i's, but this is clearly satisfied as x_i's are in virtually all practical cases just real numbers).

The most commonly used defuzzification procedure in fuzzy (logic) control (cf. Section 1.1) is certainly the *center-of-area*, also called the *center-of-gravity*, method which in essence is

$$a = \frac{\sum_{i=1}^{n} x_i \mu_A(x_i)}{\sum_{i=1}^{n} \mu_A(x_i)} \tag{2.83}$$

However, as we will clearly see in the following, the above defuzzification procedure – though very simple indeed – is not suitable for prescriptive-type considerations in this book which are optimization-related.

Therefore, we will use throughout this book (cf. Chapter 3 for a justification) an even simpler defuzzification method which simply assumes that the defuzzified value of a fuzzy value is $x_i \in X = \{x_1, \ldots, x_n\}$ for which $\mu_A(x)$ takes on its maximum values, i.e.

$$\mu_A(a) = \max_{x_i \in X} \mu_A(x) \tag{2.84}$$

with an obvious extension that if the A determined in (2.84) is not unique, then we take, say, the mean value of such equivalent a's.

In Section 4.2.1.2 we will use another, more specific defuzzification method, while for even more specialized and less traditional ones we refer the reader to Yager and Filev (1993).

2.2   FUZZY LOGIC – BASIC ISSUES

Fuzzy logic is one of the much-talked-about concepts today, both in scientific, professional and engineering discourse, and in the mass media. Unfortunately, the very concept of a *fuzzy logic* is not uniquely understood. Basically, it may be meant in (at least) three following ways:

- as a foundation of reasoning based on ambiguous, vague and imprecise statements (see, e.g., Goguen, 1969),
- as a foundation of reasoning based on ambiguous, vague and imprecise statements in which fuzzy sets theoretic tools are used (see, e.g., Zadeh, 1975a, 1979 or Zadeh and Kacprzyk, 1992), and
- as a multivalued logic with truth values in the unit interval in which the logical operations of negation, union, intersection, implication, equivalence, etc. are chosen in a special way, and have some fuzzy interpretation.

It is easy to see that the meaning of fuzzy logic in the first and second sense is similar, though the generality of the former is clearly higher, while its meaning in the third way is different.

For the purposes of this volume, we will assume the third view on fuzzy logic, and present a very limited survey of basic issues, and only those which will be relevant to our discussion as tools of fuzzy logic are explicitly used only in some of our later considerations. The interested reader is referred for more detail on various aspects of fuzzy logic to Zadeh and Kacprzyk's (1992) volume which is practically the only up-to-date and exhaustive treatise on fuzzy logic available today.

Suppose therefore that we have a statement (predicate)

$$u \text{ is } P$$

denoted, for brevity, as

$$P$$

exemplified by, say, "temperature (u) is high (P)," where u is a variable taking on its values in a universe of discourse $U = \{u\}$, and P is an imprecise term equated with a fuzzy set in U, i.e. $P = \{\mu_P(u)/u\}$.

If we have a specified value $u \in U$, then the truth of "u is P" (or of P) is denoted $\tau(P)$ and meant to be

$$\tau(u \text{ is } P) = \tau(P) = \mu_P(u)$$

for each $u \in U$.

We will now use the following general definitions of basic logical operations (in terms of their respective truth values):

The *negation* of statement P, i.e. "*not* P," which is denoted $\neg P$:

$$\tau(\neg P) = 1 - \tau(P) \tag{2.85}$$

The *intersection* of statements P and Q, i.e. "P *and* Q," which is denoted $P \cap Q$:

$$\tau(P \cap Q) = t[\tau(P), \tau(Q)] \tag{2.86}$$

where $t : [0,1] \times [0,1] \longrightarrow [0,1]$ is a t-norm defined by (2.33), page 31, and exemplified by (2.34)–(2.36). And again, the original Zadeh's definition is

$$\tau(P \cap Q) = \tau(P) \wedge \tau(Q) = \min[\tau(P), \tau(Q)] \tag{2.87}$$

The *union* of statements P and Q, i.e. "P *or* Q," which is denoted $P \cup Q$:

$$\tau(P \cup Q) = s[\tau(P), \tau(Q)] \tag{2.88}$$

where $s : [0,1] \times [0,1] \longrightarrow [0,1]$ is an s-norm defined by (2.37), page 32, and exemplified by (2.38)–(2.40). And again, the original Zadeh's definition is

$$\tau(P \cup Q) = \tau(P) \vee \tau(Q) = \max[\tau(P), \tau(Q)] \tag{2.89}$$

The implication, i.e. "if P, then Q," denoted $P \Longrightarrow Q$, may be defined in various ways among which the following definitions are commonly employed:

- the *Łukasiewicz implication*

$$\tau(P \Longrightarrow Q) = \min\{1 - \tau(P) + \tau(Q), 1\} \tag{2.90}$$

- the *Gödel implication*

$$\tau(P \Longrightarrow Q) = \begin{cases} 1 & \text{if } \tau(P) \leq \tau(Q) \\ \tau(Q) & \text{otherwise} \end{cases} \tag{2.91}$$

- the *Goguen implication*

$$\tau(P \Longrightarrow Q) = \begin{cases} 1 & \text{if } \tau(P) = 0 \\ \min\{1, \frac{\tau(Q)}{\tau(P)}\} & \text{otherwise} \end{cases} \qquad (2.92)$$

- the *Kleene–Dienes implication*

$$\tau(P \Longrightarrow Q) = \max\{1 - \tau(P), \tau(Q)\} \qquad (2.93)$$

- the *Zadeh implication*

$$\tau(P \Longrightarrow Q) = \max\{1 - \tau(P), \min\{\tau(P), \tau(Q)\}\} \qquad (2.94)$$

- the *Reichenbach implication*

$$\tau(P \Longrightarrow Q) = 1 - \tau(P) + \tau(P) \cdot \tau(Q) \qquad (2.95)$$

The *equivalence* "P is equivalent to Q" or "if P then Q and if Q then P," denoted by $P \Longleftrightarrow Q$, is defined as:

$$\tau(P \Longleftrightarrow Q) = \tau[(P \Longrightarrow Q) \cap (Q \Longrightarrow P)] \qquad (2.96)$$

and we can assume an appropriate definition of the intersection "$\cap$" (2.31), page 30, and the implication "$\Longrightarrow$" (2.90)–(2.95).

For our purposes the use of linguistic quantifiers, exemplified by *most, almost all, a few*, etc. in the context of fuzzy logic is relevant, and this will be discussed below.

2.3 FUZZY-LOGIC-BASED CALCULI OF LINGUISTICALLY QUANTIFIED STATEMENTS

In conventional logics only the two quantifiers are in principle employed, i.e. the universal quantifier *for all* and the existential quantifier *for at least one*. However, the class of quantifiers used by humans in everyday discourse is much richer as it contains many linguistic quantifiers as: (for) *almost none, a few, a half, about 3/4, almost all, most*, etc. Linguistic expressions exemplified by "most Swedes are blond," "only a few large cars are inexpensive," etc. are easily comprehensible to humans but cannot be treated by conventional logical tools.

Although attempts to enlarge the class of quantifiers which might be used beyond the usual "for all" and "for at least one" have been undertaken for some time, and even some "generalized" quantifiers have been proposed, it has first been fuzzy logic to provide simple and constructive formal tools for dealing with linguistic quantifiers of the type mentioned above.

Below we will present the fundamentals of how to deal with linguistic quantifiers using fuzzy logic, and we will concentrate on those aspects which are relevant to our discussion, i.e. on fuzzy-logic-based calculi of linguistically quantified statements. For more details on more general aspects of linguistic quantifiers and fuzzy logic we refer the reader to the source paper (Zadeh, 1983a).

A *linguistically quantified statement* is exemplified by "most experts are convinced" and may be generally written as

$$Qy\text{'s are } F \tag{2.97}$$

where Q is a linguistic quantifier (e.g., most), $Y = \{y\}$ is a set of objects (e.g., experts) and F is a property (e.g., convinced).

Importance B may also be added to the linguistically quantified statement (2.97) yielding

$$QBy\text{'s are } F \tag{2.98}$$

exemplified by "most (Q) of the important (B) experts $(y\text{'s})$ are convinced (F)."

For our purposes, the problem is to find the (degree of) truth of such linguistically quantified statements (2.97) and (2.98), denoted $\tau(Qy\text{'s are } F)$ in the former case and $\tau(QBy\text{'s are } F)$ in the latter case, knowing the truth of the statements "y is F," denoted $\tau(y \text{ is } F)$, for all $y \in Y$. Evidently, all these degrees of truth (truths) will be meant as real numbers from the unit interval.

We now present the two calculi of linguistically quantified statements which will provide a formal means to calculate the above-mentioned degrees of truth, the so-called algebraic (or consensory) and substitution (or competitive) calculi.

2.3.1 Algebraic or consensory calculus

In this classical calculus proposed by Zadeh (1983a) in the source papers on fuzzy linguistic quantifiers, the linguistic quantifier Q is assumed to be a fuzzy set defined in $[0, 1]$, and hence such linguistic quantifiers will be called here *fuzzy linguistic quantifiers* since they are equated with fuzzy sets. For instance, a fuzzy linguistic quantifier $Q =$ "most" may be given as

$$\mu_{''\text{most}''}(x) = \begin{cases} 1 & \text{for } x > 0.8 \\ 2x - 0.6 & \text{for } 0.3 \leq x \leq 0.8 \\ 0 & \text{for } x < 0.3 \end{cases} \tag{2.99}$$

which may be meant as that if less than 30% of the objects considered possess some property, then it is sure that not most of those objects possess it, if more than 80% of these object possess the property, then it is sure that most of these objects possess it, and for the cases in between (i.e. between 30% and 80%), the fact that most of the objects possess the property is true (sure) to an extent, from 0 to 1, the more the percentage the higher the truth.

Notice that such a fuzzy linguistic quantifier is an example of a *relative quantifier* because it concerns the percentage of the objects (a relative measure). The same argument can be applied for *absolute quantifiers* exemplified by *about 5, much more than 7*, etc. However, since in our context only the relative quantifiers are relevant, we will consider in the sequel such quantifiers only. For details on absolute quantifiers we refer the reader to Kacprzyk (1985–1986c) or Zadeh (1983a).

Particularly important in our context are the *non-decreasing fuzzy linguistic quantifiers* defined as

$$x' > x'' \implies \mu_Q(x') \geq \mu_Q(x''), \qquad \text{for each } x', x'' \in [0, 1] \tag{2.100}$$

They reflect, roughly speaking, an attitude of the type "the more the better" which is appropriate in the context of fuzzy control as we will see later. $Q = $ "most" given by (2.99) is clearly a non-decreasing quantifier.

Property F is defined as a fuzzy set in the set of objects Y, and if $Y = \{y_1, \ldots, y_p\}$, then we assume that $\tau(y_i$ is $F) = \mu_F(y_i), i = 1, \ldots, p$.

The (degree of) truth of the linguistically quantified statement (2.97), $\tau(Qy$'s is $F)$, is now calculated using the non-fuzzy cardinalities, the so-called $\sum$Count's (2.18), page 27, of the respective fuzzy sets in the following two steps (Zadeh, 1983a):

Step 1. Calculate

$$r = \frac{\sum\text{Count}(F)}{\sum\text{Count}(Y)} = \frac{1}{p}\sum_{i=1}^{p}\mu_F(y_i) \tag{2.101}$$

Step 2. Calculate

$$\tau(Qy\text{'s are } F) = \mu_Q(r) \tag{2.102}$$

In the case with importance, i.e. (2.98), importance is defined as a fuzzy set $B = $ "important" in the set of objects Y, such that $\mu_B(y_i) \in [0, 1]$ is a degree of importance of object y_i: from 0 for definitely unimportant to 1 for definitely important, through all intermediate values (such that the higher the more important).

We rewrite first the linguistically quantified statement "QBy's are F" as "$Q(B$ and $F)y$'s are B" which clearly implies the following counterparts of the two steps (2.101) and (2.102), respectively:

Step 1. Calculate

$$r' = \frac{\sum\text{Count}(B \text{ and } F)}{\sum\text{Count}(B)} = \frac{\sum_{i=1}^{p} t[\mu_B(y_i), \mu_F(y_i)]}{\sum_{i=1}^{p} \mu_B(y_i)} \tag{2.103}$$

Step 2. Calculate

$$\tau(QBy\text{'s are } F) = \mu_Q(r') \tag{2.104}$$

where $t(.,.)$ is a t-norm defined by, say, (2.34)–(2.36), page 31.

In the most common case when the t-norm is assumed to be the minimum, the above calculation steps become:

Step 1. Calculate

$$r' = \frac{\sum\text{Count}(B \text{ and } F)}{\sum\text{Count}(B)} = \frac{\sum_{i=1}^{p} [\mu_B(y_i) \wedge \mu_F(y_i)]}{\sum_{i=1}^{p} \mu_B(y_i)} \tag{2.105}$$

Step 2. Calculate

$$\tau(QBy\text{'s are } F) = \mu_Q(r') \tag{2.106}$$

Example 2.20 Let $Y = \{X, V, Z\}$, $F = $ "convinced" $= 0.1/X + 0.6/V + 0.8/Z$, $Q = $ "most" be given by (2.99), t is "$\wedge$," and $B = $ "important" $= 0.2/X + 0.5/V + 0.6/Z$.

Then, $r = 0.5$ and $r' = 0.8$, and

$$\tau(\text{"most experts are convinced"}) = 0.4$$
$$\tau(\text{"most of the important experts are convinced"}) = 1$$

□

Finally, we can also use in the above framework the fuzzy cardinalities of fuzzy sets, the so-called FECounts (2.24), page 28, but they will not be used in this book since they lead to more complicated formulas which are not suitable for optimization-related setting in all our discussions.

2.3.2 Substitution or competitive calculus

This alternative calculus of linguistically quantified statements was proposed by Yager (1983b).

As previously, the set of objects is $Y = \{y_1, \ldots, y_p\}$ and the property F is defined as a fuzzy set in Y. A statement "y_i is F" is denoted by P_i, and its (degree of) truth is $\tau P_i = \tau(y_i \text{ is } F) = \mu_F(y_i)$, $i = 1, \ldots, p$.

We introduce the set $V = \{v\} = 2^{\{P_1, \ldots, P_p\}}$, where v is a generic element of the set V, "$P_{k1}, \ldots, P_{km}$" which is meant as the statements "P_{k1} and $\ldots$ and P_{km}"; in the following we will use v and "P_{k1} and $\ldots$ and P_{km}" interchangeably.

Each v, or its corresponding "P_{k1} and $\ldots$ and P_{km}," is seen to be true to the degree

$$\mu_T(v) =$$
$$= \tau(P_{k1} \text{ and } \ldots \text{ and } P_{km}) = \tau(P_{k1})t \ldots t\tau(P_{km}) =$$
$$= t_{i=1}^m \tau(P_{ki}) = \mu_F(y_{k1})t \ldots t\mu_F(y_{km}) = t_{i=1}^m \mu_F(y_{ki}) \qquad (2.107)$$

where t is a t-norm defined by, say, (2.34)–(2.36), page 31, notably the minimum, "$\wedge$"; i.e. (2.107) defines a fuzzy set T in V.

The fuzzy linguistic quantifier Q is now defined as a fuzzy set in V. For instance, if $p = 3$, i.e. if $= \{P_1 \text{ and } P_2 \text{ and } P_3, P_1 \text{ and } P_2, P_1 \text{ and } P_3, P_2 \text{ and } P_3, P_1, P_2, P_3\}$, then $Q =$ "most" may be defined as

$$\mu_{\text{"most"}}(v) = \begin{cases} 1 & \text{for } v \in \{P_1 \text{ and } P_2 \text{ and } P_3\} \\ 0.7 & \text{for } v \in \{P_1 \text{ and } P_2, P_1 \text{ and } P_3, P_2 \text{ and } P_3\} \\ 0.3 & \text{for } v \in \{P_1, P_2, P_3\} \end{cases} \qquad (2.108)$$

which should be meant as follows: if v "contains" all the statements, then it certainly contains ($= 1$) "most" of them, if it contains two out of three statements, then it contains "most" of the statements to degree 0.7, i.e. medium, and if it contains one out of three statements, then it contains "most" of the statements to degree 0.3, i.e. low.

Note that in this case the fuzzy linguistic quantifier is defined for "k out of n" statements, i.e. for discrete quantities, as opposed to the definition of a fuzzy linguistic quantifier in Zadeh's (1983a) calculus [cf. (2.99), page 49] which is defined as some abstract quantity, not related to the real situation of an integer number of statements. This may be relevant in practice when a fuzzy linguistic quantifier needs to be elicited from a user who may not have enough experience and knowledge on fuzzy sets, mathematics, logic, etc.

Clearly, the fuzzy linguistic quantifier $Q =$ "most" given by (2.108) may be equivalently written as $Q =$ "most" $= 1/(P_1 \text{ and } P_2 \text{ and } P_3) + 0.7/(P_1 \text{ and } P_2) + 0.7/(P_1 \text{ and } P_3) + 0.7/(P_2 \text{ and } P_3) + 0.3/P_1 + 0.3/P_2 + 0.3/P_3$.

In our context, however, the fuzzy linguistic quantifier Q may be defined in the following way.

First, we define the *length* of v, or of its corresponding "$P_{k1} \ldots P_{km}$" as

$$| v | = | P_{k1} \ldots P_{km} | = m \qquad (2.109)$$

that is, as the number of terms "P_{ki}."

The membership function of Q is defined as a function of $| v |$, i.e. $\mu_Q(| v |)$. In the above example of $Q =$ "most", we have: $| P_1 \text{ and } P_2 \text{ and } P_1 | = 3$, $| P_1 \text{ and } P_2 | = | P_1 \text{ and } P_3 | = | P_2 \text{ and } P_3 | = 2$, and $| P_1 | = | P_2 | = | P_3 | = 1$.

Thus, (2.108) is equivalent to

$$\mu_{\text{"most"}} = \begin{cases} 1 & \text{for } | v | = 3 \\ 0.7 & \text{for } | v | = 2 \\ 0.3 & \text{for } | v | = 1 \end{cases} \qquad (2.110)$$

Particularly important in our control context are the so-called *monotonic* fuzzy linguistic quantifiers defined as

$$\mu_Q(v' \text{ and } v') \geq \mu_Q(v') \vee \mu_Q(v'), \qquad \text{for each } v', v' \in V \qquad (2.111)$$

the essence of which is also "the more the better."

In terms of the lengths of v' and v'' the definition of a monotonic fuzzy linguistic quantifier (2.111) becomes

$$\mu_Q(| v' \text{ and } v' |) \geq \mu_Q(| v' |) \vee \mu_Q(| v'' |), \qquad \text{for each } v', v' \in V \qquad (2.112)$$

Note that the monotonic fuzzy linguistic quantifiers as defined above by (2.111) and (2.112) are close in meaning to the non-decreasing fuzzy linguistic quantifiers in the context of the algebraic calculus defined as (2.100), though they are not exactly the same.

Evidently, $Q =$ "most" defined as above, i.e. by (2.108) and (2.110), is monotonic. We now have

$$\tau(Qy\text{'s are } F) = s_{v \in V} t[\mu_Q(v), \mu_T(v)] \qquad (2.113)$$

and in the case of importance, B, we obtain

$$\tau(QBy\text{'s are } F) = s_{v \in V} \{ t[\mu_Q(v), (t_{i=1}^m (\mu_B(y_{ki}) \Longrightarrow \mu_F(y_{ki})))] \} \qquad (2.114)$$

For instance, in a commonly used case when t is "$\wedge$," s is "$\vee$," and the implication $a \Longrightarrow b = (1 - a) \vee b$, we obtain

$$\tau(QBy\text{'s are } F) =$$

$$= \max_{v \in V} \{ \mu_Q(v) \wedge \bigwedge_{i=1}^m ([1 - \mu_B(y_{ki})] \vee \mu_F(y_{ki})) \} \qquad (2.115)$$

Example 2.21 For the same data as in Example 2.20, i.e. $Y = \{X, V, Z\}$, $F =$ "convinced" $= 0.1/X + 0.6/V + 0.8/Z$, t is "$\wedge$," and $B =$ "important" $= 0.2/X + 0.5/V + 0.6/Z$, and – which is the only obvious difference – $Q =$ "most" is given by (2.110), we obtain:

$$\tau(\text{"most experts are convinced"}) = 0.6$$
$$\tau(\text{"most of the important experts are convinced"}) = 0.8$$

and these results are clearly different than in the case of the algebraic method (cf. Example 2.20). □

For more details on the substitution method, we refer the reader to Yager (1983b) or Kacprzyk and Yager (1984a, b).

2.3.3 Fuzzy linguistic quantifiers and the ordered weighted averaging (OWA) operators

In the previous section we have provided the reader with some fuzzy-logic-based tools to handle fuzzy linguistic quantifiers. In particular, we have presented two calculi of linguistically quantified statements. These calculi may serve the purpose of a linguistic-quantifier-based aggregation of pieces of evidence such that, say, *a few, much more than a half, most*, etc. of such pieces are taken into account during the aggregation process.

It is natural that it might be very useful to have some simple yet general aggregation operator which would make it possible to aggregate pieces of evidence in various ways, including those based on linguistic quantifiers. Such an aggregation may be provided by Yager's (1988) ordered weighted averaging (or OWA, for short) operators.

An *ordered weighted averaging* (OWA) operator of dimension n is a mapping

$$F : [0,1]^n \longrightarrow [0,1] \tag{2.116}$$

if associated with F is a weighting vector

$$W = [w_1, \ldots, w_n]^T \tag{2.117}$$

such that:

1. $w_i \in [0,1]$, for all $i = 1, \ldots, n$,
2. $\sum_{i=1}^{n} w_i = 1$, and

$$F(a_1, \ldots, a_n) = W^T B = \sum_{j=1}^{n} w_j b_j \tag{2.118}$$

where b_j is the j-th largest element in the set $\{a_1, \ldots, a_n\}$, and $B = [b_1, \ldots, b_n]$. B is called an ordered argument vector if for each $b_i \in [0,1]$, $j > i$ implies $b_i \geq b_j$, $i = 1, \ldots, n$.

Example 2.22 Suppose that $W^T = [0.2, 0.3, 0.1, 0.4]$ and $A = [0.6, 1.0, 0.3, 0.5]$, i.e. $B = [1.0, 0.6, 0.5, 0.3]$. Then

$$F(a_1, \ldots, a_4) = F(0.6, 1.0, 0.3, 0.5) = W^t B =$$
$$= [0.2, 0.3, 0.1, 0.4] \times [1.0, 0.6, 0.5, 0.3] = 0.55$$

$\square$

The OWA operators have some interesting properties (cf. Yager, 1988 or Kacprzyk and Yager, 1990 and Yager and Kacprzyk, 1997) exemplified by:

- *Commutativity*: the indexing of the arguments a_i, $i = 1, \ldots$, is irrelevant, i.e. $\{\bar{a}_1, \ldots, \bar{a}_n\}$ is a permutation of the set $\{a_1, \ldots, a_n\}$, then $F(\bar{a}_1, \ldots, \bar{a}_n) = F(a_1, \ldots, a_n)$.

- *Monotonicity*: if $a_i \geq a'_i$, for all $i = 1, \ldots, n$, then $F(a_1, \ldots, a_n) \geq F(a'_1, \ldots, a'_n)$.
- *Idempotency*: $F(a, \ldots, a) = a$.
- And

$$\max_{i=1,\ldots,n} a_i \geq F(a_1, \ldots, a_n) \geq \min_{i=1,\ldots,n} a_i$$

It is now easy to see that by assuming different forms of the weighting vector W we can obtain different types of aggregation operators. Namely, for more interesting cases we obtain the following:

- if

$$w_i = \begin{cases} 1 & \text{for } i = 1 \\ 0 & \text{otherwise} \end{cases}$$

then we obtain the max-type aggregation

$$F^*(a_1, \ldots, a_n) = \max_{i=1,\ldots,n} a_i = a_1, \vee \ldots \vee a_n \qquad (2.119)$$

as the highest element of the set $\{a_1, \ldots, a_n\}$ only is chosen,

- if

$$w_i = \begin{cases} 1 & \text{for } i = n \\ 0 & \text{otherwise} \end{cases}$$

then we obtain the min-type aggregation

$$F_*(a_1, \ldots, a_n) = \min_{i=1,\ldots,n} a_i = a_1, \wedge \ldots \wedge a_n \qquad (2.120)$$

as the lowest element of the set $\{a_1, \ldots, a_n\}$ only is chosen,

- if

$$w_i = \frac{1}{n}, \qquad \text{for all } i = 1, \ldots, n$$

then we obtain the normal average (arithmetic mean) aggregation

$$F_A(a_1, \ldots, a_n) = \frac{1}{n} \sum_{i=1}^{n} a_1 \qquad (2.121)$$

- if

$$w_i = \begin{cases} 1 & \text{for } i = k, 1 \leq k \leq n \\ 0 & \text{otherwise} \end{cases}$$

then we obtain

$$F(a_1, \ldots, a_n) = b_k \qquad (2.122)$$

i.e. the k-th largest element of the set $\{a_1, \ldots, a_n\}$ only is chosen.

Thus, from the above more representative examples of diverse aggregation modes we can immediately see that the OWA operators can provide us with a very general, yet simple aggregation operator covering a whole range of cases, from the minimum type aggregation on the one extreme to the maximum type aggregation on the other extreme. And cases in between may clearly be accounted for by a proper choice of the weighting vector.

Clearly, the minimum type aggregation may be viewed to correspond to the universal quantifier *for all* as the "goodness" of all the aggregated elements matters (and the worst one determines the value of aggregate), while the maximum type aggregation may be viewed to correspond to the existential quantifier *for at least one* as the "goodness" of only the best element matters.

Therefore, we conclude that by an appropriate choice of the weighting vector W, between $W = [1, 0, \ldots, 0]$, as in the maximum type aggregation, and $W = [0, \ldots, 0, 1]$, as in the minimum type aggregation, we can obtain an aggregation operator corresponding to "intermediate" linguistic quantifiers as, e.g., *at least a half*, *most, almost all*, etc. Therefore, an OWA operator may be viewed to provide in general a (fuzzy) linguistic-quantifier-based aggregation.

Suppose therefore that Q denotes such a linguistic quantifier, and we are interested in an aggregation based on such a linguistic quantifier, denoted $F_Q(.)$. It is evident that the main problem is to find the weighting vector W corresponding to the particular linguistic quantifier Q.

First, suppose that the fuzzy linguistic quantifier is meant in the sense of Zadeh (1983a), i.e. (cf. Section 2.3.1) $\mu_Q : [0, 1] \longrightarrow [0, 1]$, $\mu_Q(x) \in [0, 1]$, and we restrict our attention to the so-called *regular non-decreasing monotone* quantifiers such that

1. $\mu_Q(0) = 0$,
2. $\mu_Q(1) = 1$, and
3. if $x > y$, then $\mu_Q(x) \geq \mu_Q(y)$.

Such fuzzy linguistic quantifiers are exemplified by *most, almost all*, etc., and only they are relevant to our discussion as they reflect the attitude "the more the better."

For regular non-decreasing monotone quantifiers Yager (1988) generates the weighting vector $W = [w_1, \ldots, w_n]^T$ as follows:

$$w_i = \mu_Q(i) - \mu_Q(i - 1), \qquad i = 1, \ldots, n \qquad (2.123)$$

and since, by definition, $\mu_Q(0) = 0$ and $\mu_Q(1) = 1$, then $w_1 + \cdots + w_n = 1$, so that the weighting vector W^T generated in the above way constitutes the legitimate weights of the OWA operator.

Example 2.23 Suppose that $Q =$ "most" is defined as (2.99), page 49, i.e.

$$\mu_{\text{"most"}}(x) = \begin{cases} 1 & \text{for } x \geq 0.8 \\ 2x - 0.6 & \text{for } 0.3 < x < 0.8 \\ 0 & \text{for } x \leq 0.3 \end{cases}$$

Then, if $n = 5$, we determine $W = [w_1, \ldots, w_5]^T$ as follows:

$$w_1 = \mu_Q(0.2) - \mu_Q(0) = 0 - 0 = 0$$
$$w_2 = \mu_Q(0.4) - \mu_Q(0.2) = 0.2 - 0 = 0.2$$
$$w_3 = \mu_Q(0.6) - \mu_Q(0.4) = 0.6 - 0.2 = 0.4$$
$$w_4 = \mu_Q(0.8) - \mu_Q(0.6) = 1 - 0.6 = 0.4$$
$$w_5 = \mu_Q(1) - \mu_Q(0.8) = 1 - 1 = 0$$

□

The above procedure for determining the weighting vector W is simple and intuitively appealing. However, one should bear in mind that it does not provide in general the same result of aggregation as that using the original fuzzy logic based aggregations via Zadeh's (or Yager's) calculus of linguistically quantified statements (cf. Section 2.3), but its approximation. It will be used in this book. As to other procedures for determining the weighting vector W, the interested reader may consult Yager and Kacprzyk's (1996) volume.

For our purposes an important issue is how to aggregate (with a fuzzy linguistic quantifier, of course) pieces of evidence with importance assigned as this is related to discounting, i.e. the assignment of different weights to the particular control stages. Unfortunately, these issues will be first clarified in Chapter 3 where we will discuss multistage control under fuzziness.

Suppose therefore that, as before, we have some vector $A = [a_1, \ldots, a_n]$ of pieces of evidence to be aggregated (by using an OWA operator), and a vector of importances of the particular piece of evidence $V = [v_1, \ldots, v_n]$ such that $v_i \in [0, 1]$ is the importance of a_i, $i = 1, \ldots, n$ (notice that $v_1 + \cdots + v_n$ need not be 1 in general). Suppose now that we have a fuzzy linguistic quantifier Q, i.e. we have its corresponding weighting vector $W = [w_1, \ldots, w_n]^T$ determined via (2.123).

Evidently, since in addition to the original definition of an OWA operator (2.116)–(2.118) we have now importances V, we cannot directly use these formulas. A natural approach would be now to define an OWA operator under importances as follows.

An *ordered weighted averaging* operator of dimension n with *importance qualification*, denoted OWA_I, is a mapping

$$F_I : [0, 1]^n \longrightarrow [0, 1] \tag{2.124}$$

if associated with F_I is a weighting vector

$$\overline{W}_I = [\overline{w}_1, \ldots, \overline{w}_n]^T \tag{2.125}$$

such that:

1. $\overline{w}_i \in [0, 1]$, for all $i = 1, \ldots, n$;
2. $\sum_{i=1}^n \overline{w}_i = 1$,

and

$$F_I(a_1, \ldots, a_n) = \overline{W}^T \cdot B = \sum_{j=1}^n \overline{w}_j b_j \tag{2.126}$$

where b_j is the j-th largest element in the set $\{a_1, \ldots, a_n\}$, and $B = [b_1, \ldots, b_n]$. B is called an ordered argument vector if for each $b_i \in [0, 1]$, $j > i$ implies $b_i \geq b_j$, $i = 1, \ldots, n$.

Therefore, the addition of importances I is basically a transformation of the weighting vector of the OWA operator, from W into $\overline{W}$. This is evidently a simple, straightforward approach which has been proposed by Yager (1996) and will be used here. However, we should bear in mind that the issue of OWA operators with importance qualification is still somewhat of an open problem that needs further research.

The essence of Yager's (1996) proposal to determine the new weights $\overline{w}_i$'s is as follows. First, as in the source OWA operator, we order the pieces of evidence a_i, $i = 1, \ldots, n$, in descending order so that we obtain the vector B such that b_j is the j-th largest element of the set $\{a_1, \ldots, a_n\}$. Next, we denote by u_j the importance of this piece of evidence b_j, i.e. of that a_i which is the j-th largest; $i, j = 1, \ldots, n$. Finally, the new weights of the transformed weighting vector $\overline{W}$ are defined as

$$\overline{w}_j = \mu_Q \left(\frac{\sum_{k=1}^{i} u_k}{\sum_{k=1}^{n} u_k} \right) - \mu_Q \left(\frac{\sum_{k=1}^{i-1} u_k}{\sum_{k=1}^{n} u_k} \right) \qquad (2.127)$$

Example 2.24 Suppose that the pieces of evidence to be aggregated are $A = [a_1, a_2, a_3, a_4] = [0.7, 1, 0.5, 0.6]$, i.e. $a_1 = 0.7$, $a_2 = 1$, $a_3 = 0.5$, and $a_4 = 0.6$, and the importances associated with them are $V = [v_1, v_2, v_3] = [1, 0.6, 0.5, 0.9]$, i.e. the importance of a_1 is 1, of a_2 is 0.6, of a_3 is 0.5, and of a_4 is 0.9.

Suppose that the fuzzy linguistic quantifier is assumed to be in a very simple form given as

$$\mu_Q(x) = x, \qquad \text{for all } x \in [0, 1] \qquad (2.128)$$

i.e. it is the so-called *unitor fuzzy quantifier* which may be viewed to correspond to "most."

Then, we order the a_i's in descending order obtaining $B = [b_1, b_2, b_3, b_4]$ such that $b_1 = a_2 = 1$, $b_2 = a_1 = 0.7$, $b_3 = a_4 = 0.6$, and $b_4 = a_3 = 0.5$. Hence the importances $U = [u_1, u_2, u_3, u_4]$ associated with $B = [b_1, b_2, b_3, b_4]$ are $u_1 = 0.6$, $u_2 = 1$, $u_3 = 0.9$, and $u_4 = 0.5$, i.e. $u_1 + \cdots + u_4 = 3$.

Now, using (2.127), we calculate the new weighting vector $\overline{W} = [\overline{w}_1, \overline{w}_2, \overline{w}_3, \overline{w}_4]$ to be

$$\overline{w}_1 = \mu_Q(\tfrac{0.6}{3}) - \mu_Q(\tfrac{0}{3}) = 0.04 - 0 = 0.04$$
$$\overline{w}_2 = \mu_Q(\tfrac{1.6}{3}) - \mu_Q(\tfrac{0.6}{3}) = 0.28 - 0.04 = 0.24$$
$$\overline{w}_3 = \mu_Q(\tfrac{2.5}{3}) - \mu_Q(\tfrac{1.6}{3}) = 0.69 - 0.28 = 0.41$$
$$\overline{w}_4 = \mu_Q(\tfrac{3}{3}) - \mu_Q(\tfrac{2.5}{3}) = 1 - 0.69 = 0.31$$

Note that $w_1 + \cdots + w_4 = 1$, and also other requirements on the weighting vector of the OWA operator are fulfilled [cf. (2.118)].

Thus

$$F_I(A) = \sum_{i=1}^{4} \overline{w}_i b_i =$$
$$= \; 0.04 \cdot 1 + 0.24 \cdot 0.7 + 0.41 \cdot 0.6 + 0.31 \cdot 0.5 = 0.609$$

It may be very illustrative to solve the same example for a different vector of the pieces of evidence to be aggregated.

So, suppose now that the pieces of evidence to be aggregated are $A = [a_1, a_2, a_3, a_4] = [0.6, 0.3, 0.9, 1]$, i.e. $a_1 = 0.6$, $a_2 = 0.3$, $a_3 = 0.9$, and $a_4 = 1$, and the importances associated with them are the same as before, i.e. $V = [v_1, v_2, v_3] = [1, 0.6, 0.5, 0.9]$, i.e. the importance of a_1 is 1, of a_2 is 0.6, of a_3 is 0.5, and of a_4 is 0.9.

Thus, we order the a_i's in descending order obtaining $B = [b_1, b_2, b_3, b_4]$ such that $b_1 = a_4 = 1$, $b_2 = a_3 = 0.9$, $b_3 = a_1 = 0.6$, and $b_4 = a_2 = 0.3$. Hence the importances $U = [u_1, u_2, u_3, u_4]$ associated with $B = [b_1, b_2, b_3, b_4]$ are now $u_1 = 0.9$, $u_2 = 0.5$, $u_3 = 1$, and $u_4 = 0.6$, i.e. $u_1 + \cdots + u_4 = 3$ as before.

Now, by (2.127) we calculate the new vector of weights $\overline{W} = [\overline{w}_1, \overline{w}_2, \overline{w}_3, \overline{w}_4]$ to be

$$\overline{w}_1 = \mu_Q(\tfrac{0.9}{3}) - \mu_Q(\tfrac{0}{3}) = 0.09 - 0 = 0.09$$
$$\overline{w}_2 = \mu_Q(\tfrac{1.4}{3}) - \mu_Q(\tfrac{0.9}{3}) = 0.22 - 0.09 = 0.13$$
$$\overline{w}_3 = \mu_Q(\tfrac{2.4}{3}) - \mu_Q(\tfrac{1.4}{3}) = 0.64 - 0.22 = 0.42$$
$$\overline{w}_4 = \mu_Q(\tfrac{3}{3}) - \mu_Q(\tfrac{2.4}{3}) = 1 - 0.64 = 0.36$$

so that, as before, $\overline{w}_1 + \cdots + \overline{w}_4 = 1$.
Thus

$$
\begin{aligned}
F_I(A) = \sum_{i=1}^{4} \overline{w}_i b_i &= \\
&= 0.09 \cdot 1 + 0.13 \cdot 0.9 + 0.42 \cdot 0.6 + 0.36 \cdot 0.3 = 0.567
\end{aligned}
$$

Notice that the new weights $\overline{w}_i$'s are different for the two cases.   □

The ordered weighted averaging (OWA) operators may be used in some control problems discussed in this book, notably in those discussed in Section 4.1.5, and also in some applications presented in Chapter 8, notably in Section 8.1.

2.4   DETERMINISTIC, STOCHASTIC AND FUZZY SYSTEMS UNDER CONTROL

As we may remember from Section 1.2, a crucial element in our multistage fuzzy control models is a *dynamic system under control* which – for the purposes of this book – is meant to only represent the time evolution, i.e. the transition from the state (equated throughout this book for simplicity with the output) at the current time (control stage) to the state at the next time under a control (equated with input) at the current time. Such a state transition (a relationship between these states and control) is assumed to be known here.

The nature of that transition may evidently have a different character. In the simplest case, a *deterministic* relationship may be known, i.e. if we are in a given (nonfuzzy) state and apply a particular (nonfuzzy) control, then we know exactly the next (nonfuzzy) state. Next, a *stochastic* relationship may be known, i.e. if we are in a given (nonfuzzy) state and apply a particular (nonfuzzy) control, then we only know the probabilities of the particular resulting (nonfuzzy) states. Finally, a *fuzzy* relationship may be known. In this case there are a couple of possible situations as, e.g., we may know that from a nonfuzzy state under a nonfuzzy control the system proceeds to a fuzzy state, or the states and/or controls, and the state transitions may be fuzzy, given by, say, a fuzzy relation. In principle, we will mean the fuzzy system under control in the latter sense.

2.4.1   Deterministic system under control

Assume that the *state space* is $X = \{s_1, \ldots, s_n\}$ and the *control space* is $U = \{c_1, \ldots, c_m\}$. They are assumed to be finite throughout the book. If the state transition of a system under control is governed by the function

$$f : X \times U \longrightarrow X \tag{2.129}$$

such that

$$x_{t+1} = f(x_t, u_t), \qquad t = 0, 1, \ldots \tag{2.130}$$

where $x_t, x_{t+1} \in X$ are the states at the control stage (time) t and $t+1$, respectively, and $u_t \in U$ is the control at control stage t, the the system under control is *time-invariant deterministic*.

If, on the other hand, the state transitions are governed by

$$f_t : X \times U \longrightarrow X, \qquad t = 0, 1, \ldots \tag{2.131}$$

such that

$$x_{t+1} = f_t(x_t, u_t), \qquad t = 0, 1, \ldots \tag{2.132}$$

then the system under control is *time-varying deterministic*.

In principle, in our next discussions we will assume that the deterministic system under control is time-invariant, and is called a deterministic system, for brevity. However, virtually all our arguments may be extended to time-varying deterministic systems under control in a straightforward manner.

In the case of finite state space X and control space U, as is assumed here, the state transition equation (2.130) describing the dynamics of the (time-invariant) deterministic system may be conveniently written in matrix form as shown in Example 2.25.

Example 2.25 If $X = \{s_1, s_2, s_3\}$ and $U = \{c_1, c_2\}$, then (2.130) may be written in matrix form as, e.g.,

$$x_{t+1} = \quad \begin{array}{c|ccc} & x_t = s_1 & s_2 & s_3 \\ \hline u_t = c_1 & s_1 & s_2 & s_3 \\ c_2 & s_1 & s_3 & s_1 \end{array} \tag{2.133}$$

to be read as follows: if the current state is $x_t = s_1$ and the control applied is c_1, then the state attained is $x_{t+1} = s_1$, ..., if the current state is $x_t = s_3$ and the control applied is c_2, then the state attained is $x_{t+1} = s_1$, etc. □

The above matrix representation of a deterministic system under control will be extensively used in this book, in particular in examples.

2.4.2   Stochastic system under control

In this case we also assume that the state space and control space are finite and given as $X = \{s_1, \ldots, s_n\}$ and $U = \{c_1, \ldots, c_m\}$, respectively.

If the state transitions of a system under control are governed by the function

$$p : X \times U \times X \longrightarrow [0, 1] \tag{2.134}$$

such that

$$p(x_{t+1} \mid x_t, u_t) \in [0, 1], \qquad t = 0, 1, \ldots \tag{2.135}$$

is a *conditional probability function* which specifies the probability of attaining the state $x_{t+1} \in X$ from the state $x_t \in X$ and under the control $u_t \in U$, the the system under control is *time-invariant stochastic*.

On the other hand, if the state transitions of a system under control are governed by the function

$$p_t : X \times U \times X \longrightarrow [0,1] \qquad\qquad (2.136)$$

such that

$$p_t(x_{t+1} \mid x_t, u_t) \in [0,1], \qquad t = 0,1,\ldots \qquad\qquad (2.137)$$

is a *conditional probability function* which specifies the probability, that is valid at the particular control stage t, of attaining the state $x_{t+1} \in X$ from the state $x_t \in X$ and under the control $u_t \in U$, the the system under control is *time-varying stochastic*.

Similarly as in the case of a deterministic system, we will also assume in the sequel for simplicity that the stochastic systems under control are time-invariant, called for brevity stochastic systems, and virtually all our discussions can be directly extended to time-varying stochastic systems.

And, again, in the case of finite state space and control space a convenient representation of a stochastic system is the matrix form shown in Example 2.26.

Example 2.26 If $X\{s_1, s_2, s_3\}$ and $U = \{c_1, c_2\}$, then the conditional probability (2.135) may be written in matrix form as, e.g.,

$$p(x_{t+1} \mid x_t, u_t) =$$

	$x_{t+1} = s_1$	s_2	s_3
$u_t = c_1$ $x_t = s_1$	0.1	0.2	0.7
s_2	0	0.5	0.5
s_3	0.2	0.6	0.2

	$x_{t+1} = s_1$	s_2	s_3
$u_t = c_2$ $x_t = s_1$	1	0	0
s_2	0.6	0.3	0.1
s_3	0.1	0.9	0

$$(2.138)$$

which should be read as follows: if $x_t = s_1$ and $u_t = c_1$, then the probability of attaining $x_{t+1} = s_1$ is 0.1, the probability of attaining $x_{t+1} = s_2$ is 0.2, and the probability of attaining $x_{t+1} = s_3$ is 0.7, ..., if $x_t = s_3$ and $u_t = c_2$, then the probability of attaining $x_{t+1} = s_1$ is 0.1, the probability of attaining $x_{t+1} = s_2$ is 0.9 while the probability of attaining $x_{t+1} = s_3$ is 0. Note that, evidently, the probabilities in each row of the tables (matrices) in (2.138) sum up to 1. □

2.4.3   Fuzzy system under control

We assume here again the state space and control space to be finite, $X = \{s_1, \ldots, s_n\}$ and $U = \{c_1, \ldots, c_m\}$, respectively. The (fuzzy) state of the system under control is now characterized by a fuzzy set X_t defined in X whose membership function is $\mu_{X_t}(x_t)$. The control at control stage t may be either nonfuzzy, i.e. $u_t \in U$, or fuzzy, i.e. characterized by a fuzzy set U_t defined in U with the membership function $\mu_{U_t}(u_t)$.

If the state transitions of a system under control are governed by the function

$$F : \mathcal{L}(X) \times \mathcal{L}(U) \longrightarrow \mathcal{L}(X) \tag{2.139}$$

such that

$$X_{t+1} = F(X_t, U_t), \qquad t = 0, 1, \ldots \tag{2.140}$$

where $\mathcal{L}(.)$ is the family of all the fuzzy sets defined in the respective space, then the system is *time-invariant fuzzy with fuzzy control*.

On the other hand, if the state transitions of a system under control are governed by the function

$$F_t : \mathcal{L}(X) \times \mathcal{L}(U) \longrightarrow \mathcal{L}(X) \tag{2.141}$$

such that

$$X_{t+1} = F_t(X_t, U_t), \qquad t = 0, 1, \ldots \tag{2.142}$$

then the system is *time-varying fuzzy with fuzzy control*.

If the control is nonfuzzy, i.e. $u_t \in U$, then the state transition equation (2.140) becomes

$$X_{t+1} = F(X_t, u_t), \qquad t = 0, 1, \ldots \tag{2.143}$$

and the system under control is *time-invariant fuzzy with nonfuzzy control*.

In the case of a nonfuzzy control, (2.142) becomes

$$X_{t+1} = F_t(X_t, u_t), \qquad t = 0, 1, \ldots \tag{2.144}$$

and the system under control is *time-varying fuzzy with nonfuzzy control*.

Finally, one can imagine that the fuzzy system under control is of the "Takagi-Sugeno-type" [cf. Section 1.3, or Sugeno and Yasukawa, 1993], specified as

IF $X_t = $ NS AND $U_t = $ NS THEN $x_{t+1} = f_{1,1}(x_t, u_t)$
IF $X_t = $ NS AND $U_t = $ NM THEN $x_{t+1} = f_{1,2}(x_t, u_t)$
$\ldots$
IF $X_t = $ NS AND $U_t = $ PB THEN $x_{t+1} = f_{1,5}(x_t, u_t)$
$\ldots$
IF $X_t = $ NM AND $U_t = $ NS THEN $x_{t+1} = f_{2,1}(x_t, u_t)$
$\ldots$
IF $X_t = $ ZE AND $U_t = $ ZE THEN $x_{t+1} = f_{3,1}(x_t, u_t)$
$\ldots$
IF $X_t = $ PM AND $U_t = $ ZE THEN $x_{t+1} = f_{4,3}(x_t, u_t)$
$\ldots$
IF $X_t = $ PB AND $U_t = $ NS THEN $x_{t+1} = f_{5,1}(x_t, u_t)$
$\ldots$
IF $X_t = $ PB AND $U_t = $ PB THEN $x_{t+1} = f_{5,5}(x_t, u_t)$

but this will not be used in this book.

In the following we will consider in principle, for simplicity, time-invariant fuzzy systems, and it will be explicitly stated whether or not we will discuss the systems with nonfuzzy or fuzzy controls. One will be able to extend the discussion for the respective time-varying fuzzy systems in virtually all cases.

As may be recalled from Section 2.1.4, the fuzzy state transition equations (2.140) and (2.142)–(2.144) are equivalent to a fuzzy relation in $X \times U \times X$ (a conditioned fuzzy set) whose membership function is $\mu_{X_{t+1}}(x_{t+1} \mid x_t, u_t)$ specifying the membership grade of x_{t+1}. The above fuzzy state transition equation may also be represented in a matrix form, similar to (2.133) and (2.138), and details will be shown in Example 2.27 to be given below.

The temporal evolution – or, equivalently, the state transitions – of the fuzzy system under control is now governed by the following equations [which are clearly equivalent to the max–min composition of fuzzy relations (2.53), page 36]:

- for the fuzzy system with fuzzy control

$$\mu_{X_{t+1}}(x_{t+1}) =$$
$$= \max_{x_t \in X} \max_{u_t \in U} [\mu_{X_t}(x_t) \wedge \mu_{U_t}(u_t) \wedge \mu_{X_{t+1}}(x_{t+1} \mid x_t, u_t)] \quad (2.145)$$

for each $x_{t+1} \in X$ and $t = 0, 1, \ldots$;
- for the fuzzy system with nonfuzzy control

$$\mu_{X_{t+1}}(x_{t+1}) =$$
$$= \max_{x+t \in X} [\mu_{X_t}(x_t) \wedge \mu_{X_{t+1}}(x_{t+1} \mid x_t, u_t)] \quad (2.146)$$

for each $x_{t+1} \in X$ and $t = 0, 1, \ldots$.

One should be aware of the following extremely important (and dangerous!) effect related to the use of the state transition equations (2.145) and (2.146). Suppose that we have a fuzzy system with fuzzy control so that its state transitions are governed by (2.145). Since in all practically relevant cases the system evolves over multiple control stages, $t = 0, 1, \ldots$, we consecutively obtain:

- at $t = 0$

$$\mu_{X_1}(x_1) =$$
$$= \max_{x_0 \in X} \max_{u_0 \in U} [\mu_{X_0}(x_0) \wedge \mu_{U_0}(u_0) \wedge \mu_{X_1}(x_1 \mid x_0, u_0)] \quad (2.147)$$

- at $t = 1$

$$\mu_{X_2}(x_2) =$$
$$= \max_{x_1 \in X} \max_{u_1 \in U} [\mu_{X_1}(x_1) \wedge \mu_{U_1}(u_1) \wedge \mu_{X_2}(x_2 \mid x_1, u_1)] \quad (2.148)$$

- …
- at $t = N - 1$

$$\mu_{X_N}(x_N) =$$
$$= \max_{x_{N-1} \in X} \max_{u_{N-1} \in U} [\mu_{X_{N-1}}(x_{N-1}) \wedge$$
$$\wedge \mu_{U_{N-1}}(u_{N-1}) \wedge \mu_{X_N}(x_N \mid x_{N-1}, u_{N-1})] \quad (2.149)$$

Unfortunately, due to the max–min composition employed in (2.147)–(2.149), the consecutive fuzzy states $X_1, X_2, \ldots, X_N$ become more and more fuzzy ("flatter and flatter" fuzzy numbers). In practice, after a couple of iterations they may be too fuzzy to become a meaningful point of departure for a next state transition.

This is an extremely serious problem which should be somehow solved. A straightforward, though often not fully justifiable procedure may be to sharpen (e.g., by contrast intensification) first the fuzzy state attained before taking it for the next state transition iteration.

Another approach, which will be adopted throughout this book, is to predefine some (sufficiently small) number of reference (standard) fuzzy states and controls, and redefine all state transitions of the type (2.145) in terms of these reference entities. Then, all fuzzy states and controls occurring are first approximated by their closest reference counterparts, and these are used in state transitions. Therefore, since only finite, sufficiently small sets of fuzzy states and controls are involved, the above-mentioned effect of increasing fuzziness is not present. However, it is clear that by operating on reference, not "real" fuzzy states and controls, we do solve some auxiliary problems which may be similar but not exactly the same as the source problems considered.

Both the above remedies for the fuzzification effect are efficient but should be viewed as *ad hoc* measures. There is, however, no well-founded general mathodology to deal with this extremely important issue.

Example 2.27 Let the the fuzzy system under control be with nonfuzzy control, for simplicity. Assume again that $X = \{s_1, s_2, s_3\}$ and $U = \{c_1, c_2\}$, and the fuzzy system (its state transitions) is represented in the following matrix form:

$$\mu_{X_{t+1}}(x_{t+1} \mid x_t, u_t) =$$

		$x_{t+1} = s_1$	s_2	s_3
$u_t = c_1$	$x_t = s_1$	0.2	0.7	0.9
	s_2	0.7	1	0.3
	s_3	1	0.7	0.4

		$x_{t+1} = s_1$	s_2	s_3
$u_t = c_2$	$x_t = s_1$	0.2	0.6	1
	s_2	0.5	0.8	1
	s_3	0.8	1	0.6

If now, e.g., the current state (at control stage t) is $X_t = 1/s_1 + 0.7/s_2 + 0.4/s_4$ and $u_t = c_2$, then we obtain via (2.146)

$$X_{t+1} = 0.6/s_1 + 0.7/s_2 + 0.8/s_3$$

$\square$

The representation of a fuzzy system under control as a fuzzy state transition equation, e.g. (2.140), i.e. $X_{t+1} = F(X_t, U_t)$, $t = 0, 1, \ldots$, is clearly a very general representation which can be specified in various ways as, for instance:

- directly, i.e. just as a state transition equation (2.140),

- as a set of IF–THEN rules of the type (2.62), page 38, i.e.

$$\begin{cases} \text{IF } X_t = S_1 \text{ AND } U_t = C_1 \text{ THEN } X_{t+1} = S_{k1} \\ \text{ELSE } \ldots \text{ ELSE} \\ \text{IF } X_t = S_n \text{ AND } U_t = C_m \text{ THEN } X_{t+1} = S_{kw} \end{cases} \qquad (2.150)$$

which is equivalent to the following fuzzy relation [cf. (2.62)]

$$R = S_1 \times C_1 \times S_{k1} + \cdots + S_n \times C_m \times S_{kw} \qquad (2.151)$$

where $S_1, S_{k1}, \ldots, S_n, S_{kw}$ are particular values (fuzzy sets) of X_t and X_{t+1}, $C_1, \ldots, C_m$ are particular values (fuzzy sets) of U_t, "$\times$" is the Cartesian product of two fuzzy sets (2.56), page 37, and "+" is the union of fuzzy sets (2.32), page 30, and

- a neural network whose input layer represents X_t and U_t and output layer represents X_{t+1}, and weights encode the fuzzy state transition equation (cf. Kosko, 1992b).

All these possible representations of fuzzy state transitions provide only the information specified by (2.140), so they are equivalent from our point of view, and will not be separately discussed – we will assume in the following that dynamics of the fuzzy system under control is given by its fuzzy state transition equation (2.140), page 61.

For a more general discussion of fuzzy systems, or more generally fuzzy modeling, we refer the reader to the fundamental work by Sugeno and Yasukawa (1993).

2.4.3.1   Remarks on the identification of fuzzy systems under control

In this book we assume as a point of departure that a model of a system under control is known, to be more specific as a state transition equation. However, emphasis in this book is more on algorithmic, optimization-oriented elements in a new, prescriptive, model-based approach to fuzzy control which is being advocated.

Therefore, though the derivation (identification) of those models of systems under control is a *sine qua non*, and a prerequisite, for the approach presented, we will give below only some more relevant aspects of model identification referring the interested reader to the literature, which is unfortunately not rich.

Since the identification of the deterministic (cf. Section 2.4.1) and stochastic (cf. Section 2.4.2) systems is the same as in non-fuzzy control, we will not consider these cases here, and the interested reader can find details in numerous books on control, identification or parameter estimation which are available from virtually all major publishers. We will discuss below only basic issues related to the identification of fuzzy systems under control given by (2.140), page 61, i.e.

$$X_{t+1} = F(X_t, U_t), \qquad t = 0, 1, \ldots \qquad (2.152)$$

where X_t, X_{t+1} are fuzzy states at control stages t and $t+1$, respectively, and U_t is a fuzzy control at t.

In general, the above fuzzy state transition equation (2.152) is usually given a set of IF–THEN rules relating linguistically specified values of the state and control at t,

X_t and U_t, with that of the resulting state at $t + 1$, X_{t+1}. These linguistic values are represented by their equivalent fuzzy sets (numbers).

Then, such a set of IF–THEN rules is transformed, using the relational representation of IF–THEN rules (which are clearly fuzzy conditional statements – cf. Section 2.1.4) into a fuzzy relation $\mu_{X_{t+1}}(x_{t+1} \mid x_t, u_t)$ – cf. Section 2.4.3.

On the other hand, the relation between X_t, U_t and X_{t+1} can be clearly represented by a neural network. However, this relation is also between their linguistic values represented by fuzzy sets (numbers).

The identification of all such fuzzy models involves the following three basic steps:

- determination of the model's structure,
- estimation of the model's parameters from a data set provided for identification, and
- validation of the model.

The structure of the model is in our case equivalent to the "length" of IF–THEN rules employed in the former case, and the structure (number of the hidden layers, number of nodes in the input, output and hidden layers, etc.) of the neural network in the latter case.

The identification of parameters is performed by using historic data. Usually, we form first (linguistic) IF–THEN rules describing various combinations of X_t, U_t and X_{t+1} which occurred in the past. Each such rule is transformed into its equivalent (partial) fuzzy relation, and all such fuzzy relations are aggregated into a final fuzzy relation which represents the state transition of a fuzzy system under control (2.152), i.e. $X_{t+1} = F(X_t, U_t)$, $t = 0, 1, \ldots$.

One should be aware that while forming the above-mentioned (linguistic) IF–THEN rules, with linguistic (fuzzy) terms standing for the values of X_t, U_t and X_{t+1}, an extremely important problem is a proper choice of fuzzy sets (e.g., triangular fuzzy numbers) representing such terms. This is similar to the tuning of fuzzy controllers, and may be done by some *ad hoc* trial-and-error techniques, by using genetic algorithms as recently advocated, etc.

For more details on the identification of models described by fuzzy relations we refer the interested reader to, e.g., Pedrycz (1993, 1995a, 1996).

The identification of a fuzzy system given as a neural network amounts to the derivation of weights from past data. And, again, a proper choice of fuzzy sets (numbers) standing for the linguistic values of X_t, U_t and X_{t+1} is of utmost importance, and can be performed as mentioned above. The derivation of weights is the same as for any other neural network, and the interested reader may consult numerous books from all major publishers as, e.g., Kosko (1992b).

The validation of the fuzzy models is then done as always by testing them against some (real) data sets.

In general, the identification of fuzzy models of dynamic systems, which has been given above, is a non-trivial, extremely important issue. Unfortunately, there seems to be no systematic, generally accepted methodology in this respect, and *ad hoc* techniques are widely used.

This completes our brief discussion of those aspects of dynamic systems under control, deterministic, stochastic and fuzzy, which will be needed for our further discussions.

3

A General Setting for Multistage Control under Fuzziness

The purpose of this chapter is to provide the reader with a brief introduction to Bellman and Zadeh's (1970) general approach to decision making under fuzziness, originally termed *decision making in a fuzzy environment*, a simple yet extremely powerful framework within which virtually all fuzzy models related to decision making, optimization and control have been dealt with. This framework will also be employed throughout this book. We will start its presentation with a more general analysis, and then proceed to control-related extensions.

3.1 DECISION MAKING IN A FUZZY ENVIRONMENT – BELLMAN AND ZADEH'S APPROACH

Since decision making is omnipresent in any human activity, it is quite clear that not much later after the concept of a fuzzy set was introduced as a tool for a formal description and handling of imprecise concepts, relations, etc., a next rational step was an attempt to devise a general framework for dealing with decision making under fuzziness. Needless to say that since control is evidently quite closely related to decision making, such a framework would also serve as a basis for control under fuzziness.

Such a general approach appeared in the famous Bellman and Zadeh's (1970) article, and it was termed "decision making in a fuzzy environment." And, indeed, it had proved to be extremely successful and general, probably one of a few such universal tools in science, and had since then served as a point of departure for an enormous variety of fuzzy models and approaches. This approach is also the core of all the models considered in this book. Hence, the model will be now dealt with in some detail.

3.1.1 The concept of a fuzzy goal, fuzzy constraint, and fuzzy decision

In Bellman and Zadeh's (1970) setting the imprecision (fuzziness) of the environment within which the decision making (or control) process proceeds is modeled by the introduction of the so-called *fuzzy environment* which consists of fuzzy goals, fuzzy constraints, and a fuzzy decision.

A formal definition of these elements of the fuzzy environment starts with the assumption of some set of possible *options* (or alternatives, variants, choices, decisions,

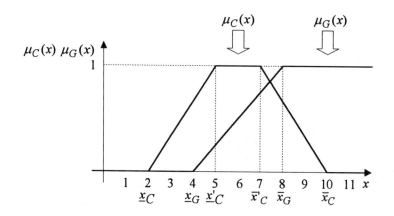

Figure 14   Fuzzy goal, G, "x should be *much larger than* 5" and fuzzy constraint, C, "x should be *about 6*"

...) denoted by $X = \{x\}$ where x means a generic element of X. The set X contains all the possible (relevant, feasible, ...) values, courses of action, etc. in the situation considered. For the time being no particular restriction on X will be assumed such as its finiteness, countability, etc. The only commonsense type requirement would be that X be rich enough to make the analysis significant.

The *fuzzy goal* is now defined as a fuzzy set G in the set of options X, characterized by its membership function $\mu_G : X \longrightarrow [0,1]$ such that $\mu_G(x) \in [0,1]$ specifies the grade of membership of a particular option $x \in X$ in the fuzzy goal G.

The *fuzzy constraint* is similarly defined as a fuzzy set C in the set of options X, characterized by its membership function $\mu_C : X \longrightarrow [0,1]$ such that $\mu_C(x) \in [0,1]$ specifies the grade of membership of a particular option $x \in X$ in the fuzzy constraint C.

For example, suppose that $X = R$, the set of real numbers. Then the fuzzy goal "x should be *much larger* than 5" may be represented by a fuzzy set whose membership, $\mu_G(x)$, function is shown in Figure 14. On the other hand, the fuzzy constraint "x should be *more or less* 6" may be represented by a fuzzy set whose membership function, $\mu_C(x)$, is also shown in Figure 14.

An important issue is how the fuzzy goal and fuzzy constraint are to be interpreted. Their definitions, which are virtually the same, do indicate an intrinsic analogy. And, indeed, such an analogy exists but this will be discussed a little later. For now we will present an interpretation of the fuzzy goal only.

In some, mostly earlier, works as, e.g., in Bellman and Zadeh (1970) or Tanaka, Okuda and Asai (1974), the following view on the essence of the fuzzy goal is advocated.

Suppose that

$$f : X \longrightarrow R \tag{3.1}$$

is a conventional performance (objective) function which associates with each option $x \in X$ a real number $f(x) \in R$. We assume that the function f is bounded, i.e.

$f(x) \leq M < \infty$, for each $x \in X$. By M we denote

$$M = \max_{x \in X} f(x) \qquad (3.2)$$

Then the membership function of the fuzzy goal G can be defined as a normalized performance function f, i.e.

$$\mu_G(x) = \frac{f(x)}{M} = \frac{f(x)}{\max_{x \in X} f(x)}, \qquad \text{for each } x \in X \qquad (3.3)$$

A fuzzy goal (and, as we will see later, a fuzzy constraint as well) may be, however, viewed from a different perspective that is presumably more convenient for our discussion in this book. Namely, we will rather interpret the fuzzy goal in terms of *satisfaction levels*. This is facilitated by the representation of the fuzzy goal's membership function in a piecewise linear form as assumed here. Then the piecewise linear membership function of a fuzzy goal shown in Figure 14 should be understood as follows: if the value of x attained is at least $\overline{x}_G$ (equal 8), which is the *satisfaction level* of x, i.e. for $x \geq \overline{x}_G$, then $\mu_G(x) = 1$ which means that we are fully satisfied with the x attained. On the other hand, if the x attained does not exceed $\underline{x}_G$ (equal 5), which is the lowest possible value of x, then $\mu_G(x) = 0$ which means that we are fully dissatisfied with such a value of x or, in other words, this value is impossible. For the intermediate values, $\underline{x}_G < x < \overline{x}_G$, we have $0 < \mu_G(x) < 1$ which means that our satisfaction as to a particular value of x is intermediate, i.e. between the full satisfaction and the full dissatisfaction, and – evidently – the closer x to $\overline{x}$, the higher $\mu_G(x)$, and hence the higher the satisfaction. The interpretation of a fuzzy constraint is analogous.

One can therefore view the concept of a fuzzy goal and constraint as meant above as some fuzzification of the famous Simon's satisfaction rule whose essence is that an option may result in either a satisfactory or unsatisfactory outcome; in our case we have a gradual rather than abrupt transition from the (full) satisfaction to the (full) dissatisfaction.

It is now easy to see that the above interpretation provides a "common denominator" for the fuzzy goal and fuzzy constraint. They may be treated in an analogous way which, as we will see soon, is one of merits of Bellman and Zadeh's (1970) approach. Such an analogy between the fuzzy constraint and fuzzy goal (performance function) is clearly not new. For instance, the Lagrange multipliers or penalty functions used in optimization do essentially play such a role. This intrinsic similarity is, however, made explicit, without any "trickery," in the approach considered.

The above-mentioned identity of handling a fuzzy goal G and a fuzzy constraint C suggest the following general formulation of the decision making problem in a fuzzy environment:

$$\text{``Attain } G \underline{\text{ and}} \text{ satisfy } C\text{''} \qquad (3.4)$$

which should be meant as to determine a decision (an option or a set of options) which simultaneously fulfills the fuzzy goal and fuzzy constraint; evidently, such a decision should belong to those available, or perhaps to those relevant or feasible.

The fuzziness of the fuzzy goal and fuzzy constraint implies the above decision, which is called a fuzzy decision, to be a fuzzy set defined in the set of options which results from some aggregation of the two fuzzy sets: the fuzzy goal and fuzzy constraint; this aggregation is equivalent to the intersection of two fuzzy sets (2.31), page 30, that corresponds to the "and" connective.

In a formal way, if G is a fuzzy goal and C is a fuzzy constraint, both defined as fuzzy sets in the set of options $X = \{x\}$, the *fuzzy decision* D is a fuzzy set defined also in the set of options X resulting from an aggregation $\star : [0,1] \times [0,1] \longrightarrow [0,1]$ of G and C, that is

$$D = G \star C \tag{3.5}$$

or, in terms of membership functions,

$$\mu_D(x) = \mu_G(x) \star \mu_C(x), \qquad \text{for each } x \in X \tag{3.6}$$

The aggregation "$\star$" is evidently some operation on two fuzzy sets. Therefore, there immediately arises a question as to which one is appropriate. Basically, if we take as a point of departure the general formulation of the decision making problem (3.4), i.e. "attain G <u>and</u> satisfy C," this operation should correspond to the "and" connective in the definition of the intersection of two fuzzy sets (2.31), page 30. Therefore, all aggregation operators related to the intersection as, e.g., a t-norm exemplified by (2.34)–(2.36), page 31, or one of more complex (though presumably more adequate) families of aggregation operations exemplified by (2.42), (2.46) and (2.48), pages 33–33, may also be used; other aggregation operations are also possible and will be occasionally employed in this book.

It should be observed, however, that although many more complex definitions are possible here, only the simple ones are relevant in our context of control since the fuzzy decision is used as a performance function that is to be maximized. Needless to say that for an efficient maximization, in nontrivial problems, a simple form of the aggregation function is crucial.

Let us start with the most important type of the fuzzy decision that is related to Zadeh's (1965a) standard definition of the intersection of two fuzzy sets (2.31), page 30, that has also been accepted by an overwhelming majority of his followers.

The *min-type fuzzy decision* is defined as

$$D = G \cap C \tag{3.7}$$

or, in terms of the membership functions,

$$\mu_D(x) = \mu_G(x) \wedge \mu_C(x), \qquad \text{for each } x \in X \tag{3.8}$$

where "$\wedge$" is the minimum operation, i.e. $a \wedge b = \min(a, b)$.

Example 3.1 Suppose that the fuzzy goal G is "x should be much larger than 5," and the fuzzy constraint C is "x should be about 6," as in Figure 14, and also as in Figure 15.

The (membership function of) min-type fuzzy decision is given in bold line, and should be interpreted as follows. The set of possible options is the interval $[5, 10]$ because $\mu_D(x) > 0$, for $5 \leq x \leq 10$. The other options, i.e. $x < 5$ and $x > 10$ are impossible since $\mu_D(x) = 0$. However, not all the options belonging to the interval $[5, 10]$ are equally satisfactory (or preferable). The value of $\mu_D(x) \in [0, 1]$ may be

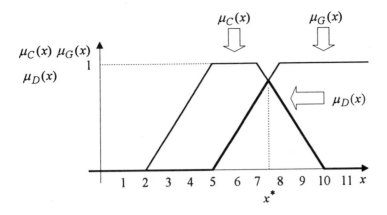

Figure 15 Fuzzy goal, fuzzy constraint, fuzzy decision, and the optimal (maximizing) decision

meant as the degree of satisfaction from the choice of a particular $x \in X$, from 0 for full dissatisfaction (impossibility of x) to 1 for full satisfaction, through all intermediate values; thus, the higher the value of $\mu_D(x)$, the higher the satisfaction from x. □

Note that in Figure 15, $\mu_D(x) < 1$ which means that there is no option which fully satisfies both the fuzzy goal and fuzzy constraint. In other words, there is a discrepancy or conflict between the goal and constraint.

We have therefore a concept of a fuzzy decision, that is a (fuzzy) solution to the decision making problem considered (3.4). In practice, however, if we wish to implement such a solution, we need to find a nonfuzzy solution. The above interpretation of the fuzzy decision's membership function $\mu_D(x)$ as the degree to which a particular option x is satisfactory as a solution to the problem (3.4) immediately suggests that the best (nonfuzzy) choice in this case would be the one corresponding to the highest value of $\mu_D(x)$.

The *maximizing decision* is defined as an $x^* \in X$ such that

$$\mu_D(x^*) = \max_{x \in X} \mu_D(x) \tag{3.9}$$

and an example may be found in Figure 15 where $x^* = 7.5$.

In the following we will basically use the min-type fuzzy decision, and – if not otherwise stated – it will be called just the fuzzy decision.

We should, however, be aware that the determination of a nonfuzzy (maximizing) decision from a fuzzy decision is basically the problem of defuzzification, and the simple defuzzification of the type 3.9) is clearly not a perfect solution, and its simplicity is the only advantage. The issue of defuzzification has attracted much attention in recent years in relation to a huge worldwide interest in fuzzy logic control where this defuzzification is of major concern (cf. Driankov, Hellendoorn and Reinfrank, 1993; Yager and Filev, 1994).

Among better known approaches to the defuzzification that are advocated in the literature, a particular role is played by the so-called *center-of-gravity* (or *center-of-*

area) method presented in Section 2.1.9 for which

$$x^* = \frac{\sum_{i=1}^{n} x_i \mu_D(x_i)}{\sum_{i=1}^{n} \mu_D(x_i)} \tag{3.10}$$

where, evidently, it was assumed – as always in practice, and throughout this book – that the fuzzy decision is defined in a finite set $X = \{x_1, \ldots, x_n\}$, i.e. $D = \mu_D(x_1)/x_1 + \cdots + \mu_D(x_n)/x_n$.

However, as will be seen in virtually all the following chapters, the fact that the fuzzy decision is to be maximized does practically preclude the use of all those more complicated (though maybe more adequate) defuzzification techniques [such as (3.10)], and the maximizing decision (3.9) is the only practically feasible choice, and it will be in principle sought. Clearly, and $x^* \in X$ given by (3.9) need not be unique.

As to the other types of the fuzzy decision, the rationale behind them is mainly an inability of the min-type fuzzy decision to reflect any interplay or compensation between the values of $\mu_G(x)$ and $\mu_C(x)$ as this fuzzy decision always chooses the smaller of the two values. Moreover, a min-type fuzzy decision has a clearly safety-first and pessimistic character.

Among other operations that may help overcome the above inherent disadvantage of the min-type fuzzy decision, the *t*-norm (2.33), page 31, is particularly relevant. So, it may well serve our purpose to define another type of fuzzy decision.

The *t-norm-type fuzzy decision* is defined as

$$\mu_D(x) = \mu_G(x) \, t \, \mu_C(x), \qquad \text{for each } x \in X \tag{3.11}$$

And among the *t*-norms a particular role is played by the *product-type fuzzy decision* which is defined as

$$\mu_D(x) = \mu_G(x) \cdot \mu_C(x), \qquad \text{for each } x \in X \tag{3.12}$$

Evidently, the value of $\mu_D(x)$ for the product-type fuzzy decision is never higher than for the min-type fuzzy decision.

One can evidently use other *t*-norms [cf. (2.36), page 32] but, as we have already mentioned, only the above two fuzzy decisions: the minimum-type (3.7) and product-type (3.12) are simple enough for optimization-type procedures employed in the following.

Both the min-type and product-type fuzzy decisions tacitly assume that the importance of the fuzzy goal and fuzzy constraint is the same. If this is not the case, the most straightforward type of fuzzy decision that makes it possible to take into account that different importance is the *weighted-sum-type* or *convex-combination-type fuzzy decision* defined as

$$\mu_D(x) = w\mu_G(x) + (1 - w)\mu_C(x), \qquad \text{for each } x \in X \tag{3.13}$$

where $w \in [0, 1]$ is an importance coefficient with values ranging from 0 for when the satisfaction of the fuzzy constraint only matters, to 1 for when that of the fuzzy goal only matters, through all intermediate values.

Yager (1992b) presented another approach to the inclusion of different importances associated with the fuzzy goals and fuzzy constraints. Basically, if $w_G \in [0, 1]$ is the importance of the fuzzy goal G, and $w_C \in [0,1]$ is the importance of the fuzzy constraint C, then the min-type fuzzy decision (3.7) with importance qualification may take on one of the following forms (cf. Yager, 1992b):

$$\mu_D(x) = [(1 - w_C) \vee \mu_C(x)] \wedge [(1 - w_G) \vee \mu_G(x)] \qquad (3.14)$$

$$\mu_D(x) = [(1 - w_C) + w_c\mu_C(x)] \wedge [(1 - w_G) + w_g\mu_G(x)] \qquad (3.15)$$

$$\mu_D(x) = [\mu_C(x)]^{w_C} \wedge [\mu_G(x)]^{w_G} \qquad (3.16)$$

and, as mentioned before, only the simplest form (3.16) is practically relevant in our context in view of optimization-type procedures employed.

A similar line of reasoning may be applied for the product-type fuzzy decision (3.12), and the counterparts of (3.14)–(3.16) may be devised.

Moreover, a similar role may be played by the following simple expression:

$$\mu_D(x) = w_C\mu_C(x) \wedge w_G\mu_G(x) \qquad (3.17)$$

since $w_C, w_G \leq 1$; such a simple idea will be used in Chapter 8.

All the types of fuzzy decision defined above reflect the essence of the problem formulation (3.4), page 69, i.e. "attain G <u>and</u> satisfy C." Sometimes, however, it may be more appropriate to assume that the problem formulation is

$$\text{"Attain } G \text{ \underline{or} satisfy } C\text{"} \qquad (3.18)$$

instead of "...<u>and</u>..." in (3.4). This leads to fuzzy decisions that are related to the union of the two fuzzy sets (2.32), page 30, which corresponds to the connective "or" – cf. (2.38)–(2.40), page 32, for various definitions of the union.

The most important of them is the *max-type fuzzy decision* defined as

$$\mu_D(x) = \mu_G(x) \vee \mu_C(x), \qquad \text{for each } x \in X \qquad (3.19)$$

where "$\vee$" is the maximum operation, i.e. $a \vee b = \max(a, b)$.

And again, the max-type fuzzy decision does not make it possible to reflect any compensation or interplay between the values of $\mu_G(x)$ and $\mu_C(x)$ as it always chooses the higher one. This may be possible – to some extent – by using, e.g., an s-norm (t-conorm), i.e. (2.37), page 32, exemplified by (2.39) or (2.40).

Therefore, the *s-norm-type fuzzy decision* is defined as

$$\mu_D(x) = \mu_G(x)s\mu_C(x), \qquad \text{for each } x \in X \qquad (3.20)$$

Among the s-norm-type fuzzy decisions one may mention, e.g., the *algebraic-sum-type fuzzy decision* which is defined as

$$\mu_D(x) = \mu_G(x) + \mu_C(x) - \mu_G(x)\mu_G(x), \qquad \text{for each } x \in X \qquad (3.21)$$

The importance qualification may proceed for this kind of fuzzy decision similarly as in the case of the min-type fuzzy decision, i.e. by using (3.14)–(3.17). For instance, for (3.17) and the max-type fuzzy decision, we obtain

$$\mu_D(x) = (1 - w_C)\mu_C(x) \vee (1 - w_G)\mu_G(x) \qquad (3.22)$$

since $w_C, w_G \leq 1$.

We have therefore discussed major types of a fuzzy decision. Observe again that, since all our problem formulations will be concerned with optimization, only "simpler" types will be applicable. Notably, we will use the min-type fuzzy decision (3.7), page 70, throughout the book, and only mention how other types may be employed.

Some other types of the fuzzy decision will also be presented in Section 3.1.2 while considering the case of multiple fuzzy goals and constraints.

3.1.2 Multiple fuzzy goals and/or fuzzy constraints

In the previous section we discussed the case of just one fuzzy goal and fuzzy constraint. This has best served the purpose of presenting the philosophy and essence of the Bellman and Zadeh (1970) approach. However, in all nontrivial practical problems we face multiple fuzzy goals and fuzzy constraints, and – moreover – such a case is crucial for our next discussion. The extension of the framework presented in the previous section to the case of multiple fuzzy goals and fuzzy constraints may be carried out, fortunately enough, in quite a straightforward manner.

Suppose therefore that we have $n > 1$ fuzzy goals, $G_1, \ldots, G_n$, and $m > 1$ fuzzy constraints, $C_1, \ldots, C_m$, all defined as fuzzy sets in X.

The fuzzy decision can be defined analogously as in the case of one fuzzy goal and one fuzzy constraint. And we obtain the following definitions that parallel those introduced in the previous section.

Namely, in the general case, the *fuzzy decision* is defined as

$$\mu_D(x) = \mu_{G_1}(x) \star \ldots \mu_{G_n}(x) \star$$
$$\star \mu_{C_1}(x) \star \ldots \star \mu_{C_m}(x), \qquad \text{for each } x \in X \qquad (3.23)$$

And again, the most important is the *min-type fuzzy decision* defined as

$$\mu_D(x) = \mu_{G_1}(x) \wedge \ldots \mu_{G_n}(x) \wedge$$
$$\wedge \mu_{C_1}(x) \wedge \ldots \wedge \mu_{C_m}(x), \qquad \text{for each } x \in X \qquad (3.24)$$

The *t-norm-type fuzzy decision* is defined as

$$\mu_D(x) = \mu_{G_1}(x) t \ldots \cdot \mu_{G_n}(x) t$$
$$t \mu_{C_1}(x) t \ldots t \mu_{C_m}(x), \qquad \text{for each } x \in X \qquad (3.25)$$

And among the *t*-norms a particular role is played by the *product-type fuzzy decision* which is defined as

$$\mu_D(x) = \mu_{G_1}(x) \cdot \ldots \mu_{G_n}(x) \cdot$$
$$\cdot \mu_{C_1}(x) \cdot \ldots \cdot \mu_{C_m}(x), \qquad \text{for each } x \in X \qquad (3.26)$$

and, evidently, the value of $\mu_D(x)$ for the product-type fuzzy decision is never higher than for the min-type fuzzy decision.

One can analogously define other *t*-norm-type fuzzy decisions, in principle for all other *t*-norms exemplified by (2.36), page 32, but they will not be used in this book.

The *weighted-sum-type* or *convex-combination-type fuzzy decision* is defined as

$$\mu_D(x) = \sum_{i=1}^{n} \alpha_i \mu_{G_i}(x) + \sum_{j=1}^{m} \beta_j \mu_{C_j}(x), \qquad \text{for each } x \in X \qquad (3.27)$$

where $\alpha_i, \beta_j \in [0,1]; i = 1, \ldots, n; j = 1, \ldots, m$; and $\sum_{i=1}^{n} \alpha_i + \sum_{j=1}^{m} \beta_j = 1$.
The *max-type fuzzy decision* is defined as

$$\mu_D(x) = \mu_{G_1}(x) \vee \ldots \vee \mu_{G_n}(x) \vee$$
$$\vee \mu_{C_1}(x) \vee \ldots \vee \mu_{C_m}(x), \qquad \text{for each } x \in X \qquad (3.28)$$

And, generally, the *s-norm-type fuzzy decision* is defined as

$$\mu_D(x) = \mu_{G_1}(x) s \ldots s \mu_{G_n}(x) s$$
$$s \mu_{C_1}(x) s \ldots s \mu_{C_m}(x), \qquad \text{for each } x \in X \qquad (3.29)$$

and, similarly as for the *t*-norm-type fuzzy decision, one can in principle use any other *s*-norm exemplified by (2.40), page 32.

All remarks made in the previous section concerning the case of one fuzzy goal and one fuzzy constraint remain valid here. This is also true for the importance qualification. To give an example, in the case of the min-type fuzzy decision (3.24), and importances associated with the particular fuzzy goals, $w_{G_1}, \ldots, w_{G_n} \in [0,1]$, and importances associated with the particular fuzzy constraints, $w_{C_1}, \ldots, w_{C_m} \in [0,1]$, then the min-type fuzzy decision becomes

$$\mu_D(x) = w_{G_1} \mu_{G_1}(x) \wedge \ldots \wedge w_{G_n} \mu_{G_n}(x) \wedge$$
$$\wedge w_{C_1} \mu_{C_1}(x) \wedge \ldots \wedge w_{C_m} \mu_{C_m}(x), \qquad \text{for each } x \in X \qquad (3.30)$$

and similarly for other *t*-norm-type fuzzy decisions (3.25), exemplified by the product-type fuzzy decision (3.26).

On the other hand, analogously as in the case of (3.22), for the max-type fuzzy decision (3.28), the addition of importances associated with the particular fuzzy goals, $w_{G_1}, \ldots, w_{G_n} \in [0,1]$, and importances associated with the particular fuzzy constraints, $w_{C_1}, \ldots, w_{C_m} \in [0,1]$, yields

$$\mu_D(x) = (1 - w_{G_1}) \mu_{G_1}(x) \vee \ldots \vee (1 - w_{G_N}) \mu_{G_n}(x) \vee$$
$$\vee (1 - w_{C_1}) \mu_{C_1}(x) \vee \ldots \vee (1 - w_{C_m}) \mu_{C_m}(x), \qquad \text{for each } x \in X \quad (3.31)$$

However, in the case of multiple fuzzy goals and fuzzy constraints the simplicity of the fuzzy decision becomes even more relevant. In fact, we will again employ the min-type fuzzy decision (3.24) [or (3.30) with importances] throughout the book, mentioning only how other types may be used.

All the above types of a fuzzy decision do explicitly reflect the willingness to attain "all" the fuzzy goals and satisfy "all" the fuzzy constraints. This is certainly a very strict and rigid condition. A milder requirement, exemplified by insisting on the attainment of just "most" of the fuzzy goals and the satisfaction of "almost all" of the fuzzy constraints may be more realistic and adequate in many cases. This will be also relevant for our fuzzy control models and will be discussed in Section 4.1.5.

In the following, for all the types of fuzzy decision introduced above, we will assume as an optimal decision – similarly as in the case of one fuzzy goal and one fuzzy constraint – the *maximizing decision* (3.9), page 71, i.e.

$$\mu_D(x^*) = \max_{x \in X} \mu_D(x)$$

and such a simple defuzzification form is implied by the maximization involved in the models to be discussed later in the book. Other defuzzification methods exemplified by that of center-of-gravity (2.83), page 46, would make optimization procedures inefficient, if not impossible to employ.

3.1.3 Fuzzy goals and fuzzy constraints defined in different spaces

Our discussion of the Bellman and Zadeh (1970) approach has concerned so far the fuzzy goals and fuzzy constraints defined in the same space X. This has best served the purpose of introducing the idea of the approach. However, as we will see in the following, for the issues considered in this book, and for virtually all applications in general, an extension of the approach is needed to cover the case of fuzzy goals and fuzzy constraints that are defined as fuzzy sets in different spaces.

Suppose therefore that the fuzzy constraint C is defined as a fuzzy set in $X = \{x\}$, and the fuzzy goal G is defined as a fuzzy set in $Y = \{y\}$. Moreover, suppose that a function $f : X \longrightarrow Y$, $y = f(x)$, is known. Typically, X and Y may be sets of options and outcomes, notably causes and effects, respectively.

Now, the *induced fuzzy goal* G' in X generated by the given fuzzy goal G in Y is defined as

$$\mu_{G'}(x) = \mu_G[f(x)], \qquad \text{for each } x \in X \tag{3.32}$$

Example 3.2 Let $X = \{1, 2, 3, 4\}$, $Y = \{2, 3, \ldots, 10\}$, and $y = 2x + 1$. If now

$$G = 0.1/2 + 0.2/3 + 0.4/4 + 0.5/5 + 0.6/6 + 0.7/7 + 0.8/8 + 1/9 + 1/10$$

then

$$G' = \mu_G(3)/1 + \mu_G(5)/2 + \mu_G(7)/7 = 0.2/1 + 0.5/2 + 0.7/3 + 1/4$$

$$\square$$

The *fuzzy decision* is now defined analogously as in Section 3.1.1, i.e.

$$\mu_D(x) = \mu_{G'}(x) \star \mu_C(x), \qquad \text{for each } x \in X \tag{3.33}$$

where "$\star$" is some aggregation operator as in (3.5), page 70.

By introducing the induced fuzzy goal, it can be seen that both G' and C are defined as fuzzy sets in the same space X which is evidently a *sine qua non* to use all the aggregation operations "$\star$" to be discussed below.

And again, analogously as in Section 3.1.2, we may define the following basic types of the fuzzy decision:

- the *min-type fuzzy decision*

$$\mu_D(x) = \mu_{G'}(x) \wedge \mu_C(x), \qquad \text{for each } x \in X \tag{3.34}$$

- the *t-norm-type fuzzy decision*

$$\mu_D(x) = \mu_{G'}(x) t \mu_C(x), \qquad \text{for each } x \in X \qquad (3.35)$$

- the *product-type fuzzy decision*

$$\mu_D(x) = \mu_{G'}(x) \cdot \mu_C(x), \qquad \text{for each } x \in X \qquad (3.36)$$

- the *weighted-sum-type* or *convex-combination-type fuzzy decision*

$$\mu_D(x) =$$
$$= \alpha \mu_{G'}(x) + (1 - \alpha) \mu_C(x), \qquad \text{for each } x \in X; \alpha \in [0, 1] \qquad (3.37)$$

- the *max-type fuzzy decision*

$$\mu_D(x) = \mu_{G'}(x) \vee \mu_C(x), \qquad \text{for each } x \in X \qquad (3.38)$$

- the *s-norm-type fuzzy decision*

$$\mu_D(x) = \mu_{G'}(x) s \mu_C(x), \qquad \text{for each } x \in X \qquad (3.39)$$

- the *algebraic-sum-type fuzzy decision*

$$\mu_D(x) = \mu_{G'}(x) + \mu_C(x) - \mu_{G'}(x) \cdot \mu_C(x), \qquad \text{for each } x \in X \qquad (3.40)$$

Evidently, all remarks and comments concerning the particular fuzzy decisions are valid here too.

Finally, for $n > 1$ fuzzy goals $G_1, \ldots, G_n$ defined in Y, $m > 1$ fuzzy constraints $C_1, \ldots, C_m$ defined in X, and a function $f : X \longrightarrow Y$, $y = f(x)$, we analogously have

$$\mu_D(x) = \mu_{G'_1}(x) \star \cdots \star \mu_{G'_n}(x)$$
$$\star \mu_{C_1}(x) \star \cdots \star \mu_{C_n}(x), \qquad \text{for each } x \in X \qquad (3.41)$$

and as the aggregation operation "$\star$" may be taken, e.g., any of the operations used in (3.34)–(3.40).

Clearly, the min-type fuzzy decision given below is fundamental here:

$$\mu_D(x) = \mu_{G'_1}(x) \wedge \cdots \wedge \mu_{G'_n}(x)$$
$$\wedge \mu_{C_1}(x) \wedge \cdots \wedge \mu_{C_n}(x), \qquad \text{for each } x \in X \qquad (3.42)$$

The inclusion of importances associated with the particular fuzzy goals and fuzzy constraints proceeds analogously as in Section 3.1.2, and will not be repeated.

In all the types of fuzzy decision, an optimal decision is assumed, analogously as in the case of one fuzzy goal and one fuzzy constraint, to be the *maximizing decision* defined as (3.9), i.e.

$$\mu_D(x^*) = \max_{x \in X} \mu_D(x)$$

though we can clearly assume another defuzzification method, notably the center of gravity (2.83), page 46, but for practical reasons the above maximizing decision will be used since optimization will be involved in all our models.

Finally, let us mention that Piskunov (1992a, b) presents a disussion of inconsistencies in Bellman and Zadeh's (1970) model. Basically, he states that the fuzzy goals and fuzzy constraints are defined in different universes of discourse, so they may be methodologically and semantically different. Therefore, he proposes to view the fuzzy goal as an absolute measure of the level of satisfaction of the decision maker. Next, two general attitutes are possible which are reflected by the resulting two problem formulations:

- to attain the best possible satisfaction of the fuzzy goal(s) and to find decisions (feasible, in the sense of the the set of fuzzy constraints) that make this possible, and
- to attain a best possible compromise between the satisfaction of the fuzzy goals(s) and constraint(s).

Though Piskunov's (1992a, b) ideas are certainly relevant and point out some relevant issues related to Bellman and Zadeh's (1970) approach, one may view the situation as follows: Bellman and Zadeh's (1970) approach is certainly not perfect but this is a natural consequence of its inherent simplicity. It is therefore, first, a perfect point of departure for an analysis, and second, it is perfect for our purposes since in the control models to be considered an important element is always some kind of optimization, and this precludes in practice the use of any more sophisticated formal representations, if we wish to develop constructive procedures that would be able to solve nontrivial problems. Hence, Bellman and Zadeh's (1970) framework presented in this chapter will be employed throughout this book.

We are now in a position to proceed to the presentation of multistage decision making (control) under fuzziness (fuzzy goals and fuzzy constraint) in Bellman and Zadeh's setting. This will constitute the core of all the following discussions.

3.2   MULTISTAGE DECISION MAKING (CONTROL) IN A FUZZY ENVIRONMENT

The decision-making problem is now of a dynamic type, and – in our context – is better referred to as *control*. We will therefore extensively use here, and later in this book, control-related notation and terminology. Mainly, decisions will be referred to as *controls*, the discrete time moments at which decisions are to be made – as *control stages*, and the input–output (or cause–effect) relationship – as a *system under control*.

To formally delimit the class of control problems considered, we will present first a simple model and then indicate some of its possible extensions.

The essence of multistage control in a fuzzy environment (under fuzziness) has already been schematically presented in Figure 5, page 14, and we show this figure again, in Figure 16, for convenience of the reader.

First, suppose that the control space is $U = \{u\} = \{c_1, \ldots, c_m\}$ and the state space is $X = \{x\} = \{s_1, \ldots, s_n\}$. For simplicity, which will not, however, limit the generality of our discussion, the control is equated with the input, and the state with the output. See also that the control and state spaces are assumed finite here and throughout this book which is mainly motivated by practical requirements as, in any

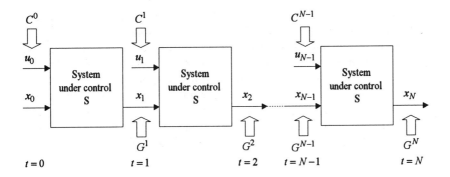

Figure 16 Essence of the multistage control in a fuzzy environment (under fuzziness)

case, even if we assumed infinite control and state spaces, we would need to discretize them (i.e. replace by finite ones) for computations. Moreover, this will simplify most of our discussion.

The *control process* proceeds basically as follows. In the beginning we are in some initial state $x_0 \in X$. We apply a control $u_0 \in U$ which is subjected to a fuzzy constraint $\mu_{C^0}(u_0)$. We attain a state $x_1 \in X$ via a known cause–effect relationship, i.e. a state transition equation of the system under control governing the system under control S; a fuzzy goal $\mu_{G^1}(x_1)$ is imposed on x_1. Next, we apply a control u_1 which is subjected to a fuzzy constraint $\mu_{C^1}(u_1)$, and attain a fuzzy state x_2 on which a fuzzy goal $\mu_{G^2}(x_2)$ is imposed, etc.

We consider here, and in the following, *open-loop control*. Unfortunately, not much is known about closed-loop (feedback) control in a fuzzy environment in the optimal-control-type Bellman and Zadeh (1970) setting considered here.

Suppose now that the system under control is deterministic and its temporal evolution is governed by a *state transition equation*

$$f : X \times U \longrightarrow X \tag{3.43}$$

such that

$$x_{t+1} = f(x_t, u_t), \qquad t = 0, 1, \ldots \tag{3.44}$$

where $x_t, x_{t+1} \in X = \{s_1, \ldots, s_n\}$ are the states at control stages t and $t + 1$, respectively, and $u_t \in U = \{c_1, \ldots, c_m\}$ is the control at control stage t.

Equation (3.44) represents a time-invariant deterministic system under control. In a more general case of a time-varying system, the system's temporal evolution is given as $f_t : X \times U \longrightarrow X$ such that

$$x_{t+1} = f_t(x_t, u_t), \qquad t = 0, 1, \ldots \tag{3.45}$$

i.e. the state transition function f_t itself depends on a particular control stage t; we will use in the following the time-invariant system (3.44), and the use of the time-varying system (3.45) would not change the reasoning.

The temporal evolution of the system under control results therefore in a *trajectory* $(x_0, u_0, x_1, u_1, x_2, u_2, \ldots, x_{t-1}, u_{t_1}, x_t, \ldots)$.

At each control stage t, the control applied $u_t \in U$ is subjected to a *fuzzy constraint* $\mu_{C^t}(u_t)$, and on the state attained $x_{t+1} \in X$ a *fuzzy goal* is imposed; $t = 0, 1, \ldots$.

The *initial state* is $x_0 \in X$ and is assumed to be known, and given in advance. The *termination time* (planning, or control, horizon), i.e. the maximum number of control stages, is denoted by $N \in \{1, 2, \ldots\}$, and may be finite or infinite.

The *performance* (goodness) of the multistage control process under fuzziness is evaluated by the fuzzy decision $D(x_0)$ which, being an aggregate of the subsequent (at the subsequent control stages) fuzzy constraints and goals, evaluates how well these are satisfied by controls applied and by the resulting states attained, respectively.

Therefore, the fuzzy decision is

$$D(x_0) = C^0 \star G^1 \star \cdots \star C^{N-1} \star G^N \tag{3.46}$$

or, in terms of membership functions,

$$\mu_D(u_0, \ldots, u_{N-1} \mid x_0) =$$
$$= \mu_{C^0}(u_0) \star \mu_{G^1}(x_1) \star \cdots \star \mu_{C^{N-1}}(u_{N-1}) \star \mu_{G^N}(x_N) \tag{3.47}$$

where the consecutive attained states x_t are expressed by the trajectory $(x_0, u_0, x_1, u_1, x_2, \ldots, x_{t-1}, u_{t-1})$ via the consecutive application of the state transition equation (3.44), i.e.

$$\begin{cases} x_1 = f(x_0, u_0) \\ x_2 = f(x_1, u_1) = f(f(x_0, u_0), u_1) \\ \cdots\cdots\cdots\cdots\cdots\cdots\cdots\cdots\cdots\cdots\cdots\cdots\cdots\cdots \\ x_N = f(x_{N-1}, u_{N-1}) = f(f(\ldots(f(x_0, u_0), u_1), \ldots, u_{N-2}), u_{N-1} \end{cases} \tag{3.48}$$

The fuzzy decision (3.47) is therefore assumed to be a decomposable [in the sense of (2.57), page 37] fuzzy set in $\underbrace{U \times \cdots \times U}_{n}$, and hence we use here and later the notation of the type (3.47), i.e. in which the particular terms depend only on one particular control.

The most relevant and popular is again the min-type fuzzy decision

$$\mu_D(u_0, \ldots, u_{N-1} \mid x_0) =$$
$$= \mu_{C^0}(u_0) \wedge \mu_{G^1}(x_1) \wedge \ldots \wedge \mu_{C^{N-1}}(u_{N-1}) \wedge \mu_{G^N}(x_N) \tag{3.49}$$

In most cases, however, a slightly simplified form of the fuzzy decision (3.47) is used, namely it is assumed that all the subsequent fuzzy controls, $u_0, u_1, \ldots, u_{N-1}$, are subjected to the fuzzy constraints, $\mu_{C^0}(u_0), \mu_{C^1}(u_1), \ldots, \mu_{C^{N-1}}(u_{N-1})$, while the fuzzy goal is just imposed on the final state x_N, $\mu_{G^N}(x_N)$. In such a case the fuzzy decision becomes

$$D(x_0) = C^0 \star \cdots \star C^{N-1} \star G^N \tag{3.50}$$

or, in terms of membership functions,

$$\mu_D(u_0, \ldots, u_{N-1} \mid x_0) = \mu_{C^0}(u_0) \star \cdots \star \mu_{C^{N-1}}(u_{N-1}) \star \mu_{G^N}(x_N) \tag{3.51}$$

And, again, the most relevant for our discussions is the min-type fuzzy decision

$$\mu_D(u_0, \ldots, u_{N-1} \mid x_0) = \mu_{C^0}(u_0) \wedge \ldots \wedge \mu_{C^{N-1}}(u_{N-1}) \wedge \mu_{G^N}(x_N) \tag{3.52}$$

Such a simplified form of a fuzzy decision will be assumed in most parts of this book, but this will not limit the generality of discussion as one may readily assume the "extended" fuzzy decision (3.49).

An important issue is importance qualification of the particular control stages the essence of which is that some weights $w_0, w_1, \ldots, w_{N-1} \in [0,1]$ are assigned to $t = 0, 1, \ldots, N-1$, respectively. In practice, this takes the form of *discounting* whose very meaning is that the importance of what happens earlier is greater than of what happens later. The rationale behind this is simple. On the one hand, it is motivated by a generally accepted attitude that now is more important than in future. On the other hand, the models consider here "start" from now, i.e. from $t = 0$, but determine what to do in the future, i.e. the next optimal controls $u_0^*, u_1^*, \ldots, u_{N-1}^*$. However, this planning is full of uncertainty as, e.g., the behavior of the system under control cannot be exactly predicted now for the future.

The discounting can be added to the min-type fuzzy decision in the following simple ways:

- for the fuzzy decision with the fuzzy constraints and fuzzy goals imposed at all the consecutive control stages, i.e. for (3.49):

$$
\begin{aligned}
\mu_D(u_0, \ldots, u_{N-1} \mid x_0) &= \\
&= b^0[\mu_{C^0}(u_0) \wedge \mu_{G^1}(x_1)] \wedge \ldots \\
&\quad \ldots \wedge b^{N-1}[\mu_{C^{N-1}}(u_{N-1}) \wedge \mu_{G^N}(x_N)] = \\
&= \bigwedge_{t=0}^{N-1} b^t[\mu_{C^t}(u_t) \wedge \mu_{G^{t+1}}(x_{t+1})]
\end{aligned}
\tag{3.53}
$$

where $b > 1$, and

- for the fuzzy decision with the fuzzy constraints imposed at $t = 0, 1, \ldots, N-1$ and the fuzzy goal imposed at $t = N$, i.e. for (3.52):

$$
\begin{aligned}
\mu_D(u_0, \ldots, u_{N-1} \mid x_0) &= \\
&= b^0 \mu_{C^0}(u_0) \wedge b^1 \mu_{C^0}(u_0) \wedge \ldots \\
&\quad \ldots \wedge b^{N-1}[\mu_{C^{N-1}}(u_{N-1}) \wedge \mu_{G^N}(x_N)]
\end{aligned}
\tag{3.54}
$$

The multistage control problem in a fuzzy environment is now formulated as to find an optimal sequence of controls $u_0^*, \ldots, u_{N-1}^*$, $u_t^* \in U$, $t = 0, 1, \ldots, N-1$, such that:

$$
\mu_D(u_0^*, \ldots, u_{N-1}^* \mid x_0) = \max_{u_0, \ldots, u_{N-1} \in U} \mu_D(u_0, \ldots, u_{N-1} \mid x_0)
\tag{3.55}
$$

where $\mu_D(u_0, \ldots, u_{N-1} \mid x_0)$ is given by one of those introduced above, both with and without discounting.

In the class of control problems considered it is usually more convenient to express the solution, i.e. the controls to be applied, in the form of a *control policy function*, or, briefly, a *policy function* or even *policy* which is defined, for the control stage t, as

$$
a_t : X \longrightarrow U, \qquad t = 0, 1, \ldots
\tag{3.56}
$$

such that

$$
u_t = a_t(x_t), \qquad t = 0, 1, \ldots
\tag{3.57}
$$

that is, the control to be applied at control stage t is expressed as a function of the state at stage t.

Such a simple definition is, however, not always applicable, and then the policy is to be defined in the following extended form:

$$a_t : X \times U \times X \times \cdots \times U \times X \longrightarrow U, \qquad t = 0, 1, \ldots \qquad (3.58)$$

such that

$$u_t = a_t(x_0, u_0, x_1, \ldots, u_{t-1}, x_t), \qquad t = 0, 1, \ldots \qquad (3.59)$$

i.e. the control to be applied at the current stage t depends not only on the current state x_t but also on the whole past trajectory.

In practice, however, the above definition (3.58) usually takes the following simpler form:

$$a_t : [0, 1] \times X \longrightarrow U, \qquad t = 0, 1, \ldots \qquad (3.60)$$

such that

$$u_t = a_t[w_t(x_0, u_0, x_1, \ldots, u_{t-1}), x_t], \qquad t = 0, 1, \ldots \qquad (3.61)$$

where

$$w_t : X \times U \times X \times \cdots \times X \times U \longrightarrow [0, 1], \qquad t = 0, 1, \ldots \qquad (3.62)$$

is some function "subsuming" the past trajectory; therefore, in this case the control at the current stage t depends on the current state and a "summary" of the past (up to stage $t - 1$) trajectory.

In our next discussion we will not refer to the above different policy functions by different names but we will always clearly indicate which one of the policy types will be employed.

We have now an extremely important concept of a *stationary policy* defined as

$$a : X \longrightarrow U \qquad (3.63)$$

such that

$$u_t = a(x_t), \qquad t = 0, 1 \ldots \qquad (3.64)$$

i.e. the current control to be applied is related to the current state always (at all control stages) in the same way.

And, analogously as for (3.58), sometimes we need to employ the stationary policy meant as

$$a : X \times U \times X \times \cdots \times U \times X \longrightarrow U \qquad (3.65)$$

such that

$$u_t = a(x_0, u_0, x_1, \ldots, u_{t-1}, x_t), \qquad t = 0, 1, \ldots \qquad (3.66)$$

and, analogously as for (3.60), we may often be able to use a simplified form

$$a : [0, 1] \times X \longrightarrow U \qquad (3.67)$$

such that

$$u_t = a[w(x_0, u_0, x_1, \ldots, u_{t-1}), x_t], \qquad t = 0, 1, \ldots \qquad (3.68)$$

where

$$w : X \times U \times X \times \cdots \times X \times U \longrightarrow [0, 1], \qquad t = 0, 1, \ldots \qquad (3.69)$$

is some function "subsuming" the past trajectory; therefore, in this case the control depends on the current state and a "summary" of the past trajectory.

A *control strategy*, or briefly a *strategy*, is now defined as a sequence of (control) policies

$$A = (a_0, a_1, \ldots, a_{N-1}) \tag{3.70}$$

and the *stationary strategy* is

$$a_N = a = \underbrace{(a, a, \ldots, a)}_{N} \tag{3.71}$$

and note that we use the same notation for the strategy (a_N) and policy (a) which should not lead to confusion.

Evidently, for an infinite termination time, the strategy (actually, a stationary strategy as it makes no sense to consider non-stationary strategies in such a case) is denoted by

$$a_\infty = (a, a, \ldots) \tag{3.72}$$

Thus, in terms of the strategy and policies the fuzzy decision (3.51), page 80, becomes

$$\mu_D(A \mid x_0) =$$
$$= \mu_{C^0}(a_0(x_0)) \star \cdots \star \mu_{C^{N-1}}(a_{N-1}(x_{N-1})) \star \mu_{G^N}(x_N) \tag{3.73}$$

and the problem (3.55), page 81, becomes here that of finding an *optimal strategy* $A^* = a_0^*, \ldots, a_{N-1}^*)$ such that

$$\mu_D(A^* \mid x_0) = \max_A \mu_D(A \mid x_0) \tag{3.74}$$

with the following ordering "$\succeq$" between the strategies

$$A' \succeq A'' \iff \mu_D(A' \mid x_0) \geq \mu_D(A'' \mid x_0), \qquad \text{for each } x_0 \in X \tag{3.75}$$

i.e. the strategy A' is said to be better or equal to the strategy A'' if and only if the right-hand-side inequality in (3.75) holds for each initial state.

Evidently, *an optimal strategy* A^* satisfies (3.75) for each strategy A, i.e. $A^* \succeq A$, for each A.

We have therefore a basic formulation of multistage control in a fuzzy environment. One may readily see that its various extensions are possible. The most important aspects of the problem formulation are as follows:

- the type of termination time,
- the type of system under control, and
- the type of fuzzy decision (aggregation operation).

These aspects influence to the greatest extent the specifics of the formulation and solution of the problem. Therefore, the division of the next chapters will proceed according to, and in the order of, the above main aspects.

We will divide the book into chapters dealing with particular types of the termination time, and these chapters will be subdivided into parts dealing with the particular types of system under control. In addition, the cases of particular types of fuzzy decision will be discussed.

The following types of termination time will be discussed:

- fixed and specified in advance,
- implicitly given by entering a termination set of states,
- fuzzy, and
- infinite.

The following types of system under control will be considered:

- deterministic,
- stochastic, and
- fuzzy.

As to the type of fuzzy decision, our discussion will basically proceed for the basic case of the min-type fuzzy decision as it is the most important and most widely used. In most cases we will mention the cases of other more important types of fuzzy decision as, e.g., the product type, t-norm type, weighted average type, max type, etc.

4

Control Processes with a Fixed and Specified Termination Time

This chapter presents the analysis of the basic class of multistage optimal control problems considered in this book, namely with a fixed and specified termination time given as a predefined number of control stages. This case will be a point of departure for virtually all problems considered in the following, and will be therefore discussed in more detail.

As already indicated in Section 1.2, the analysis will be carried out for the following three systems under control:

- deterministic,
- stochastic, and
- fuzzy.

Our discussion will basically proceed for the min-type fuzzy decision (3.24), page 74, though its results will be valid for other types of fuzzy decisions too, notably for those based on t-norms (3.25) [as, e.g., the product-type one (3.26)]. In most cases it will be quite clear as to which property of a fuzzy decision (its underlying operation) is relevant for the applicability of a particular algorithm. We will also discuss this in more detail.

4.1 CONTROL OF A DETERMINISTIC SYSTEM

As we have already indicated in Section 2.4, this is the basic system under control whose analysis provides a point of departure for other systems under control.

We assume that the temporal evolution (dynamics) of the deterministic system under control is given by the following state transition equation [i.e. by (2.130), page 59]

$$x_{t+1} = f(x_t, u_t), \qquad t = 0, 1, \ldots \qquad (4.1)$$

where: $x_t, x_{t+1} \in X = \{s_1, \ldots, s_n\}$ is the *state* (equated for simplicity with the output throughout this book) at control stage (time) t and $t + 1$, respectively, and $u_t \in U = \{c_1, \ldots, c_m\}$ is the *control* (input) at control stage (time) t. The initial state is $x_0 \in X$, and the (finite!) termination time N is fixed and specified in advance, and given as a positive integer number of control stages after which the control process should terminate.

At each control stage t, the control $u_t \in U$ is subjected to a *fuzzy constraint*, $\mu_{C^t}(u_t)$, and on the final (i.e., at the final control stage N) state $x_N \in X$ a *fuzzy goal*, $\mu_{G^N}(x_N)$, is imposed. This assumption that a fuzzy goal is imposed on the final state only is basically made for simplicity, and does not limit the generality of analysis in virtually all the cases to be considered. Therefore, we could well assume that the fuzzy goals are imposed on all the consecutively attained states, i.e. $\mu_{G^1}(x_1), \ldots, \mu_{G^{N-1}}(x_{N-1}), \mu_{G^N}(x_N)$.

Both the control and state are now assumed to be nonfuzzy, and the fuzziness is in the environment in which the control process is to proceed, i.e. in the fuzzy constraints on the controls and fuzzy goals on the states.

As to the performance requirement for the control process, we require the controls applied at the consecutive control stages to best fulfill the fuzzy constraints, and the states attained to best satisfy the fuzzy goals.

The fuzzy decision, which measures the degree to which the above requirement is met, is therefore [cf. (3.49, page 80]

$$\mu_D(u_0, \ldots, u_{N-1} \mid x_0) = \mu_{C^0}(u_0) \wedge \ldots \wedge \mu_{C^{N-1}}(u_{N-1}) \wedge \mu_{G^N}(x_N) \tag{4.2}$$

where x_N is expressed by $x_0, u_0, u_1, \ldots, u_{N-1}$ via (4.1), i.e. $x_N = f(x_{N-1}, u_{N-1}) = f[f(x_{N-2}, u_{N-2}), u_0]$, etc.

The problem is to find an *optimal sequence of controls* – $u_0^*, \ldots, u_{N-1}^*$; $u_t^* \in U$, $t = 0, \ldots, N-1$ – which best satisfies the above fuzzy constraints and fuzzy goals [i.e. which maximizes the fuzzy decision (4.2) [cf. (3.55), page 81], i.e.

$$\mu_D(u_0^*, \ldots, u_{N-1}^* \mid x_0) =$$
$$= \max_{u_0, \ldots, u_{N-1}} \mu_D(u_0, \ldots, u_{N-1} \mid x_0) =$$
$$= \max_{u_0, \ldots, u_{N-1}} [\mu_{C^0}(u_0) \wedge \ldots \wedge \mu_{C^{N-1}}(u_{N-1}) \wedge \mu_{G^N}(x_N)] \tag{4.3}$$

This problem can be solved using the following two basic traditional techniques:

- dynamic programming, and
- branch-and-bound,

and also using the following two new ones:

- a genetic algorithm, and
- a neural network.

These techniques will be discussed in the following sections.

4.1.1   *Solution by dynamic programming*

The application of dynamic programming for the solution of problem (4.3) was proposed in the seminal paper of Bellman and Zadeh (1970). For clarity, it is better to begin the presentation of this approach by slightly rewriting the problem to be solved (4.3): find an optimal sequence of controls $u_0^*, \ldots, u_{N-1}^*$ such that

$$\mu_D(u_0^*, \ldots, u_{N-1} \mid x_0) = \max_{u_0, \ldots, u_{N-1}} [\mu_{C^0}(u_0) \wedge \ldots$$
$$\ldots \wedge \mu_{C^{N-1}}(u_{N-1}) \wedge \mu_{G^N}(f(x_{N-1}, u_{N-1}))] \tag{4.4}$$

It is easy to see that the very structure of this problem (4.4) makes the application of dynamic programming possible. Namely, the last two right-hand-side terms, i.e.

$$\mu_{C^{N-1}}(u_{N-1}) \wedge \mu_{G^N}(f(x_{N-1}, u_{N-1}))$$

depend only on control u_{N-1} and not on any previous controls.

The maximization over the sequence of controls $u_0, \ldots, u_{N-1}$ in (4.4) can be therefore divided into two phases:

- the maximization over the control sequence: $u_0, \ldots, u_{N-2}$, and
- the maximization over u_{N-1},

which can be written as

$$\mu_D(u_0^*, \ldots, u_{N-1}^* \mid x_0) =$$
$$= \max_{u_0, \ldots, u_{N-2}} \{\mu_{C^0}(u_0) \wedge \ldots \wedge \mu_{C^{N-2}}(u_{N-2}) \wedge$$
$$\wedge \max_{u_{N-1}} [\mu_{C^{N-1}}(u_{N-1}) \wedge \mu_{G^N}(f(x_{N-1}, u_{N-1}))]\} \tag{4.5}$$

And further, continuing the same line of reasoning, since the next term $\mu_{C^{N-2}}(u_{N-2})$ depends only on the control u_{N-2}, then (4.5) can be rewritten as

$$\mu_D(u_0^*, \ldots, u_{N-1}^* \mid x_0) =$$
$$= \max_{u_0, \ldots, u_{N-3}} (\mu_{C^0}(u_0) \wedge \ldots \wedge \mu_{C^{N-3}}(u_{N-3}) \wedge \max_{u_{N-2}} \{\mu_{C^{N-2}}(u_{N-2}) \wedge$$
$$\wedge \max_{u_{N-1}} [\mu_{C^{N-1}}(u_{N-1}) \wedge \mu_{G^N}(f(x_{N-1}, u_{N-1}))]\}) \tag{4.6}$$

This backward iteration, which reflects the essence of dynamic programming, can be repeated for $u_{N-3}, u_{N-4}, \ldots, u_0$ (i.e. for the control stages $N - 3, N - 4, \ldots, 0$) which leads to the following set of dynamic programming recurrence equations

$$\begin{cases} \mu_{G^{N-i}}(x_{N-i}) = \max_{u_{N-i}} [\mu_{C^{N-i}}(u_{N-i}) \wedge \mu_{G^{N-i+1}}(x_{N-i+1})] \\ x_{N-i+1} = f(x_{N-i}, u_{N-i}), \qquad i = 0, 1, \ldots, N \end{cases} \tag{4.7}$$

where $\mu_{G^{N-i}}(x_{N-i})$ may be regarded as a fuzzy goal at control stage $t = N - i$ induced by the fuzzy goal at $t = N - i + 1$, $i = 0, 1, \ldots, N$.

The optimal sequence of control sought, $u_0^*, \ldots, u_{N-1}^*$, is given by the successive maximizing values of u_{N-i}, $i = 1, \ldots, N$ in (4.7). Each such a maximizing value, u_{N-i}^* is obtained as a function of x_{N-i} which is evident if we look at the left-hand side of the first equation in the set (4.7).

Clearly, the solution of the problem considered by solving the set of recurrence equations (4.7) reflects the famous Bellman's *optimality principle* which states that in an optimal sequence of controls from the initial to the final control stage, $u_0^*, \ldots, u_{N-1}^*$, its part from the $t = K - 1$ to the final control stage $t = N - 1$, i.e. $u_{K-1}^*, \ldots, u_{N-1}^*$, must itself be optimal.

We obtain therefore an *optimal control policy* (or *optimal policy*, for short)

$$a_{N-i}^* : X \longrightarrow U, \qquad i = 1, \ldots, N \tag{4.8}$$

such that

$$u^*_{N-i} = a^*_{n-i}(x_{N-i}), \qquad i = 1, \ldots, N \qquad (4.9)$$

An optimal solution to problem (4.3), $u^*_0, \ldots, u^*_{N-1}$, exists if there is at least one control sequence $u_0, \ldots, u_{N-1}$ for which $\mu_D(u_0, \ldots, u_{N-1} \mid x_0) > 0$.

A simple example from the source paper of Bellman and Zadeh (1970) would be a perfect illustration of the algorithm presented above.

Example 4.1 Let the state space be $X = \{s_1, s_2, s_3\}$, the control space be $U = \{c_1, c_2\}$, the termination time be $N = 2$, and the fuzzy constraints and fuzzy goal be, respectively:

$$C^0 = 0.7/c_1 + 1/c_2$$
$$C^1 = 1/c_1 + 0.8/c_2$$
$$G^2 = 0.3/s_1 + 1/s_2 + 0.8/s_3$$

Suppose that the dynamics of the deterministic system under control is governed by the state transition equation of type (4.1), represented by the following state transition table:

$$x_{t+1} = \quad
\begin{array}{c|ccc}
 & x_t = s_1 & s_2 & s_3 \\
\hline
u_t = c_1 & s_1 & s_3 & s_1 \\
c_2 & s_2 & s_1 & s_3
\end{array}
\qquad (4.10)$$

to be read that, e.g., if the current state is $x_t = s_2$ and the control applied is c_2, then the state attained is $x_{t+1} = s_1$.

First, using the fuzzy dynamic programming recurrence equations (4.7) for $i = 1$, we obtain $G^1 = 0.6/s_1 + 0.8/s_2 + 0.6/s_3$, and the corresponding optimal control policy

$$a^*_1(s_1) = c_2 \quad a^*_1(s_2) = c_1 \quad a^*_1(s_3) = c_2$$

Next, the subsequent iteration of (4.7), for $i = 2$, yields $G^0 = 0.8/s_1 + 0.6/s_2 + 0.6/s_3$ and the corresponding optimal control policy

$$a^*_0(s_1) = c_2 \quad a^*_0(s_2) \in \{c_1, c_2\} \quad a^*_1(s_3) \in \{c_1, c_2\}$$

Therefore, for instance, if we start at $t = 0$ from $x_0 = s_1$, then $u^*_0 = a^*_0(s_1) = c_2$ and we obtain $x_1 = s_2$. Next, at $t = 1$, $u^*_1 = a^*_1(s_2) = c_1$ and $\mu_D(u^*_0, u^*_1 \mid s_1) = \mu_D(c_2, c_1 \mid s_1) = 0.8$. $\qquad \square$

4.1.1.1   Dynamic programming recurrence equations for other types of fuzzy decision

It is easy to see that the very structure of the problem considered (4.3), i.e. with the min-type fuzzy decision, as well as the line of reasoning adopted that has led to the use of dynamic programming for the solution of the problem via the set of recurrence equations of type (4.7), implies that a similar procedure is applicable for some other types of the fuzzy decision presented in Section 3.1.2.

For illustration we will only show below the sets of recurrence equations of the product type (3.26), page 74, weighted-sum type (3.27), and max type (3.28) fuzzy decisions. The example given for the min-type fuzzy decision can be easily solved for these other types of fuzzy decision, for example making comparisons possible.

The sets of dynamic programming recurrence equations for the above-mentioned fuzzy decisions are as follows:

- for the product-type fuzzy decision

$$\begin{cases} \mu_{G^{N-i}}(x_{N-i}) = \max_{u_{N-i}}[\mu_{C^{N-i}}(u_{N-i}) \cdot \mu_{G^{N-i+1}}(x_{N-i+1})] \\ x_{N-i+1} = f(x_{N-i}, u_{N-i}); \qquad i = 0, 1, \ldots, N \end{cases} \quad (4.11)$$

- for the weighted-sum-type fuzzy decision

$$\begin{cases} \mu_{\overline{G}^N}(x_N) = r_N \cdot \mu_{G^N}(x_N) \\ \mu_{\overline{G}^{N-i}}(\quad N-i \\ \qquad = \max_{u_{N-i}}[r_N \cdot \mu_{C^{N-i}}(u_{N-i}) + \mu_{\overline{G}^{N-i+1}}(x_{N-i+1})] \\ x_{N-i+1} = f(x_{N-i}, u_{N-i}); \qquad i = 0, 1, \ldots, N \end{cases} \quad (4.12)$$

where $\mu_{\overline{G}^N}(.)$ is some auxiliary fuzzy goal at a respective control stage which makes the use of the weighted sum possible,

- for the max-type fuzzy decision

$$\begin{cases} \mu_{G^{N-i}}(x_{N-i}) = \\ \qquad = \max_{u_{N-i}}[\mu_{C^{N-i}}(u_{N-i}) \vee \mu_{G^{N-i+1}}(x_{N-i+1})] \\ x_{N-i+1} = f(x_{N-i}, u_{N-i}); \qquad i = 0, 1, \ldots, N \end{cases} \quad (4.13)$$

Needless to say, by solving each of the above sets of dynamic programming recurrence equations we obtain an optimal control at control stage $N - i$, u_{N-i}^*, in the form of an optimal control policy a_{N-i}^* such that $u_{N-i}^* = a_{N-i}^*(x_{N-i}), i = 1, \ldots, N$.

4.1.1.2 Remarks on other formulations of fuzzy dynamic programming

As we have already mentioned, Chang (1969a, b) was probably the first to extend dynamic programming to accommodate fuzziness, when he proposed its application to find approximate optimal solutions of partially known systems in optimal terminal control. Basically, his model involved the following main elements:

- the state vector at stage $t = 0, 1, \ldots, N - 1$, x_t,
- the decision vector at stage $t = 0, 1, \ldots, N - 1$, u_t,
- the transition function governing the dynamics of the system under control

$$x_{t+1} = f(x_t, u_y, q_t)$$

where q_t is an unknown parameter, and
- the return function at stage $t = 0, 1, \ldots, N - 1$, $g(x_t, u_t, q_t)$.

The problem is to find a sequence of optimal controls, $u_0^*, u_1^*, \ldots, u_{N-1}^*$, to move the process to a terminal set of states $X_N \in X$ with the minimal total return S_N, i.e. such that

$$\min_{u_0, u_1, \ldots, u_{N-1}} \{S_N = R(X_N) + \sum_{t=0}^{N-1} g(x_t, u_t, q_t)\} \quad (4.14)$$

where $R(X_N)$ is some return associated with the reaching of X_N.

If now ρ is a control policy (relating the control to a current state), then fuzziness in the model is represented by the membership function $\mu_M(\rho, x_0, R)$ which constitutes a measure of the degree to which policy ρ yields a return $S_N \leq R$ when starting from an initial state x_0; this estimate is therefore fuzzy.

It may easily be noticed that in Chang's (1969a, b) approach the fuzziness is introduced in a different way than in Bellman and Zadeh's (1970) approach. In this book we will use only the latter.

4.1.2 Solution by branch-and-bound

The very nature of dynamic programing, which stems from its underlying Bellman optimality principle, implies that the solution process (consecutive iterations of the set of recurrence equations (4.7) proceeds "backwards," i.e. from the control stage $t = N - 1$ to the control stage $t = 0$. This is evidently not changed in its substance in that by an appropriate change of control stage indices we receive "forward" dynamic programming recurrence equations [cf. Nemhauser (1966) or any book on dynamic programming] in which these control stage indices increase in the course of the algorithm, i.e. from $t = 0$ to $t = N - 1$.

This backward solution process may be viewed to be in some way contradicting the very essence of real decision making processes in which, evidently, we start at an initial state x_0 (at control stage $t = 0$), apply control u_0, attain state x_1, apply control u_1, attain state x_2, ..., and, finally, apply control u_{N-1} and attain state x_N.

Fortunately enough, it turns out that a forward solution by *branch-and-bound* technique can be devised in this case too. It was proposed by Kacprzyk (1978a, 1979) solving fuzzy control and multistage decision making problems [cf. a more general setting discussed in Chang and Pavlidis (1977)].

The branch-and-bound technique will be widely used throughout this book as it is simple and efficient. In presenting its applications in the particular problem classes we will assume the familiarity with the basic idea and principle of branch-and-bound, but even the discussion in this section will make it possible to understand the idea.

The branch-and-bound procedure starts in our context from the initial state x_0. We apply control u_0 and proceed to state x_1. Next, we apply control u_1 and proceed to state x_2, etc. Finally, being in state x_{N-1}, we apply control u_{N-1} and attain state x_N. This process, for a finite control space (as we assume throughout the book) may be best presented as a decision tree (cf. Figure 17) whose nodes are the particular states attained, and with whose edges the controls applied are associated. A simple example may illustrate this, as well as clarify many of the issues presented next.

Example 4.2 Suppose that the basic problem specifications are as in Example 4.1, page 88, i.e. the state space is $X = \{s_1, s_2, s_3\}$, the control space is $U = \{c_1, c_2\}$, the termination time is $N = 2$, and the state transition are given by (4.10), i.e.

$$x_{t+1} = \quad \begin{array}{c|ccc} & x_t = s_1 & s_2 & s_3 \\ \hline u_t = c_1 & s_1 & s_3 & s_1 \\ c_2 & s_2 & s_1 & s_3 \end{array}$$

Then the decision tree is as in Figure 17. The nodes of the decision tree are associated with the particular $x_1, x_2 \in \{s_1, s_2, s_3\}$ obtained via the state transition equation given by the above table, while the edges are assigned the values of controls applied at the consecutive control stages $t = 0, 1$, i.e. $u_0, u_1 \in \{c_1, c_2\}$.

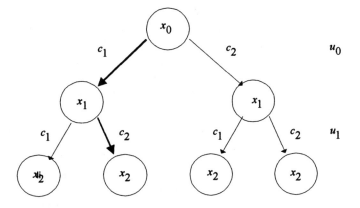

Figure 17 Example of a decision tree

To each path in the decision tree, i.e. to each sequence of nodes end edges from x_0 to x_2 in Figure 17 representing the consecutive controls applied and states attained, there corresponds some value of the fuzzy decision

$$\mu_D(u_0, u_1 \mid x_0) = \mu_{C^0}(u_0) \wedge \mu_{C^1}(u_1) \wedge \mu_{G^2}(x_2) \tag{4.15}$$

For instance, for the path shown by the bold arrows in Figure 17, i.e. to that corresponding to $u_0 = c_1$ and $u_1 = c_2$, the value of (4.15) will be equal to

$$\mu_D(c_1, c_2 \mid x_0) = \mu_{C^0}(c_1) \wedge \mu_{C^1}(c_2) \wedge \mu_{G^2}(x_2) =$$
$$= \mu_{C^0}(c_1) \wedge \mu_{C^1}(c_2) \wedge \mu_{G^2}[f(f(x_0, c_1), c_2)] \tag{4.16}$$

An optimal sequence of control u_0^*, u_1^* is now sought which maximizes the fuzzy decision (4.15). In this simple case one can well perform the full enumeration as there are only four possible paths from x_0 to x_2:

$$(x_0, c_1, x_1, c_1, x_2)$$
$$(x_0, c_1, x_1, c_2, x_2)$$
$$(x_0, c_2, x_1, c_1, x_2)$$
$$(x_0, c_2, x_1, c_2, x_2)$$

and no more sophisticated algorithm is needed for the solution of the problem in such a simple case.

An optimal solution is clearly $u_0^* = c_1$ and $u_1^* = c_2$.   □

Due to the finiteness of the control space assumed throughout the book, even the problem formulation in a general case given by (4.3), page 86, i.e. to find an optimal sequence of controls $u_0^*, \ldots, u_{N-1}^*$ such that

$$\mu_D(u_0^*, \ldots, u_{N-1}^* \mid x_0) =$$
$$= \max_{u_0, \ldots, u_{N-1}} \mu_D(u_0, \ldots, u_{N-1} \mid x_0) =$$
$$= \max_{u_0, \ldots, u_{N-1}} [\mu_{C^0}(u_0) \wedge \ldots \wedge \mu_{G^N}(x_N)]$$

is equivalent to finding a path in a (finite) decision tree of the type shown in Figure 17.

The solution based on full enumeration, i.e., on the analysis of all the possible paths from the initial to the final state, is in principle possible. However, it is quite evident that for any nontrivial case with a nontrivial, realistic dimension of the control space and the number of control stages, such a procedure is highly inefficient. An implicit (partial) enumeration search should therefore replace the above "blind" full enumeration search.

Basically, it should consist in the neglecting as early as possible (at as low as possible a control stage) of nonpromising paths, and the traversing of a possibly small set of promising paths only.

Since the decision tree is traversed from the initial state x_0 through the consecutive edges (controls) and nodes (states), then the very essence of the above-mentioned implicit enumeration mechanism may be viewed to be roughly equivalent to the answer of the following question:

> If we currently (at the current control stage) arrive at some node (state), then to which node (out of those traversed so far) should we most rationally (to proceed further along the currently most promising path) add next edges (controls)?

> Or, in other words, at a specified moment of the decision tree traversal, what is the best continuation of the search process?

The answer to this crucial question is in fact the essence of the branch-and-bound procedure. To present this idea, let us first denote

$$\begin{cases} v_0 = \mu_{C^0}(u_0) \\ \dots \\ v_k = \mu_{C^0}(u_0) \wedge \dots \wedge \mu_{C^k}(u_k) = v_{k-1} \wedge \mu_{C^k}(u_k) \\ \dots \\ v_{N-1} = \mu_{C^0}(u_0) \wedge \dots \wedge \mu_{C^{N-1}}(u_{N-1}) = v_{N-2} \wedge \mu_{C^{N-1}}(u_{N-1}) \\ v_N = \mu_{C^0}(u_0) \wedge \dots \wedge \mu_{C^{N-1}}(u_{N-1}) \wedge \mu_{G^N}(x_N) = \\ \quad = v_{N-1} \wedge \mu_{G^N}(x_N) = \mu_D(u_0, \dots, u_{N-1} \mid x_0) \end{cases} \qquad (4.17)$$

Let us now discuss a very important implication of the use of "$\wedge$" (the minimum operator): if we consider some sequence of controls $u_0, \dots, u_k$, $0 < k < N - 1$, then we have, for each $k < w \leq N - 1$, that

$$v_k \geq v_w = v_k \wedge \mu_{C^{k+1}}(u_{k+1}) \wedge \dots \wedge \mu_{C^w}(u_w) \qquad (4.18)$$

because, due to "$\wedge$," by "adding" to v_k any further terms we cannot increase the value of v_w.

In particular, there also holds

$$v_k \geq v_N = \mu_D(u_0, \dots, u_{N-1} \mid x_0) \qquad (4.19)$$

Let us therefore assume that we are at the k-th control stage, and have traversed so far some nodes and edges (from x_0 to x_k). Now we have to most rationally choose a node, among those attained so far, to which we should add the edges (apply the controls). Evidently, since we are willing to find at the end a trajectory corresponding to the maximal value of $\mu_D(u_0, \ldots, u_{N-1} \mid x_0)$, the most rational choice (from the point of view of the current situation!) is to choose the most promising node, i.e., the one which corresponds to the greatest value of v_i, $i-1, \ldots, k$. The other nodes cannot lead (at that particular moment!) to any optimal solution since they cannot obviously yield any higher value of v_i if we add next edges (apply next controls).

The above property of the min-type fuzzy decision makes it possible to devise a branch-and-bound algorithm in which the branching is through the controls applied at the consecutive control stages, and the bounding is via the values of the particular v_k's, $k = 0, \ldots, N$.

A formal description of this branch-and-bound algorithm is not necessary here as, first, we generally assume in this book that the reader is familiar with the fundamentals of such basic solution techniques as dynamic programming or branch-and-bound which. On the other hand, the branching and bounding used in this algorithm is simple and transparent.

Some comments on a simple implementation of this branch-and-bound algorithm used here may be of help to the readers who may want to use the algorithm. This implementation is shown as a block diagram in Figure 18. Some symbols shown in this block diagram need explanation; these concern mostly tables (arrays) used to store temporary data and results. All the tables (or vectors if there is just one row) occurring in the block diagram have the same format as table VX whose rows are:

1	2	3	...	$k+3$	...	$N+3$
v_k	k	u_0	...	u_k	...	x_{k+1}

to be understood as follows: as a result of the k-th control stage (k in column 2) the state x_{k+1} (see column $N+3$) was attained through the sequence of consecutive controls $u_0, \ldots, u_k$ (see columns $2-k+3$) with $v_k = \mu_{C^0}(u_0) \wedge \ldots \wedge \mu_{C^k}(u_k)$ (see column 1); a row of table VX represents therefore a particular trajectory from x_0 to the current x_k (the intermediate states are not included as they are not relevant).

The rows of table VX are ordered from the highest to the lowest value of v_k, that is

$$VX(i_i, 1) \geq VX(i_2, 1), \qquad \text{for each } i_1 > i_2$$

so that at the top of the table (the first row) there is always data on the most promising (at the particular control stage!) sequence of controls applied so far.

All the other tables and vectors occurring in the block diagram of the algorithm have the same format of rows. Table OPT contains the optimal results for $k = N$, i.e., the sequences of controls for N control stages (for the whole control process considered) whose corresponding value of v_N is the highest; these are the optimal solutions sought. Note that the algorithm may be implemented to yield both an optimal solution (i.e., the first optimal solution found) and all the optimal solutions; in the first case we may often arrive at a solution in a short time, while in the second case it may take much more time to complete the algorithm (but the results obtained are evidently much more prolific), all the optimal solutions are determined, and a further choice of the "desired" optimal solution is possible using, e.g., some more profound analysis.

Moreover, though the implementation of the branch- and-bound algorithm shown in Figure 18 is meant to yield an optimal solution(s), it may often be desirable to obtain a suboptimal solution but in a shorter time. In such a case the algorithm

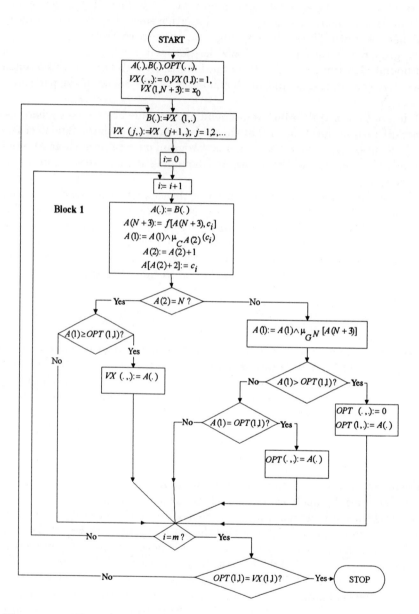

Figure 18   Block diagram of the branch-and-bound algorithm for the control of a deterministic system with a fixed and specified termination time

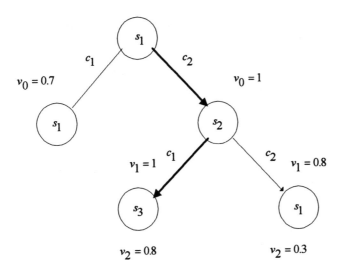

Figure 19   Decision tree with the optimal solution for Example 4.3

can be easily modified in a couple of ways as, e.g., by eliminating the backtracking while considering the promising nodes to add controls or by limiting the breadth and depth of analysis while looking for promising nodes [i.e., how many candidate nodes to consider at a particular level (control stage) and how many past control stages to consider]. These are quite straightforward modifications of the original procedure and will not be considered here. Note also that other heuristic rules can well be employed to modify the procedure so that its efficiency is higher.

To illustrate the procedure we will consider now the same simple example used in Section 4.1.1 for the case of dynamic programming, i.e. Example 4.1 on page 88.

Example 4.3 For clarity let us briefly repeat the formulation of the problem. Suppose that, as in Example 4.1, the state space is $X = \{s_1, s_2, s_3\}$, the control space is $U = \{c_1, c_2\}$, the termination time is $N = 2$, and the fuzzy constraints and fuzzy goal are, respectively:

$$C^0 = 0.7/c_1 + 1/c_2$$
$$C^1 = 1/c_1 + 0.8/c_2$$
$$G^2 = 0.3/s_1 + 1/s_2 + 0.8/s_3$$

Dynamics of the system under control is governed by the state transition equation of type (4.1) represented by the following state transition table:

$$x_{t+1} =$$

		$x_t = s_1$	s_2	s_3
$u_t = c_1$		s_1	s_3	s_1
c_2		s_2	s_1	s_3

(4.20)

The solution obtained via branch-and-bound is shown graphically in Figure 19, page 95, which should be read as follows: at $t = 0$ in the initial state $x_0 = s_1$ we can apply control c_1 and attain $v_0 = 0.7$ [cf. (4.17)] or c_2 and attain $v_0 = 1$. We choose the latter case as the more promising, hence we proceed to $x_1 = s_2$ and apply control

c_1 attaining $v_1 = 1$ or c_2 attaining $v_1 = 0.8$. The former is more promising, so we assume $x_2 = s_3$ attaining $v_2 = 0.8$. As there is no other higher v_k attained so far, this is the optimal solution sought; its corresponding trajectory is shown in bold arrows in Figure 19. Evidently, this optimal solution is the same as the one obtained using dynamic programming (cf. Example 4.1, page 88).   □

The branch-and-bound procedure is also applicable for some other types of the fuzzy decision (cf. Section 4.1.1.1), notably for the product-type fuzzy decision (4.11), that is

$$\mu_D(u_0, \ldots, u_{N-1} \mid x_0) = \mu_{C^0}(u_0) \cdot \ldots \cdot \mu_{C^{N-1}}(u_{N-1}) \cdot \mu_{G^N}(x_N)$$

In this case we obtain the following set of equations similar to (4.17)

$$\begin{cases} v_0 = \mu_{C^0}(u_0) \\ \ldots \\ v_k = \mu_{C^0}(u_0) \cdot \ldots \cdot \mu_{C^k}(u_k) = v_{k-1} \cdot \mu_{C^k}(u_k) \\ \ldots \\ v_{N-1} = \mu_{C^0}(u_0) \cdot \ldots \cdot \mu_{C^{N-1}}(u_{N-1}) = v_{N-2} \cdot \mu_{C^{N-1}}(u_{N-1}) \\ v_N = \mu_{C^0}(u_0) \cdot \ldots \cdot \mu_{C^{N-1}}(u_{N-1}) \cdot \mu_{G^N}(x_N) = \\ \quad = v_{N-1} \cdot \mu_{G^N}(x_N) = \mu_D(u_0, \ldots, u_{N-1} \mid x_0) \end{cases} \quad (4.21)$$

and then, analogously as for the min-type fuzzy decision, there holds, for each $k < w \leq N - 1$,

$$v_k \geq v_w = v_k \cdot \mu_{C^{k+1}}(u_{k+1}) \cdot \ldots \cdot \mu_{C^{N-1}}(u_w) \geq$$
$$\geq v_N = \mu_D(u_0, \ldots, u_{N-1} \mid x_0) \quad (4.22)$$

The block diagram of the branch-and-bound algorithm for the case of the product-type fuzzy decision is obviously the same as that for the min-type fuzzy decision (Figure 18), with the obvious replacement of "∧" by "·" in the respective blocks.

Evidently, the branch-and-bound algorithm is not applicable for all types of fuzzy decision since not all of them satisfy the branch-and-bound's underlying properties of types (4.18) and (4.19). For instance, for the weighted-sum-type (3.27) and max-type (3.28) fuzzy decisions such properties evidently do not hold so that the branch-and-bound algorithm cannot be employed.

Finally, let us take a note that a different branch-and-bound algorithm proposed by Esogbue, Theologidu and Guo (1992) will be presented in Section 8.2.3. That algorithm is also applicable for non-monotonic objective functions (fuzzy decisions), but the bounding procedure is much more complex. For virtually all the problems (fuzzy decisions) considered in this book the branch-and-bound algorithm described in this section is applicable. One can also mention the work of Isik and Ammar (1992) in which some search methods of branch-and-bound type in decision trees have been developed, also for fuzzy evaluations of paths traversed.

4.1.3 Solution by a genetic algorithm

From the previous sections we have learned that the solution of the multistage control problem considered (4.3), page 86, may be really difficult for practical problems of a non-trivial size, in spite of being relatively simple conceptually. This has been particularly true for the dynamic programming formulation discussed in Section 4.1.1, plagued by its inherent "curse of dimensionality."

The branch-and-bound formulation discussed in Section 4.1.2 has proven more efficient in practice although its efficiency is also limited by the combinatorial nature of the problem solved.

A natural consequence of all the above would therefore be to try to employ something different, which would constitute a more general solution tool, possibly easily implementable. Such a tool may be, e.g., a so-called genetic algorithm, and its use for solving the problem considered (4.3), page 86, which has been proposed by Kacprzyk (1995a, b), will be presented below.

First, we will provide the reader with a short introduction to genetic algorithms, concentrating on those aspects which will be of use for our discussion here and later in this section.

In general, any problem-solving process (including the multistage fuzzy control problems discussed here!) can be perceived as a search through the space of potential solutions to end up with an "optimal," or maybe a best possible one, e.g., for the particular length of the search process allowed. This should all be performed "intelligently."

Genetic algorithms are a prominent example of such techniques. They are stochastic algorithms whose search methods "mirror" some phenomena underlying natural processes, notably evolution. Among such phenomena, *genetic inheritance* and the Darwinian *survival of the fittest* play a particular role.

In this section we will provide a short overview of genetic algorithms. However, its basic purpose would rather be to fix the terminology used in this book in relation to genetic algorithms, and present their basic principles of operation. This would make our analysis practically self-contained as we will use only genetic algorithms in their basic versions, without all the sophisticated additional mechanisms. Thus, since genetic algorithms have already become a standard topic included in many textbook, then the reader has either already been exposed to them or can easily find a proper reference in a vast literature on this subject published by all major publishers. For instance, Michalewicz's (1994) book may be recommended.

In our context, the basic terminology will be as follows. First, by an *individual* we will mean a particular solution, i.e. the particular values of controls at the consecutive control stages, $u_0, \ldots, u_{N-1}$. It is *evaluated* in our context by the fuzzy decision (4.2), page 86, which is here the *evaluation function*, called also a *fitness functions*. This function expresses the *fitness* (goodness) of the particular solutions (sequences of controls).

A set of potential solutions will be termed a *population*. We will assume that the algorithm will operate within a population of a fixed size. So, we initially assume some number of potential solutions (the initial population) which has been, e.g., randomly generated. Next, some members of the population, who serve the role of parents, will undergo *reproduction* through the so-called *crossover* and *mutation* to produce off-springs (children), i.e. some new solutions. Then, basically, the best ones (the fittest) will "survive," i.e. will be used while repeating this process. Finally, at the end of such a process one may expect to find a very good (if not optimal) solution.

The structure of a genetic algorithm may be portrayed as follows:

begin
$t =: 0$
set the initial population $P(t)$
evaluate strings in $P(t)$
while termination condition is not fulfilled do:
begin
$t := t + 1$
select current population $P(t)$ from $P(t-1)$
perform reproduction on elements of $P(t)$
calculate the evaluation function for each element of $P(t)$
end
end

The very essence of the above scheme is clear but we need to clarify some of its basic elements, namely:

- how to represent a potential solution,
- how to create (generate) an initial population,
- how to define an evaluation function that is meant to rate solutions in terms of their fitness,
- how to perform the reproduction, i.e. for our purposes how to define the crossover and mutation, and
- how to choose some parameters as, e.g., the population size, probabilities of applying some operations, the termination condition, etc.

To show the idea of a genetic algorithm in more detail, though just in its basic form suitable for our next discussion, we will describe now the essentials of the above issues using a very simple example.

First, in our context, the potential solutions are sequences of controls as, e.g., u_0, u_1, u_2 for the case of a control process with the termination time of $N = 3$. If now the possible values of the controls are $\{0, 1, 2, 3, 4, 5, 6, 7\}$, i.e. $u_0, u_1, u_2 \in \{0, 1, \ldots, 7\}$, then there may be the following solution candidates:

$$\begin{array}{ll} \text{Solution 1:} & (2, 4, 5) \\ \text{Solution 2:} & (1, 7, 6) \\ \text{Solution 3:} & (3, 2, 5) \end{array}$$

which, if the above values of controls are represented in binary notation (i.e. as binary strings), may be represented in a convenient tabular form as, respectively:

Solution 1:	0	1	0	1	0	0	1	0	1
Solution 2:	1	0	0	1	1	1	0	1	1
Solution 3:	1	1	0	0	1	0	1	0	1

Suppose now that the above three solution candidates (which are generated in some way as, e.g., randomly) is our initial population. Each solution candidate is then evaluated by using the fuzzy decision (4.2), and suppose that that its respective values are

$$e_1 = \mu_D(2,4,5 \mid .) = 0.7$$
$$e_2 = \mu_D(1,7,6 \mid .) = 0.3$$
$$e_3 = \mu_D(3,2,5 \mid .) = 0.9$$

Now, the probability of selection of the solution candidate $i \in \{1,2,3\}$ is defined as

$$p_i = \frac{e_i}{e_1 + e_2 + e_3} \tag{4.23}$$

Suppose therefore that by random selection, based on the selection probability p_i, the solutions 2 and 3 are selected out of the initial population to be the "parents." They are now subjected to the two basic operations to "produce" their offsprings (children).

The first operation is *crossover*. We generate, using a *crossover probability*, some point (bit number) in the binary string from 1 to the length of the string (9 in our case), e.g., 3. This may be conveniently shown using a vertical arrow as

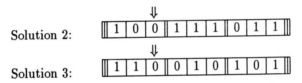

The crossover operation exchanges the respective bits between the two parents from bit 4 on which yields the following two new candidate solutions:

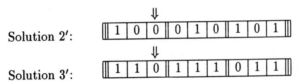

One of the two candidates produced as above, Solution 2′ or 3′, is selected at random and taken as the new individual. Suppose that we randomly select Solution 3′, i.e.

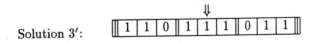

Next, using a prespecified *mutation probability*, we select a point in the above binary string, say 5, and change this bit in the string to the opposite value (i.e. 0 to 1 and vice versa), which may be shown as

– from

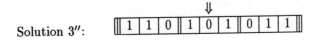

– we obtain

Solution 3″: ⇓ 1 1 0 1 0 1 0 1 1

and this new candidate solution is introduced into the new population.

The process of random selection of the parents, crossover, and mutation is repeated until the new population of the predefined size is obtained. Then, the whole process is repeated, and is continued until some termination condition is satisfied as, e.g., the maximum number of iterations (or a time limit for the process) is reached, there has not been for some time any improvement, etc.

The above process is conceptually and implementationally simple, and usually leads to good results. It may be viewed justified by mirroring some processes taking place in nature (natural selection, evolution, survival of the fittest, etc.). Due to random mechanisms widely employed, it escapes from local optima, and helps find a globally best (optimal) solution.

The scheme of a genetic algorithm given above is basic, and many modifications of both the crossover and mutation, as well as new operations, have been proposed. This is, however, beyond the scope of our discussion, and the reader is referred to, e.g., Davis (1991) or Michalewicz (1994).

Finally, let us remark that the binary representation of solutions (the so-called binary coding) has been used traditionally in genetic algorithms. It does provide a convenient conceptual framework, and facilitates theoretical analysis. However, it is easy to see that it is certainly not a necessary condition for the general genetic algorithm scheme shown in the beginning of this section.

And, indeed, one can well imagine that a real coding of solutions is used. In fact, in our context a real coding is clearly more natural as the solution candidate may be represented as a string of real numbers (the subsequent controls). Such a real coding will be employed throughout this book. We will, however, use the basic operations of crossover and mutation that will be direct derivatives from the traditional ones; for more sophisticated operation in the case of real coding the reader is referred to, e.g., Michalewicz (1994).

We are now in a position to proceed to the use of a real coded genetic algorithm in the context of multistage fuzzy control.

To present the idea of Kacprzyk's (1995a, b) genetic-algorithm-based approach to multistage fuzzy control, it may be expedient to briefly restate the problem considered.

First, the deterministic system under control is given by the state transition equation (4.1), page 85, i.e.

$$x_{t+1} = f(x_t, u_t), \qquad t = 0, 1, \ldots \tag{4.24}$$

where: $x_t, x_{t+1} \in X = \{s_1, \ldots, s_n\}$ is the *state* (output) at control stages t and $t + 1$, respectively, and $u_t \in U = \{c_1, \ldots, c_m\}$ is the *control* (input) at t. The initial state is $x_0 \in X$, and the (finite) termination time N is fixed and specified in advance.

At each control stage t, the control $u_t \in U$ is subjected to a *fuzzy constraint* $\mu_{C^t}(u_t)$, and on the final state $x_N \in X$ a *fuzzy goal* $\mu_{G^N}(x_N)$ is imposed (fuzzy goals at the subsequent t's may also be assumed, and the reasoning remains valid – cf. Section 3.2).

The evaluation (performance or fitness) function is the fuzzy decision (4.2), page 86, that is

$$\mu_D(u_0, \ldots, u_{N-1} \mid x_0) =$$
$$= \mu_{C^0}(u_0) \wedge \ldots \wedge \mu_{C^{N-1}}(u_{N-1}) \wedge \mu_{G^N}(x_N) \tag{4.25}$$

and the problem is [cf. (4.3), page 86] to find an *optimal sequence of controls*,

$u_0^*, \ldots, u_{N-1}^*$, such that

$$\mu_D(u_0^*, \ldots, u_{N-1}^* \mid x_0) =$$
$$= \max_{u_0, \ldots, u_{N-1}} \mu_D(u_0, \ldots, u_{N-1} \mid x_0) = \max_{u_0, \ldots, u_{N-1}} [\mu_{C^0}(u_0) \wedge \ldots$$
$$\ldots \wedge \mu_{C^{N-1}}(u_{N-1}) \wedge \mu_{G^N}(x_N)] \tag{4.26}$$

where $x_N = f(x_{N-1}, u_{N-1})$ by the state transition equation (4.24), and "$\wedge$" (minimum) may be replaced by another operation (cf. Section 3.2).

For solving the problem given above, we will now use a genetic algorithm. Its basic elements are meant as follows:

- the problem is represented by strings of controls $u_0, \ldots, u_{N-1}$, and we use real coding which is evidently convenient in this case;
- the evaluation (objective) function is the fuzzy decision (4.25), i.e.

$$\mu_D(u_0, \ldots, u_{N-1} \mid X_0) = \mu_{C^0}(u_0) \wedge \ldots \wedge \mu_{C^{N-1}}(u_{N-1}) \wedge \mu_{G^N}(x_N)$$

- standard random selections of elements from the consecutive populations, standard concepts of crossover and mutation (evidently applied to real coded strings), and a standard termination condition, mainly a predefined number of iterations, or iteration-to-iteration improvement lower than a threshold is used;

For the use of such a genetic algorithm in the particular problem class considered, we assume further that:

- controls are assumed as "evenly spaced" real numbers in $[0, 1]$ corresponding to $c_1, \ldots, c_m$, and
- states are defined as "evenly spaced" real numbers from $[0, 1]$ corresponding to $s_1, \ldots, s_n$.

The genetic algorithm works now basically as follows:

begin
    $t =: 0$
    set the initial population $P(t)$ which consists of
        randomly generated strings of controls
        (i.e. of randomly generated real numbers from $[0, 1]$);
    for each $u_0, \ldots, u_{N-1}$ in each
        string in the population $P(t)$,
        find the resulting x_{t+1}
        by using the state transition
        equation $x_{t+1} = f(x_t, u_t)$,
        and use the evaluation function (4.25)
        $\mu_D(u_0, \ldots, u_{N-1} \mid x_0)$ to evaluate each string in $P(t)$;
while $t <$ *maximum number of iterations* **do**
begin
    $t := t + 1$
    assign the probabilities to each string in $P(t - 1)$

which are proportional to the value of the evaluation
function for each string;
randomly (using those probabilities) generate
the new population $P(t)$;
perform crossover and mutation on the strings in $P(t)$;
calculate the value of the evaluation function (4.25) for each string in $P(t)$.
    end
end

The algorithm does evidently follow a general scheme of a genetic algorithm, without more specialized and specific mechanisms and "tricks," and has proven to be, first, conceptually simple and easily implementable (which is certainly not the case for any other solution method used, i.e. dynamic programming, branch-and-bound and the neural network based algorithm), and second, quite efficient. We will illustrate now this algorithm by a simple example.

Example 4.4 Suppose that the state space is $X = \{s_1, \ldots, s_{20}\}$, the control space is $U = \{c_1, \ldots, c_{32}\}$, the planning horizon is $N = 10$, and the initial state is $x_0 = s_1$.
The state transition equation (4.24) is given as

$$x_{t+1} = f(x_t, u_t) =$$

$x_t = s_1$	s_2	s_3	s_4	s_5	s_6	s_7	s_8	s_9	s_{10}	$\ldots$
$u_t = c_1$ $\quad$ s_1	s_1	s_2	s_3	s_4	s_5	s_6	s_7	s_8	s_9	$\ldots$
c_2 $\quad$ s_2	s_3	s_3	s_3	s_6	s_7	s_8	s_9	s_{10}	s_{11}	$\ldots$
c_3 $\quad$ s_2	s_3	s_3	s_4	s_5	s_6	s_7	s_8	s_9	s_{10}	$\ldots$
$\ldots$	$\ldots$	$\ldots$	$\ldots$	$\ldots$	$\ldots$	$\ldots$	$\ldots$	$\ldots$	$\ldots$	$\ldots$
c_{30} $\quad$ s_4	s_5	s_6	s_7	s_{10}	s_{11}	s_{12}	s_{14}	s_{15}	s_{15}	$\ldots$
c_{31} $\quad$ s_5	s_6	s_7	s_8	s_{10}	s_{11}	s_{12}	s_{14}	s_{15}	s_{16}	$\ldots$
c_{32} $\quad$ s_6	s_7	s_8	s_{10}	s_{11}	s_{12}	s_{14}	s_{15}	s_{16}	s_{17}	$\ldots$

$$\ldots$$

$\ldots$	$x_t = s_{11}$	s_{12}	s_{13}	s_{14}	s_{15}	s_{16}	s_{17}	s_{18}	s_{19}	s_{20}
$u_t = c_1$ $\quad \ldots$	s_{10}	s_{11}	s_{12}	s_{13}	s_{14}	s_{15}	s_{16}	s_{17}	s_{18}	s_{19}
c_2 $\quad \ldots$	s_{12}	s_{13}	s_{14}	s_{15}	s_{16}	s_{17}	s_{18}	s_{19}	s_{20}	s_{20}
c_3 $\quad \ldots$	s_{11}	s_{12}	s_{13}	s_{14}	s_{15}	s_{16}	s_{17}	s_{18}	s_{19}	s_{20}
$\ldots$ $\quad \ldots$	$\ldots$	$\ldots$	$\ldots$	$\ldots$	$\ldots$	$\ldots$	$\ldots$	$\ldots$		
c_{30} $\quad \ldots$	s_{16}	s_{17}	s_{19}	s_{19}	s_{20}	s_{20}	s_{20}	s_{20}	s_{20}	s_{20}
c_{31} $\quad \ldots$	s_{16}	s_{17}	s_{18}	s_{19}	s_{20}	s_{20}	s_{20}	s_{20}	s_{20}	s_{20}
c_{32} $\quad \ldots$	s_{17}	s_{18}	s_{19}	s_{20}	s_{20}	s_{20}	s_{20}	s_{20}	s_{20}	s_{20}

The fuzzy constraints and fuzzy goals at the consecutive control stages are given as trapezoid fuzzy numbers in $[0, 1]$ as shown in Figure 20, i.e. are equated with the 4-tuples (a, b, c, d) to be understood as follows. We assume that the particular states $s_1, \ldots, s_{20}$ and controls $c_1, \ldots, c_{32}$ correspond to "evenly spaced" real numbers from $[0, 1]$, with s_1 and c_1 equal to 0, and s_{20} and c_{32} equal to 1.

Then, the 4-tuple (a, b, c, d) is equivalent to $(s_{i1}, s_{i2}, s_{i3}, s_{i4})$ or $(c_{j1}, c_{j2}, c_{j3}, c_{j4})$, such that $s_{i1} \leq s_{j2} \leq s_{i3} \leq s_{i4}$ and $c_{j1} \leq c_{j2} \leq c_{j3} \leq c_{j4}$, respectively. Needless to say that though in Figure 20 the fuzzy constraints and goals are shown as functions

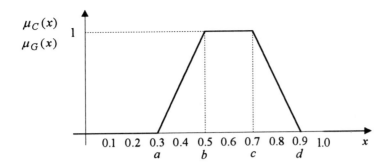

Figure 20 Trapezoid fuzzy constraints and goals for Example 4.4 – for simplicity, they are shown as being defined in $[0, 1]$ while in reality they are defined in X and U, respectively

defined in the unit interval $[0, 1]$, they are in fact defined in discrete state and control spaces.

The fuzzy constraints and goals are therefore assumed to be:

$$C^0 = (c_1, c_1, c_4, c_{32}) \qquad G^1 = (s_1, s_2, s_7, s_9)$$
$$C^1 = (c_1, c_1, c_7, c_{32}) \qquad G^2 = (s_2, s_3, s_9, s_{11})$$
$$C^2 = (c_1, c_1, c_9, c_{32}) \qquad G^3 = (s_3, s_5, s_9, s_{11})$$
$$C^3 = (c_1, c_1, c_{10}, c_{31}) \qquad G^4 = (s_4, s_7, s_{12}, s_{14})$$
$$C^4 = (c_1, c_1, c_{12}, c_{32}) \qquad G^5 = (s_5, s_8, s_{14}, s_{16})$$
$$C^5 = (c_1, c_1, c_{13}, c_{32}) \qquad G^6 = (s_6, s_{10}, s_{16}, s_{18})$$
$$C^6 = (c_1, c_1, c_{15}, c_{32}) \qquad G^7 = (s_7, s_{11} s_{16}, s_{18})$$
$$C^7 = (c_1, c_1, c_{17}, c_{32}) \qquad G^8 = (s_9, s_{14}, s_{18}, s_{20})$$
$$C^8 = (c_1, c_1, c_{18}, c_{32}) \qquad G^9 = (s_{11}, s_{16}, s_{20}, s_{20})$$
$$C^9 = (c_1, c_1, c_{20}, c_{32}) \qquad G^{10} = (s_{14}, s_{18}, s_{20}, s_{20})$$

We assume that the main parameters are:

- the population size is 250,
- the number of trials is 32,000,
- the crossover rate is 0.6, and
- the mutation rate is 0.001.

We obtain 3 best ("optimal") results (starting from $x_0 = s_1$):

$$u_0^* = c_3 \quad u_1^* = c_4 \quad u_2^* = c_4 \quad u_3^* = c_5 \quad u_4^* = c_8$$
$$u_5^* = c_{11} \quad u_6^* = c_{12} \quad u_7^* = c_{14} \quad u_8^* = c_{15} \quad u_9^* = c_{15}$$

$$u_0^* = c_2 \quad u_1^* = c_5 \quad u_2^* = c_3 \quad u_3^* = c_5 \quad u_4^* = c_8$$
$$u_5^* = c_{11} \quad u_6^* = c_{13} \quad u_7^* = c_{14} \quad u_8^* = c_{14} \quad u_9^* = c_{15}$$

$$u_0^* = c_3 \quad u_1^* = c_6 \quad u_2^* = c_3 \quad u_3^* = c_5 \quad u_4^* = c_8$$
$$u_5^* = c_{11} \quad u_6^* = c_{12} \quad u_7^* = c_{14} \quad u_8^* = c_{15} \quad u_9^* = c_{15}$$

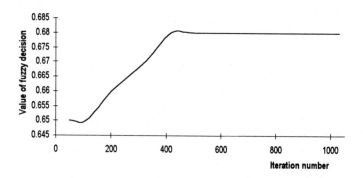

Figure 21   The value of the fuzzy decision obtained in the course of iterations
in Example 4.4

for which the corresponding value of the fuzzy decision (4.25) is

$$\mu_D(u_0^*, \ldots, u_9^* \mid s_1) = 1$$

The next best result is

$$
\begin{array}{lllll}
u_0^* = c_3 & u_1^* = c_6 & u_2^* = c_3 & u_3^* = c_5 & u_4^* = c_8 \\
u_5^* = c_{12} & u_6^* = c_{13} & u_7^* = c_{15} & u_8^* = c_{12} & u_9^* = c_{14}
\end{array}
$$

and its corresponding value of the fuzzy decision (4.25) is

$$\mu_D(u_0^*, \ldots, u_0^* \mid s_1) = 0.973684$$

while the tenth best result is

$$
\begin{array}{lllll}
u_0^* = c_2 & u_1^* = c_4 & u_2^* = c_3 & u_3^* = c_5 & u_4^* = c_8 \\
u_5^* = c_{12} & u_6^* = c_{13} & u_7^* = c_{15} & u_8^* = c_{12} & u_9^* = c_{14}
\end{array}
$$

and its corresponding value of the fuzzy decision (4.25) is

$$\mu_D(u_0^*, \ldots, u_9^* \mid s_1) = 0.973684$$

In Figure 21 we show the best values of the fuzzy decision (4.25) obtained in
the course of iterations. It may readily be seen that the optimal solution has been
attained quite early, and the 32,000 iterations assumed have not been necessary, in
fact. In general, also for many different problems solved, the algorithm has proven to
be efficient.   □

of control stages, and the size of the state and control spaces, by using any of the other techniques (dynamic programming, branch-and-bound and neural networks) would certainly be more complicated from the conceptual, implementational and computational points of view.

Evidently, though much work is still needed in regard to the use of genetic algorithms in multistage fuzzy control problems, it seems that they may be a promising tool. This may be certainly amplified by employing some evolutionary strategies (cf. Michalewicz, 1994), and the first attempt in this direction was made by Kacprzyk (1995b).

A genetic algorithm will also be used in the sequel for the control of a fuzzy system in Section 4.3.4, and – as we will see – with an even stronger motivation.

4.1.4 Solution by a neural network

In this section we will present an interesting, non-traditional approach to employ an artificial neural network for solving the fuzzy multistage control problem with the deterministic system under control and a fixed and specified termination time. This approach was proposed by Francelin and Gomide (1992, 1993) [cf. also Francelin, Gomide and Kacprzyk (1995)]. Our discussion will basically be self-contained, and we will assume that the reader is familiar with only the fundamental concepts of (artificial) neural networks. Since neural networks now attract an enormous attention among representatives of virtually all fields and domains, textbooks and monographs on this subject are available from virtually all major scientific publishers; for instance, one may consult Simpson (1989).

For convenience of the reader let us briefly repeat the problem formulation. The temporal evolution (dynamics) of the deterministic system under control is given by the state transition equation (4.1), page 85, i.e.

$$x_{t+1} = f_t(x_t, u_t), \qquad t = 0, 1, \ldots \tag{4.27}$$

where: $x_t, x_{t+1} \in X = \{s_1, \ldots, s_n\}$ is the *state* (output) at t and $t+1$, respectively, and $u_t \in U = \{c_1, \ldots, c_m\}$ is the *control* (input) at t. Note that we assume here a slightly more general case that the function f_t in the state equation given above may be different for different t's, i.e. we assume a time-varying deterministic system. The initial state is $x_0 \in X$, and the termination time $N < \infty$ is fixed and specified in advance.

At each t, $u_t \in U$ is subjected to a *fuzzy constraint*, $\mu_{C^t}(u_t)$, and on the states attained $x_{t+1} \in X$ *fuzzy goals*, $\mu_{G^{t+1}}(x_{t+1})$ are imposed. We assume here that the fuzzy goals are imposed on all the control stages, and not just on the final stage as usually done throughout our discussion thus far; evidently, this does not change the essence of the analysis.

The fuzzy decision is therefore [cf. (4.2), page 86]

$$\mu_D(u_0, \ldots, u_{N-1} \mid x_0) =$$
$$= \mu_{C^0}(u_0) \wedge \mu_{G^1}(x_1) \wedge \ldots \wedge \mu_{C^{N-1}}(u_{N-1}) \wedge \mu_{G^N}(x_N) \tag{4.28}$$

where x_{t+1} is expressed by x_t and u_t via (4.27).

The problem is [cf. (4.3), page 86] to find an *optimal sequence of controls,*

$u_0^*, \ldots, u_{N-1}^*$, such that

$$
\mu_D(u_0^*, \ldots, u_{N-1}^* \mid x_0) =
$$
$$
= \max_{u_0, \ldots, u_{N-1}} \mu_D(u_0, \ldots, u_{N-1} \mid x_0) =
$$
$$
= \max_{u_0, \ldots, u_{N-1}} [\mu_{C^0}(u_0) \wedge \mu_{G^1}(x_1) \wedge \ldots
$$
$$
\ldots \wedge \mu_{C^{N-1}}(u_{N-1}) \wedge \mu_{G^N}(x_N)] \tag{4.29}
$$

Our analysis now concerns the use of dynamic programming for solving problem (4.29) presented in Section 4.1.1 which will now be given again as it will be crucial for the description of the neural-network-based method.

We begin by rewriting the problem (4.29) to find an optimal sequence of controls $u_0^*, \ldots, u_{N-1}^*$ such that

$$
\mu_D(u_0^*, \ldots, u_{N-1} \mid x_0) =
$$
$$
= \max_{u_0, \ldots, u_{N-1}} [\mu_{C^0}(u_0) \wedge \mu_{G^1}(x_1) \wedge \ldots
$$
$$
\ldots \wedge \mu_{C^{N-1}}(u_{N-1}) \wedge \mu_{G^N}(f(x_{N-1}, u_{N-1}))] \tag{4.30}
$$

and, as we may remember from Section 4.1.1, the structure of (4.30) makes the application of dynamic programming possible.

Namely, since $\mu_{C^{N-1}}(u_{N-1}) \wedge \mu_{G^N}(f(x_{N-1}, u_{N-1}))$ depend only on u_{N-1}, then the maximization over $u_0, \ldots, u_{N-1}$ in (4.30) can be split into:

- the maximization over $u_0, \ldots, u_{N-2}$, and
- the maximization over u_{N-1},

written as

$$
\mu_D(u_0^*, \ldots, u_{N-1}^* \mid x_0) =
$$
$$
= \max_{u_0, \ldots, u_{N-2}} \{\mu_{C^0}(u_0) \wedge \mu_{G^1}(x_1) \wedge \ldots
$$
$$
\ldots \wedge \mu_{C^{N-2}}(u_{N-2}) \wedge \mu_{G^{N-1}}(x_{N-1}) \wedge
$$
$$
\wedge \max_{u_{N-1}} [\mu_{C^{N-1}}(u_{N-1}) \wedge \mu_{G^N}(f(x_{N-1}, u_{N-1}))]\} \tag{4.31}
$$

And this may be continued for u_{N-2}, u_{N-3}, etc. Such a backward iteration, which reflects the essence of dynamic programming, leads to the following set of fuzzy dynamic programming recurrence equations:

$$
\begin{cases}
\mu_{\overline{G}^{N-i}}(x_{N-i}) = \\
\quad = \max_{u_{N-i}} [\mu_{C^{N-i}}(u_{N-i}) \wedge \mu_{G^{N-i}}(x_{N-i}) \wedge \mu_{\overline{G}^{N-i+1}}(x_{N-i+1})] \\
x_{N-i+1} = f(x_{N-i}, u_{N-i}); \qquad i = 0, 1, \ldots, N
\end{cases} \tag{4.32}
$$

where $\mu_{\overline{G}^{N-i}}(x_{N-i})$ may be regarded as a fuzzy goal at control stage $t = N - i$ induced by the fuzzy goal at $t = N - i + 1$, $i = 0, 1, \ldots, N$; $\mu_{\overline{G}^N}(x_N) = \mu_{G^N}(x_N)$.

The $u_0, \ldots, u_{N-1}$ sought is given by the successive maximizing values of u_{N-i}, $i = 1, \ldots, N$ in (4.32) which are obtained as functions of x_{N-i}, i.e. as an *optimal policy*.

For the purposes of our next discussion it is convenient to rewrite the set of recurrence equations (4.32) as

$$
\begin{cases}
\mu_{\overline{G}^{N-i}}(x_{N-i}) = \\
\quad = \underbrace{\overbrace{\max_{u_{N-i}} \left[\mu_{C^{N-i}}(u_{N-i}) \wedge \mu_{G^{N-i}}(x_{N-i}) \wedge \mu_{\overline{G}^{N-i+1}}(x_{N-i+1}) \right]}^{\text{minimization at } t = N-i}}_{\text{maximization at } t = N-i} \\
x_{N-i+1} = f(x_{N-i}, u_{N-i}); \qquad i = 0, 1, \ldots, N
\end{cases}
\tag{4.33}
$$

So, proceeding backwards from the final $- t = N -$ to the initial $- t = 0 -$ control stage, at each particular control stage perform two phases: a minimization one followed by a maximization one as schematically shown in (4.33). Such a flow of computation "minimization at $t = N - 1$, maximization at $t = N - 1$, minimization at $t = N - 2$, maximization at $t = N - 2, \ldots$, minimization at $t = 0$, maximization at $t = 0$" may be modeled by a special neural network, and this was proposed by Francelin and Gomide (1992, 1993).

First, note that it cannot be a traditional neural network since we have here some "non-traditional" operations: the minimum "$\wedge$" and the maximization over a (finite) set. We need some special types of neurons which may implement these two operations. Luckily enough, such neurons may be obtained as special cases of some generalized recurrent neurons proposed by Rocha (1993).

Suppose that a neuron has n inputs, and the weighted sum of these n inputs, $a_1, \ldots, a_n$, is calculated as

$$
v = \sum_{i=1}^{n} w_i a_i
\tag{4.34}
$$

where w_i's are the weights of the synapses linking the input (pre-synaptic) neurons $n_1, \ldots, n_n$ with the post-synaptic neuron n_p.

Then, the value of v obtained is recoded into the axonic activation a_p of the post-synaptic neuron due to

$$
a_p = \begin{cases} f(v) & \text{if } v \geq \alpha \\ 0 & \text{otherwise} \end{cases}
\tag{4.35}
$$

where α is an axonic threshold and f is an encoding (transfer, squashing, $\ldots$) function.

One may also introduce two axonic thresholds, α_1 and α_2, that are usually provided by the so-called *bias neurons*. Then, we have

$$
a_p = \begin{cases} 1 & \text{if } v \geq \alpha_2 \\ f(v) & \text{if } \alpha_1 \leq v \leq \alpha_2 \\ 0 & \text{otherwise} \end{cases}
\tag{4.36}
$$

Now, looking at the transfer function (4.36) one may clearly see that we may define in quite a natural way the two types of neurons to be needed, namely the max-type and min-type neurons.

A *max-type neuron* is defined to be a neuron whose axonic threshold at time t, α_t, is defined as

$$
\alpha(t) = \begin{cases} 0 & \text{for } t = 0 \\ a_p(t-1) & \text{otherwise} \end{cases}
\tag{4.37}
$$

and

$$a_p(t) = \begin{cases} \alpha(t) & \text{if } v(t) \le \alpha(t) \\ v(t) & \text{otherwise} \end{cases} \qquad (4.38)$$

where $v(t)$ is the post-synaptic activation at time t [cf. (4.34) and (4.35)].

Thus, the output of the neuron given by (4.38) encodes the maximum of the weighted inputs, i.e.

$$a_p(t) = \max_{k=1,\ldots,t} w_k a_k \qquad (4.39)$$

while for the non-weighted case, i.e. with $w_1 = \cdots = w_t = 1$, (4.39) becomes

$$a_p(t) = \max_{k=1,\ldots,t} a_k \qquad (4.40)$$

And analogously, a *min-type neuron* is defined to be a neuron whose axonic threshold at time t, α_t, is defined as

$$\alpha(t) = \begin{cases} 1 & \text{for } t = 0 \\ a_p(t-1) & \text{otherwise} \end{cases} \qquad (4.41)$$

and

$$a_p(t) = \begin{cases} \alpha(t) & \text{if } \alpha(t) \le v(t) \\ v(t) & \text{otherwise} \end{cases} \qquad (4.42)$$

where $v(t)$ is the post-synaptic activation at time t [cf. (4.34) and (4.35)].

Thus, the output of the neuron given by (4.38) encodes the minimum of the weighted inputs, i.e.

$$a_p(t) = \min_{k=1,\ldots,t} w_k a_k \qquad (4.43)$$

while for the non-weighted case, i.e. with $w_1 = \cdots = w_t = 1$, (4.39) becomes

$$a_p(t) = \min_{k=1,\ldots,t} a_k \qquad (4.44)$$

In our next discussion we will employ a slightly modified min-type neuron, with a bias value $\bar{a}$ and with the weights $w_1 = \cdots = w_t = 1$, i.e. defined as

$$a_p(t) = \bar{a} \wedge \min_{k=1,\ldots,t} a_t \qquad (4.45)$$

and such a min-type neuron will be used in the sequel.

The inputs to the min-type neurons will basically consist of μ^{N-k}, which is the bias, being the value of the [truncated "after" $\mu_{C^{N-k}}(u_{N-k})$] fuzzy decision $\mu_D(.\,|\,.)$ (4.28), and the inputs from other neurons denoted generically as $a_1, \ldots, a_{nm}$ ($n = \text{card } X$, and $m = \text{card } U$).

Francelin and Gomide's (1992, 1993) neural network for solving fuzzy dynamic programming problems is composed of alternate layers of min-type and max-type neurons [corresponding to the minimization and maximization phases indicated in (4.33)] of the type defined above. These layers will be denoted by min-layer-k and max-layer-k which means the k-th (for $t = k$) layers of min-type and max-type neurons, respectively. The first layer is max-layer-0, and the last is min-layer-$(N-1)$.

The network's weights are not derived by training in usual manner, i.e. by feeding the network with examples, but are somehow designed, or even predetermined by the description of the problem (state transitions, fuzzy constraints and goals, etc.). So, from some points of view it may regarded as not a "real" neural network. However, on the other hand, it has a clear neural network topology, and – what is crucial for our purposes – it exhibits an inherent parallelism in its operation.

In the development of such a neural network, Francelin and Gomide (1992, 1993) employ the concepts of amounts of *transmitters, receptors,* and *controllers* (cf. Rocha, 1993) which are associated with the values of the membership function of the fuzzy constraints and goals, and state transitions.

The problem is therefore how to:

- determine the connections between min-layer-k and max-layer-k neurons, and
- determine the connections between max-layer-$(k-1)$ and min-layer-k neurons,

which will now be briefly described.

4.1.4.1 Determination of connections between the min-layer-k and max-layer-k neurons

These connections are assigned a priori, following the state transition equation (4.27). If we denote by m_k^i the i-th neuron in min-layer-k, and by M_k^j the j-th neuron in max-layer-k, then $q_R(m_k^i) = s_i$ is the amount of receptor, and $q_T(M_k^j) = c_j$ is the amount of transmitter.

Then the connections (weights) between m_k^i, the i-th neuron in min-layer-k, and M_k^j, the j-th neuron in max-layer-k, are

$$W(m_k^i, M_k^j) = \begin{cases} 1 & \text{if } f_{N-k}[q_R(m_k^i), q_T(M_k^j)] \neq \emptyset \\ 0 & \text{otherwise} \end{cases} \qquad (4.46)$$

where $W(.,.) = 1$ means that the connection exists, and $W(.,.) = 0$ that it does not exist.

4.1.4.2 Determination of connections between the max-layer-$(k-1)$ and min-layer-k neurons

Suppose that we have an m_k^i neuron (i-th min-type neuron in min-layer-k) which has the amount of receptor $q_R(m_k^i)$. This neuron receives the amount of transmitter $q_T(M_k^j)$ from the M_k^j neuron (j-th max-type neuron in max-layer-k). These two quantities activate the controller of the m_k^i neuron whose controller function is defined to be the state transition equation (4.27) at stage $N-k$, i.e.

$$q_C(m_k^i) = x_{N-k} = f_{N-k}[q_R(m_k^i), q_T(M_k^j)] \qquad (4.47)$$

This controller transmits the amount $q_C(m_k^i)$ to all the max-type neurons in M-layer-$(k-1)$, and this activates those of them, M_{k-1}^l, for which

$$q_R(M_{k-1}^l) = q_T(m_k^i) = q_C(m_k^i) \qquad (4.48)$$

The connections (weights) between the M_{k-1}^l neurons in max-layer-$(k-1)$ and the m_k^i neurons in min-layer-k are established as

$$W(M_{k-1}^l, m_k^i) = \begin{cases} 1 & \text{if } q_R(M_{k-1}^l) = q_T(m_k^i) = q_C(m_k^i) \\ 0 & \text{otherwise} \end{cases} \quad (4.49)$$

and $W(.,.) = 1$ means the existence of a connection, and $W(.,.) = 0$ the non-existence.

We have established the weights, and are in a position to derive the neural network algorithm for solving fuzzy dynamic programming problems.

The algorithm may be portrayed in the following general form:

begin
    perform the *initialization pass*
        begin
            perform the *connection pass*
            perform the *output pass-k*
        end
end

The three passes – those of initialization, connection, and pass-k – are presented below in a similar clear form as the algorithm given above.

The *initialization pass* is performed as follows:

begin
    for each M_0^j neuron in max-layer-0 **do**
        begin
            $q_R(M_0^j) = s_j \in X$
            $v(M_0^j) = \mu_{G^N}[q_R(M_0^j)]$
        end
    for each min-layer-k **do**
        begin
            for each m_k^i neuron **do**
                $q_R(m_k^i) = c_i \in U$
        end
    for each max-layer-k **do**
        begin
            for each M_k^j neuron **do**
            begin
                $q_R(M_k^j) = x_{N-k}^j \in X$
                $q_T(M_k^j) = q_R(M_k^j)$
            end
        end
end

The *connection pass* is performed as follows:

begin
Phase 1: Establishing connections between min-layer-k and max-layer-k neurons
begin

for each M_k^j neuron in max-layer-k **do**
for each m_k^i neuron in m-layer-k **do**
    begin
        $q_C(m_k^i) := f_{N-k}[q_T(M_k^j, q_R(m_k^i)]$
        $q_T(m_k^i) := q_C(m_k^i)$
        $W(m_k^i, M_k^i) := 1$
    end
Phase 2: Establishing connections between max-layer-$(k-1)$ and min-layer-k neurons
    for each neuron M_{k-1}^t in max-layer-$(k-1)$
    begin
do
        If $q_R(M_{k-1}^t) = q_t(m_k^i)$, then $W(M_{k-1}^t, m_k^i) := 1$
        Receive the output $s(M_{k-1}^t)$ from the M_{k-1}^t neuron
        $s(m_k^i) := \min\{\mu_C(q_R(m_k^i), s(M_{k-1}^t)\}$
        $a_i := a_i(M_k^j) = s(m_k^i)$
    end
end

Finally, the *output pass* is performed as follows:

for each M_k^t neuron in max-layer-k **do**
    $s(M_k^t) := \max_i\{a_i\}$

As proven by Francelin and Gomide (1992, 1993), the solution (a sequence of optimal controls at the consecutive control stages) obtained by using the neural network presented, with the above passes, is equivalent to the one yielded by the dynamic programming approach discussed in Section 4.1, and more specifically by the solution of the set of dynamic programming recurrence equations (4.32) on page 106.

To illustrate the neural network approach outlined in this section, we now present the solution of a simple example.

Example 4.5 Let the state space be $X = \{s_1, s_2, s_3\}$, the control space be $U = \{c_1, c_2\}$, the termination time be $N = 2$, and the fuzzy constraints and fuzzy goal be, respectively:

$$C^0 = 0.7/c_1 + 1/c_2$$
$$C^1 = 1/c_1 + 0.8/c_2$$
$$G^2 = 0.3/s_1 + 1/s_2 + 0.8/s_3$$

Suppose that the dynamics of the deterministic system under control is governed by the state transition equation of type (4.27) represented by the following state transition table:

$$x_{t+1} = \quad
\begin{array}{c|ccc}
 & x_t = s_1 & s_2 & s_3 \\
\hline
u_t = c_1 & s_1 & s_3 & s_1 \\
c_2 & s_2 & s_1 & s_3 \\
\end{array}
\qquad (4.50)$$

Since $N = 2$, then the neural network to be developed will have 5 layers. Moreover, evidently, $U = \{c_1, c_2\}$, $f_1 = f_2 = f$ given by (4.50), and $X = \{s_1, s_2, s_3\}$.

During the initialization pass, we obtain:

$$\begin{aligned}
q_R(M_0^i) &:= s_2^i \in X \quad i = 0, 1, 2 \\
q_R(m_1^i) &:= c_1^i \in U \quad i = 0, 1 \\
q_R(M_1^i) &:= s_1^i \in X \quad i = 0, 1, 2 \\
q_R(m_2^i) &:= c_0^i \in U \quad i = 0, 1 \\
q_R(M_2^i) &:= s_0^i \in X \quad i = 0, 1, 2
\end{aligned}$$

We set $k := 1$, and the connection pass is performed. The connections between max-layer-0 and min-layer-1 are now established. For example, for the M_1^2 and m_1^1 neurons, we have

$$q_C(m_1^1) = f[q_R(m_1^1), q_T(M_1^2)] = f(c_1, s_2) = s_3$$

which implies that $W(M_0^2, m_1^1) = 1$ because the neuron M_0^2 satisfies

$$q_R(M_0^2) = q_T(m_1^1) = q_C(m_1^1)$$

The neuron m_1^1 receives the output from the neuron M_0^2, equal $s(m_0^2) = 0.8$, and in turn its output is

$$s(m_1^1) = \mu_{C^1}(c_2) \wedge s(M_0^2) = 0.6 \wedge 0.8 = 0.6$$

The output pass is now executed. For instance, the output of the M_1^2 neuron is

$$s(M_1^2) = \max\{0.3, 0.6\} = 0.6$$

and analogously the output for the other neurons, M_1^0 and M_1^1.

We set $k = 2$, and perform the connection pass again, for each neuron M_2^j, $j = 0, 1, 2$, in max-layer-2, and for each neuron m_2^i, $i = 0, 1$, in min-layer-2. For example, for $j = 0$ and $i = 0, 1$, the following connections are established:

$$\begin{aligned}
W(M_1^0) &= 1 \\
W(M_1^1, m_2^1) &= 1
\end{aligned}$$

because

$$q_C(m_2^0) = f[q_R(m_2^0), q_T(M_2^0)] = f(c_1, s_1) = q_T(m_2^0) = q_R(M_1^0)$$

and

$$q_C(m_2^1) = q_T(m_2^1) = q_R(M_1^1) = s_2$$

The neuron m_2^0 receives therefore the output $s(M_1^0)$ from the neuron M_1^0, and its output is

$$s(m_2^0) = \mu_{C^0}(c_1) \wedge s(M_1^0) = 0.7 \wedge 0.6 = 0.6$$

and analogously

$$s(m_2^1) = \mu_{C^0}(c_2) \wedge s(M_1^0) = 1.0 \wedge 0.8 = 0.8$$

Next, the output pass is executed, and the output of the neurons in max-layer-2 are:

$$\begin{aligned}
s(M_2^0) &= \max\{0.6, 0.8\} = 0.8 \\
s(M_2^1) &= \max\{0.6, 0.6\} = 0.6 \\
s(M_2^2) &= \max\{0.6, 0.6\} = 0.6
\end{aligned}$$

The optimal solution is therefore exemplified, for $x_0 = s_3$, by

$$u_0^* = c_2 \quad u_1^8 = c_1$$

for which $\mu_D(c_1, c_2 \mid x_0) = 0.8$. $\square$

This simple example shows the essentials of the procedure well, but it should be mentioned that it may seem at first glance that the neural network approach is much more complicated that of dynamic programming (cf. Section 4.1) or branch-and-bound (cf. Section 4.1.2). This impression may be misleading, however, since for large problems the inherent parallelism of neural networks may be an advantage.

Francelin and Gomide (1992, 1993) also provide an interesting example of using the algorithm presented in long-range planning of interconnections between two power systems. We refer the interested reader to these source papers.

4.1.5   Using fuzzy linguistic quantifiers in multistage fuzzy control

In this section we will present the use of fuzzy linguistic quantifiers, to be more specific of calculi of linguistically quantified propositions presented in Section 2.3, to derive a new and interesting class of multistage fuzzy control problems. This was proposed by Kacprzyk (1983c) and Kacprzyk and Iwański (1987), and we will follow their exposition.

As in all former cases, we assume that the system under control is deterministic, and its dynamics is described by the state transition equation (4.1), page 85, i.e.

$$x_{t+1} = f(x_t, u_t), \qquad t = 0, 1, \ldots$$

where $x_t, x_{t+1} \in X\{s_1, \ldots, s_n\}$ are states at control stages t and $t + 1$, respectively, and $u_t \in U = \{c_1, \ldots, c_m\}$ is the control at stage t; $t = 0, 1, \ldots, N - 1$; $N < \infty$ is fixed and specified in advance.

At each control stage t, a fuzzy constraint $\mu_{C^t}(u_t)$ is imposed on u_t, and the resulting x_{t+1} is subjected to a fuzzy goal $\mu_{G^{t+1}}(x_{t+1})$.

The fuzzy decision [cf. (4.2), page 86] is therefore

$$\mu_D(u_0, \ldots, u_{N-1} \mid x_0) = \bigwedge_{t=0}^{N-1} [\mu_{C^t}(u_t) \wedge \mu_{G^{t+1}}(x_{t+1})] \tag{4.51}$$

and the problem [cf. (4.3), page 86] is to find an optimal sequence of controls $u_0^*, \ldots, u_{N-1}^*$ such that

$$\mu_D(u_0^*, \ldots, u_{N-1}^* \mid x_0) = \max_{u_0, \ldots, u_{N-1}} \bigwedge_{t=0}^{N-1} [\mu_{C^t}(u_t) \wedge \mu_{G^{t+1}}(x_{t+1})] \tag{4.52}$$

If we look at this problem, we can readily notice that its very essence is that at *each* control stage t, $t = 0, 1, \ldots, N - 1$, the fuzzy constraint C^t and the fuzzy goal G^{t+1} are to be satisfied. This may be written in the form of the following statements:

$$\left\{ \begin{array}{l} P_1 : \text{``}C^0 \text{ and } G^1 \text{ are satisfied''} \\ \quad \vdots \\ P_N : \text{``}C^{N-1} \text{ and } G^N \text{ are satisfied''} \end{array} \right. \tag{4.53}$$

Evidently, each C^t and G^{t+1} are satisfied to a certain extent, from 0 to 1, which depends on the control applied u_t and the state attained x_{t+1}. This may be represented by the truth of the statement P_t which is

$$\tau(P_t) = \tau(C^t \text{ and } G^{t+1} \text{ are satisfied}) = \mu_{C^t}(u_t) \wedge \mu_{G^{t+1}}(x_{t+1}) \qquad (4.54)$$

Example 4.6 Let $U = \{c_1, c_2\}$, $X = \{s_1, s_2, s_3\}$, and

$$C^t = 0.5/c_1 + 1/c_2 \qquad G^{t+1} = 0.3/s_1 + 0.7/s_2 + 1/s_3$$

If now $u_t = c_1$ and $x_{t+1} = s_2$, then $\tau(P_{t+1}) = 0.5 \wedge 0.7 = 0.5$. $\qquad\qquad\square$

The fuzzy decision (4.51) may now be rewritten as

$$
\begin{aligned}
\mu_D(u_0, \ldots, u_{N-1} \mid x_0) &= \\
&= \tau(\text{``}C^0 \text{ and } G^1 \text{ are satisfied'' and } \ldots \\
&\qquad \ldots \text{ and ``}C^{N-1} \text{ and } G^N \text{ are satisfied''}) = \\
&= \tau(P_1 \text{ and } \ldots \text{ and } P_N) = \tau(P_1) \wedge \ldots \wedge \tau(P_N) = \\
&= \bigwedge_{t=0}^{N-1} \tau(P_{t+1}) = \bigwedge_{t=0}^{N-1} [\mu_{C^t}(u_t) \wedge \mu_{G^{t+1}}(x_{t+1})]
\end{aligned} \qquad (4.55)
$$

and the problem (4.52) may be rewritten as to find $u_0^*, \ldots, u_{N-1}^*$ such that

$$
\begin{aligned}
\mu_D(u_0^*, \ldots, u_{N-1}^* \mid x_0) &= \\
&= \max_{u_0, \ldots, u_{N-1}} \tau(P_1 \text{ and } \ldots \text{ and } P_N) = \tau(P_1) \wedge \ldots \wedge \tau(P_N) = \\
&= \max_{u_0, \ldots, u_{N-1}} \bigwedge_{t=0}^{N-1} \tau(P_{t+1}) = \max_{u_0, \ldots u_{N-1}} \bigwedge_{t=0}^{N-1} [\mu_{C^t}(u_t) \wedge \mu_{G^{t+1}}(x_{t+1})] \qquad (4.56)
\end{aligned}
$$

Thus, for a particular sequence of controls $u_0, \ldots, u_{N-1}$, and a sequence of resulting states $x_0, x_1, \ldots, x_N$, the value of $\mu_D(\cdot \mid \cdot)$ given by (4.55) may be viewed as the truth value of the statement

"the fuzzy constraints and fuzzy goals are satisfied at **all** the control stages"

and the problem (4.56) is clearly to

"find an optimal sequence of controls best satisfying the fuzzy constraints and fuzzy goals at **all** the control stages."

To emphasize this *all*, let us rewrite the fuzzy decision (4.55) as

$$
\begin{aligned}
\mu_D(u_0, \ldots, u_{N-1} \mid x_0, \text{all}) &= \\
&= \tau(P_1 \text{ and } \ldots \text{ and } P_N \mid \text{all}) = \left(\bigwedge_{t=0}^{N-1} \mid \text{all}\right) \tau(P_{t+1}) = \\
&= \left(\bigwedge_{t=0}^{N-1} \mid \text{all}\right) [\mu_{C^t}(u_t) \wedge \mu_{G^{t+1}}(x_{t+1})]
\end{aligned} \qquad (4.57)
$$

and rewrite the problem (4.56) as to find $u_0^*, \ldots, u_{N-1}^*$ such that

$$
\begin{aligned}
\mu_D(u_0^*, \ldots, u_{N-1}^* \mid x_0 \mid \text{all}) = \\
= \max_{u_0, \ldots, u_{N-1}} \tau(P_1 \text{ and } \ldots \text{ and } P_N \mid \text{all}) = \\
= \max_{u_0, \ldots, u_{N-1}} \left(\bigwedge_{t=0}^{N-1} \mid \text{all} \right) \tau(P_{t+1}) = \\
= \max_{u_0, \ldots u_{N-1}} \left(\bigwedge_{t=0}^{N-1} \mid \text{all} \right) [\mu_{C^t}(u_t) \wedge \mu_{G^{t+1}}(x_{t+1})]
\end{aligned}
\tag{4.58}
$$

Note that *all* in the above notations is used just to indicate that all the control stages are to be accounted for, and has no other formal meaning.

It is now easy to see that the above requirement to satisfy the fuzzy constraints and fuzzy goals at *all* the control stages is very restrictive and often counter-intuitive. It may be fully sufficient to satisfy them at, say, an *overwhelming majority* of the control stages which is a "milder" requirement. Examples may be found in virtually all practical problems, and will be mentioned in Section 8.1.

The above idea of a milder problem formulation is the essence of Kacprzyk's (1983c), and Kacprzyk and Iwański's (1987) proposals. First, note that *all* may be viewed as a universal quantifier, i.e. *for all*. As we may remember from Section 2.3, fuzzy logic makes it possible to deal with a much wider class of quantifiers than those employed in classical logic (i.e. *for all* and *for at least one*). Notably, fuzzy logic provides tools to handle fuzzy linguistic quantifiers like *most, almost all, much more than a half*, etc.

Such linguistic quantifiers are used for the derivation of those milder problem formulations of the type

"find an optimal sequence of controls that best satisfies the fuzzy constraints and fuzzy goals at Q (e.g., most, almost all, much more than a half, etc.) control stages."

So, analogously to (4.57), the fuzzy decision of the above milder problem formulation may be written as

$$
\begin{aligned}
\mu_D(u_0, \ldots, u_{N-1} \mid x_0, Q) = \\
= \tau(P_1 \text{ and } \ldots \text{ and } P_N \mid Q) = \\
= \left(\bigwedge_{t=0}^{N-1} \mid Q \right) \tau(P_{t+1}) = \left(\bigwedge_{t=0}^{N-1} \mid Q \right) [\mu_{C^t}(u_t) \wedge \mu_{G^{t+1}}(x_{t+1})]
\end{aligned}
\tag{4.59}
$$

and the problem (4.58) may be analogously rewritten as to find $u_0^*, \ldots, u_{N-1}^*$ such that

$$
\begin{aligned}
\mu_D(u_0^*, \ldots, u_{N-1}^* \mid x_0, Q) = \\
= \max_{u_0, \ldots, u_{N-1}} \tau(P_1 \text{ and } \ldots \text{ and } P_N \mid Q) = \\
= \max_{u_0, \ldots, u_{N-1}} \left(\bigwedge_{t=0}^{N-1} \mid Q \right) \tau(P_{t+1}) =
\end{aligned}
$$

$$= \max_{u_0, \ldots u_{N-1}} \left(\bigwedge_{t=0}^{N-1} \mid Q \right) [\mu_{C^t}(u_t) \wedge \mu_{G^{t+1}}(x_{t+1})] \qquad (4.60)$$

where Q is a fuzzy linguistic quantifier.

For instance, in presumably one of the most natural cases when $Q = $ "most," the fuzzy decision (4.59) becomes

$$\mu_D(u_0, \ldots, u_{N-1} \mid x_0, \text{most}) =$$
$$= \tau(P_1 \text{ and } \ldots \text{ and } P_N \mid \text{most}) =$$
$$= \left(\bigwedge_{t=0}^{N-1} \mid \text{most} \right) \tau(P_{t+1}) = \left(\bigwedge_{t=0}^{N-1} \mid \text{most} \right) [\mu_{C^t}(u_t) \wedge \mu_{G^{t+1}}(x_{t+1})] \quad (4.61)$$

and the problem (4.60) becomes to find $u_0^*, \ldots, u_{N-1}^*$ such that

$$\mu_D(u_0^*, \ldots, u_{N-1}^* \mid x_0, \text{most}) =$$
$$= \max_{u_0, \ldots, u_{N-1}} \tau(P_1 \text{ and } \ldots \text{ and } P_N \mid \text{most}) =$$
$$= \max_{u_0, \ldots, u_{N-1}} \left(\bigwedge_{t=0}^{N-1} \mid \text{most} \right) \tau(P_{t+1}) =$$
$$= \max_{u_0, \ldots u_{N-1}} \left(\bigwedge_{t=0}^{N-1} \mid \text{most} \right) [\mu_{C^t}(u_t) \wedge \mu_{G^{t+1}}(x_{t+1})] \qquad (4.62)$$

Now, we will show how to solve the control problem with a fuzzy linguistic quantifier using the two calculi of linguistically quantified statements presented in Sections 2.3.1 and 2.3.2.

4.1.5.1   Solution via the algebraic method

Using the algebraic method presented in Section 2.3.1, we calculate first (2.101), page 50, i.e.

$$r(u_0, \ldots, u_{N-1} \mid x_0) =$$
$$= \frac{1}{N} \sum_{t=0}^{N-1} \tau(P_{t+1}) = \frac{1}{N} \sum_{t=0}^{N-1} [\mu_{C^t}(u_t) \wedge \mu_{G^{t+1}}(x_{t+1})] \qquad (4.63)$$

and then calculate (2.102), i.e.

$$\mu_D(u_0, \ldots, u_{N-1} x_0, Q) =$$
$$= \tau(P_1 \text{ and } \ldots \text{ and } P_N \mid Q) = \mu_Q[r(u_0, \ldots, u_{N-1} \mid x_0)] =$$
$$= \mu_Q \left(\frac{1}{N} \sum_{t=0}^{N-1} [\mu_{C^t}(u_t) \wedge \mu_{G^{t+1}}(x_{t+1})] \right) \qquad (4.64)$$

Example 4.7 Let $U = \{c_1, c_2\}$, $X = \{s_1, s_2, s_3\}$, $N = 3$, and

$$C^0 = 0.5/c_1 + 1/c_2 \qquad G^1 = 0.1/s_1 + 0.6/s_2 + 1/s_3$$
$$C^1 = 1/c_1 + 0.7/c_2 \qquad G^2 = 0.61/s_1 + 1/s_2 + 0.5/s_3$$
$$C^2 = 1/c_1 + 0.6/c_2 \qquad G^3 = 1/s_1 + 0.8/s_2 + 0.3/s_3$$

with the fuzzy linguistic quantifier Q = "most" given as (2.99), page 49, that is

$$\mu_{\text{"most"}}(x) = \begin{cases} 1 & \text{for } x \geq 0.8 \\ 2x - 0.6 & \text{for } 0.3 \leq x < 0.8 \\ 0 & \text{for } x < 0.3 \end{cases} \qquad (4.65)$$

Suppose now that we apply the following consecutive controls: $u_0 = c_1$, $u_1 = c_2$ and $u_2 = c_1$, and the resulting states are: $x_1 = s_2$, $x_2 = s_1$ and $x_3 = s_2$. Then

$$r(u_0, u_1, u_2 \mid x_0) =$$
$$= \frac{1}{3}[(\mu_{C^0}(c_1) \wedge \mu_{G^1}(s_2)) + (\mu_{C^1}(c_2) \wedge \mu_{G^2}(s_1)) +$$
$$+ (\mu_{C^2}(c_1) \wedge \mu_{G^3}(s_2)] = \frac{1}{3}(0.5 + 0.6 + 0.8) = 0.63$$

so that

$$\mu_D(u_0, u_1, u_2 \mid x_0) = \mu_{\text{"most"}}[r(u_0, u_1, u_2 \mid x_0)] = \mu_{\text{"most"}}(0.63) = 0.66$$

$\square$

The problem is to find an optimal sequence of controls $u_0^*, \ldots, u_{N-1}^*$ such that

$$\mu_D(u_0^*, \ldots, u_{N-1}^* \mid x_0, Q) =$$
$$= \max_{u_0, \ldots, u_{N-1}} \mu_D(u_0, \ldots, u_{N-1} \mid x_0, Q) =$$
$$= \max_{u_0, \ldots, u_{N-1}} \mu_Q[r(u_0, \ldots, u_{N-1} \mid x_0)] =$$
$$= \max_{u_0, \ldots, u_{N-1}} \mu_Q \left(\frac{1}{N} \sum_{t=0}^{N-1} [\mu_{C^t}(u_t) \wedge \mu_{G^{t+1}}(x_{t+1})] \right) \qquad (4.66)$$

It is now easy to see that it is very difficult, if not impossible, to say something about the solution of this problem for an arbitrary fuzzy linguistic quantifier Q. Fortunately, we may readily confine the class of quantifiers to be considered to the so-called *nondecreasing fuzzy quantifiers* (cf. Section 2.3.1) which fulfill [cf. (2.100), page 49]

$$r_1 > r_2 \implies \mu_Q(r_1) \geq \mu_Q(r_2) \qquad (4.67)$$

for all $r_1, r_2 \in [0, 1]$.

Clearly, only such quantifiers – in essence "the more the better," i.e. in our context "the more (at the more control stages) the fuzzy constraints and the fuzzy goals are satisfied the better – are relevant in our control context. And, luckily enough, they have some interesting properties which are important for our analysis (cf. Yager, 1983b; Kacprzyk, 1983c; Kacprzyk and Yager, 1984a, b, 1990; Kacprzyk and Iwański, 1987).

First, let a *linear quantifier*, denoted Q_L be defined as

$$\mu_{Q_L}(x) = x, \qquad \text{for each } x \in [0, 1] \qquad (4.68)$$

We now have the following important lemma.

Lemma 4.1 *For any linear quantifier* Q_L, *the solution of problem (4.66), i.e.* $u_0^*, \ldots, u_{N-1}^*$, *is given by*

$$
\begin{aligned}
\mu_D(u_0^*, \ldots, u_{N-1}^* \mid x_0, Q_L) = \\
= \max_{u_0, \ldots, u_{N-1}} r(u_0, \ldots, u_{N-1} \mid x_0) = \\
= \max_{u_0, \ldots, u_{N-1}} \frac{1}{N} \sum_{t=0}^{N-1} [\mu_{C^t}(u_t) \wedge \mu_{G^{t+1}}(x_{t+1})] = \\
= \frac{1}{N} \sum_{t=0}^{N-1} \{\mu_{C^t}(u_t^*) \wedge \mu_{G^{t+1}}[f(x_t, u_t^*)]\}
\end{aligned}
\tag{4.69}
$$

This lemma implies then an even stronger property that is given by the following proposition (Kacprzyk, 1983c).

Proposition 4.1 *If* $u_0^*, \ldots, u_{N-1}^*$ *is such that (4.69) holds, i.e. it is a solution of problem (4.66) for the linear quantifier* Q_L, *then it is also a solution of problem (4.66) for any nondecreasing quantifier* Q *[in the sense of (4.67), page 117], i.e.*

$$
\begin{aligned}
\mu_D(u_0^*, \ldots, u_{N-1}^* \mid x_0, Q) = \\
= \max_{u_0, \ldots, u_{N-1}} \mu_Q[r(u_0, \ldots, u_{N-1} \mid x_0)] = \\
= \max_{u_0, \ldots, u_{N-1}} \frac{1}{N} \sum_{t=0}^{N-1} [\mu_{C^t}(u_t) \wedge \mu_{G^{t+1}}(x_{t+1})] = \\
= \frac{1}{N} \sum_{t=0}^{N-1} \{\mu_{C^t}(u_t^*) \wedge \mu_{G^{t+1}}[f(x_t, u_t^*)]\}
\end{aligned}
\tag{4.70}
$$

Proof. First, for any $u_0, \ldots, u_{N-1}$, we have

$$
\mu_D(u_0, \ldots, u_{N-1} \mid x_0, Q) = \mu_Q[r(u_0, \ldots, u_{N-1} \mid x_0)]
$$

Suppose now that we have two sequences of controls: $u_0^a, \ldots, u_{N-1}^a$ and $u_0^b, \ldots, u_{N-1}^b$ such that $r^a = r(u_0^a, \ldots, u_{N-1}^a \mid x_0)$ and $r^b = r(u_0^b, \ldots, u_{N-1}^b \mid x_0)$. Then by the assumption of nonincreasingness of Q, we have $\mu_Q(r^a) \geq \mu_Q(r^b)$, and hence

$$
\mu_D(u_0^a, \ldots, u_{N-1}^b \mid x_0, Q) = \mu_Q[r(u_0^b, \ldots, u_{N-1}^b \mid x_0)]
$$

If now (4.69) holds, i.e. for each $u_0, \ldots, u_{N-1}$, we have

$$
\begin{aligned}
\mu_D(u_0^*, \ldots, u_{N-1}^* \mid x_0, Q_L) = \\
= \frac{1}{N} \sum_{t=0}^{N-1} \{\mu_{C^t}(u_t^*) \wedge \mu_{G^{t+1}}[f(x_t, u_t^*)]\} \geq \\
\geq \frac{1}{N} \sum_{t=0}^{N-1} \{\mu_{C^t}(u_t) \wedge \mu_{G^{t+1}}[f(x_t, u_t)]\}
\end{aligned}
$$

then, evidently, there holds $r(u_0^*, \ldots, u_{N-1}^* \mid x_0) \geq r(u_0, \ldots, u_{N-1} \mid x_0)$, for any $u_0, \ldots, u_{N-1}$ (and for any x_0, too).

This implies that

$$\mu_D(u_0^*, \ldots, lu_{N-1}^* \mid x_0, Q) \geq \mu_D(u_0, \ldots, u_{N-1} \mid x_0, Q)$$

that is, an $u_0^*, \ldots, u_{N-1}^*$ for which (4.69) holds is a solution of problem (4.66) for any nondecreasing quantifier. □

Thus, for any nondecreasing quantifier, the use of the algebraic method results in the following problem to be solved: find an optimal sequence of controls $u_0^*, \ldots, u_{N-1}^*$ such that

$$\mu_D(u_0^*, \ldots, u_{N-1}^* \mid x_0, Q) = \max_{u_0, \ldots, u_{N-1}} \frac{1}{N} \sum_{t=0}^{N-1} [\mu_{C^t}(u_t) \wedge \mu_{G^{t+1}}(x_{t+1})] \qquad (4.71)$$

which may clearly be solved by dynamic programming (cf. Section 4.1.1). The set of recurrence equations yielding an optimal solution is then

$$\begin{cases} \mu_{\underline{G}^{N-i}}(x_{N-i}) = \max_{u_{N-i}} \frac{1}{N}[\mu_{C^{N-i}}(u_{N-i}) + \\ \qquad + \mu_{G^{N-i+1}}(x_{N-i+1}) + \mu_{\underline{G}^{N-i+1}}(x_{N-i+1})] \\ x_{N-i+1} = f(x_{N-i}, u_{N-i}); \qquad i = 1, \ldots, N \end{cases} \qquad (4.72)$$

with $\mu_{\underline{G}^N}(x_N) = 0$, for each $x_N \in X$.

Example 4.8 Let the problem specifications be as in Example 4.7, page 116, i.e. $U = \{c_1, c_2\}$, $X = \{s_1, s_2, s_3\}$, $N = 3$, and

$$\begin{array}{ll} C^0 = 0.5/c_1 + 1/c_2 & G^1 = 0.1/s_1 + 0.6/s_2 + 1/s_3 \\ C^1 = 1/c_1 + 0.7/c_2 & G^2 = 0.61/s_1 + 1/s_2 + 0.5/s_3 \\ C^2 = 1/c_1 + 0.6/c_2 & G^3 = 1/s_1 + 0.8/s_2 + 0.3/s_3 \end{array}$$

with the fuzzy linguistic quantifier $Q =$ "most" given as (4.65), i.e.

$$\mu_{\text{"most"}}(x) = \begin{cases} 1 & \text{for } x \geq 0.8 \\ 2x - 0.6 & \text{for } 0.3 \leq x < 0.8 \\ 0 & \text{for } x < 0.3 \end{cases}$$

and the state transitions of the system under control be given by (4.1), i.e.

$$x_{t+1} = \begin{array}{c|ccc} & x_t = s_1 & s_2 & s_3 \\ \hline u_t = c_1 & s_1 & s_3 & s_3 \\ c_2 & s_2 & s_1 & s_2 \end{array}$$

By solving the set of recurrence equations (4.72) subsequently for $i = 1, 2, 3$, we respectively obtain

$$\underline{G}^2 = 0.66/s_1 + 0.53/s_2 + 0.46/s_3$$

and the optimal policy at control stage $t = 2$ is

$$a_2^*(s_1) = c_1 \quad a_2^*(s_2) = c_2 \quad a_2^*(s_3) = c_2$$

Next

$$\underline{G}^1 = 0.75/s_1 + 0.65/s_2 + 0.74/s_3$$

and the optimal policy at control stage $t = 1$ is

$$a_1^*(s_1) = c_1 \quad a_1^*(s_2) \in \{c_1, c_2\} \quad a_1^*(s_3) = c_2$$

Finally, we obtain

$$\underline{G}^1 = 0.75/s_1 + 0.62/s_2 + 0.75/s_3$$

and the optimal policy at control stage $t = 0$ is

$$a_0^*(s_1) \in \{c_1, c_2\} \quad a_0^*(s_2) = c_2 \quad a_0^*(s_3) = c_2$$

$\square$

Thus, on the one hand, the use of the algebraic method is simple since the control problem may be solved by dynamic programming for a large class of (plausible, to say the least) fuzzy linguistic quantifiers. On the other hand, however, the fact that for a large class of quantifiers the solution amounts to solving the same set of recurrence equations may seem to be somewhat counter-intuitive, though one should bear in mind that the solution obtained is just one of the possible solutions, i.e. *a solution*, and not *the solution*.

4.1.5.2   Solution by the substitution method

Now we will solve the control problem considered (4.66) using another calculus of linguistically quantified propositions, the so-called substitution method (cf. Section 2.3.2).

First, we will recall for convenience of the reader the fuzzy decision (4.59) which is

$$\begin{aligned}
\mu_D(u_0, \ldots, u_{N-1} \mid x_0, Q) &= \\
&= \tau(P_1 \text{ and } \ldots \text{ and } P_N \mid Q) = \\
&= \left(\bigwedge_{t=0}^{N-1} \mid Q \right) \tau(P_{t+1}) = \left(\bigwedge_{t=0}^{N-1} \mid Q \right) [\mu_{C^t}(u_t) \wedge \mu_{G^{t+1}}(x_{t+1})] \qquad (4.73)
\end{aligned}$$

where Q is a fuzzy linguistic quantifier, and the P_t's are the following statements [cf. (4.55), page 114]

$$\begin{cases} P_1 : \text{``}C^0 \text{ and } G^1 \text{ are satisfied''} \\ \vdots \\ P_N : \text{``}C^{N-1} \text{ and } G^N \text{ are satisfied''} \end{cases}$$

with their respective truth values

$$\tau(P_{t+1}) = \mu_{C^t}(u_t) \wedge \mu_{G^{t+1}}(x_{t+1}), \qquad t = 0, 1, \ldots, N-1$$

The problem considered (4.52), page 113, is therefore to find $u_0^*, \ldots, u_{N-1}^*$ such that

$$\mu_D(u_0^*, \ldots, u_{N-1}^* \mid x_0, Q) = \max_{u_0, \ldots, u_{N-1}} \mu_D(u_0, \ldots, u_{N-1} \mid x_0, Q) \qquad (4.74)$$

with $\mu_D(\cdot \mid \cdot, \cdot)$ given by (4.59).

We need to derive now a slightly transformed problem formulation which will be more convenient for the presentation.

First, we introduce the following set:

$$
\begin{aligned}
V = \{v\} = \\
= \quad & \{P_1, P_2, \ldots, P_N, P_1 \text{ and } P_2, P_1 \text{ and } P_3, \ldots, P_1 \text{ and } P_N, \\
& P_2 \text{ and } P_3, \ldots, P_{N-1} \text{ and } P_N, \ldots, P_1 \text{ and } P_2 \text{ and } P_3, \ldots \\
& \ldots, P_{N-2} \text{ and } P_{N-1} \text{ and } P_N, \ldots, P_1 \text{ and } P_2 \text{ and } \ldots \text{ and } P_N\} \quad (4.75)
\end{aligned}
$$

whose elements represent the fulfillment of fuzzy constraints and fuzzy goals at all the possible combinations of control stages ranging from all the particular single stages, all the pairs of stages, all the triples of stages, $\ldots$, to all the stages.

Example 4.9 If we have three control stages, i.e. $N = 3$, then

$$
V = \{P_1, P_2, P_3, P_1 \text{ and } P_2, P_1 \text{ and } P_3, P_2 \text{ and } P_3, P_1 \text{ and } P_2 \text{ and } P_3\} \quad (4.76)
$$

$\square$

Clearly, since the order of P_t's in the particular v's is irrelevant, we will only list in the V's in our examples, say, P_1 and P_2 and not both P_1 and P_2 and P_2 and P_1.

A quantifier is now defined as a fuzzy set in V, characterized by its membership function $\mu_Q(v)$ which indicates to what degree v satisfies the meaning of Q [cf. (2.108), page 51].

Example 4.10 If $N = 3$ and V is as in (4.76) in Example 4.9, then the traditional universal quantifier $Q =$ "all" is obviously given as

$$
\mu_{\text{"all"}}(v) = \begin{cases} 1 & \text{for } v \in \{P_1 \text{ and } P_2 \text{ and } P_3\} \\ 0 & \text{otherwise} \end{cases} \quad (4.77)
$$

because only the fulfillment of fuzzy constraints and fuzzy goals at *all* the control stages is taken into account.

The existential quantifier, $Q =$ "at least one," is clearly represented by

$$
\mu_{\text{"atleastone"}}(v) = 1, \qquad \text{for each } v \in V \quad (4.78)
$$

And for some other (fuzzy) quantifiers that might be of relevance in our context, $Q =$ "most" may be defined for instance as

$$
\mu_{\text{"most"}}(v) = \begin{cases} 1 & v \in \{P_1 \text{ and } P_2\} \\ 0.7 & v \in \{P_1 \text{ and } P_2, P_1 \text{ and } P_3, P_2 \text{ and } P_3\} \\ 0.2 & v \in \{P_1, P_2, P_3\} \end{cases} \quad (4.79)
$$

for each $v \in V$.

$\square$

Now, for each $v = \{P_{k1} \text{ and } P_{k2} \text{ and } \ldots \text{ and } P_{kp}\} \in V$ we calculate the value of

$$
\begin{aligned}
\mu_T(v) = \tau(P_{k1} \text{ and } P_{k2} \text{ and } \ldots \text{ and } P_{kp}) = \\
= \quad & \tau(P_{k1}) \wedge \tau(P_{k2}) \wedge \ldots \wedge \tau(P_{kp}) = \\
= \quad & [\mu_{C^{k1-1}}(u_{k1-1}) \wedge \mu_{G^{k1}}(x_{k1})] \wedge [\mu_{C^{k2-1}}(u_{k2-1}) \wedge \mu_{G^{k2}}(x_{k2})] \wedge \ldots \\
& \ldots \wedge [\mu_{C^{kp-1}}(u_{kp-1}) \wedge \mu_{G^{kp}}(x_{kp})] \quad (4.80)
\end{aligned}
$$

which specifies the truth value of the following statement:

"the fuzzy constraints and fuzzy goals are satisfied at the control
stages $t = k1 - 1$, $t = k2 - 1$, ..., and $t = kp - 1$."

Example 4.11 If, say, $v \in \{P_2 \text{ and } P_3\}$, then

$$\mu_T(P_2 \text{ and } P_3) = \tau(P_2) \wedge \tau(P_3) = [\mu_{C^1}(u_1) \wedge \mu_{G^2}(x_2)] \wedge [\mu_{C^2}(u_2) \wedge \mu_{G^3}(x_3)]$$

$\square$

The truth of the statement

"the fuzzy constraints and fuzzy goals are satisfied at Q control
stages"

is now expressed (cf. Section 2.3.2) as the possibility that $v \in V$ both satisfies the
meaning of the fuzzy linguistic quantifier Q and is true [to the degree expressed by
$\mu_T(v)$ given by (4.80)], i.e.

$$\mu_D(u_0, \ldots, u_{N-1} \mid x_0, Q) =$$
$$= \tau(P_1 \text{ and } \ldots \text{ and } P_N \mid Q) = \left(\bigwedge_{t=1}^{N} \mid x_0, Q \right) \tau(P_t) =$$
$$= \max_{v \in V} [\mu_Q(v) \wedge \mu_T(v)] \tag{4.81}$$

Example 4.12 For the same data as in Example 4.7 and with $Q = $ "most" given by
(4.79), if we apply $u_0 = c_1$, $u_1 = c_2$, $u_2 = c_1$, and attain $x_1 = s_2$, $x_2 = s_1$, $x_3 = s_2$,
then

$$\mu_T(P_1 \text{ and } P_2 \text{ and } P_3) =$$
$$= [\mu_{C^0}(c_1) \wedge \mu_{G^1}(s_2)] \wedge [\mu_{C^1}(c_2) \wedge \mu_{G^2}(s_1)] \wedge$$
$$\wedge [\mu_{C^2}(c_1) \wedge \mu_{G^3}(s_2)] =$$
$$= 0.5 \wedge 0.6 \wedge 0.7 \wedge 0.6 \wedge 1 \wedge 0.8 = 0.5$$

$$\mu_T(P_1 \text{ and } P_2) =$$
$$= [\mu_{C^0}(c_1) \wedge \mu_{G^1}(s_2)] \wedge [\mu_{C^1}(c_2) \wedge \mu_{G^2}(s_1)] =$$
$$= 0.5 \wedge 0.6 \wedge 0.7 \wedge 0.6 = 0.5$$

$$\mu_T(P_1 \text{ and } P_3) =$$
$$= [\mu_{C^0}(c_1) \wedge \mu_{G^1}(s_2)] \wedge [\mu_{C^2}(c_1) \wedge \mu_{G^3}(s_2)] =$$
$$= 0.5 \wedge 0.6 \wedge 1 \wedge 0.8 = 0.5$$

$$\mu_T(P_2 \text{ and } P_3) =$$
$$= [\mu_{C^1}(c_2) \wedge \mu_{G^2}(s_1)] \wedge [\mu_{C^2}(c_1) \wedge \mu_{G^3}(s_2)] =$$
$$= 0.7 \wedge 0.6 \wedge 1 \wedge 0.8 = 0.6$$

$$\mu_T(P_1) \;=\; \mu_{C^0}(c_1) \wedge \mu_{G^1}(s_2) = 0.5 \wedge 0.6 = 0.5$$

$$\mu_T(P_2) \;=\; \mu_{C^1}(c_2) \wedge \mu_{G^2}(s_1) = 0.7 \wedge 0.6 = 0.6$$

$$\mu_T(P_3) \;=\; \mu_{C^2}(c_1) \wedge \mu_{G^3}(s_2) = 1 \wedge 0.8 = 0.8$$

Therefore

$$\mu_D(c_1, c_2, c_1 \mid x_0, \text{``most''}) =$$
$$= \max_{v \in V}[\mu_{\text{``most''}}(v) \wedge \mu_T(v)] =$$
$$= [\mu_{\text{``most''}}(P_1 \text{ and } P_2 \text{ and } P_3) \wedge \mu_T(P_1 \text{ and } P_2 \text{ and } P_3)] \vee$$
$$\vee [\mu_{\text{``most''}}(P_1 \text{ and } P_2) \wedge \mu_T(P_1 \text{ and } P_2] \vee$$
$$\vee [\mu_{\text{``most''}}(P_1 \text{ and } P_3) \wedge \mu_T(P_1 \text{ and } P_3)] \vee$$
$$\vee [\mu_{\text{``most''}}(P_2 \text{ and } P_3) \wedge \mu_T(P_2 \text{ and } P_3)] \vee$$
$$\vee [\mu_{\text{``most''}}(P_1) \wedge \mu_T(P_1)] \vee [\mu_{\text{``most''}}(P_2) \wedge \mu_T(P_2)] \vee$$
$$\vee [\mu_{\text{``most''}}(P_3) \wedge \mu_T(P_3)] = 0.6$$

It may be seen that for the conventional problem formulation (4.58), i.e. for $Q = \text{``all''}$, we have

$$\mu_D(c_1, c_2, c_1 \mid x_0) =$$
$$= \mu_D(c_1, c_2, c_1 \mid x_0, \text{``all''}) = \mu_T(P_1 \text{ and } P_2 \text{ and } P_3) = 0.5$$

$\square$

The control problem considered is now to find an optimal sequence of control $u_0^*, \ldots, u_{N-1}^*$ such that

$$\mu_D(u_0^*, \ldots, u_{N-1}^* \mid x_0, Q) =$$
$$= \max_{u_0, \ldots, u_{N-1}} \mu_D(u_0, \ldots, u_{N-1} \mid x_0, Q) =$$
$$= \max_{u_0, \ldots, u_{N-1}} \max_{v \in V}[\mu_Q(v) \wedge \mu_T(v)] \qquad (4.82)$$

It may readily be seen that this problem consists in fact of two problems:

- the calculation of

$$\mu_D(u_0, \ldots, u_{N-1} \mid x_0, Q) = \max_{v \in V}[\mu_Q(v) \wedge \mu_T(v)]$$

for a fixed control sequence $u_0, \ldots, u_{N-1}$, and
- the optimization of the control sequence, i.e. the determination of $u_0^*, \ldots, u_{N-1}^*$ such that

$$\mu_D(u_0^*, \ldots, u_{N-1}^* \mid x_0, Q) =$$
$$= \max_{u_0, \ldots, u_{N-1}} \mu_D(u_0, \ldots, u_{N-1} \mid x_0, Q) =$$
$$= \max_{u_0, \ldots, u_{N-1}} \max_{v \in V}[\mu_Q(v) \wedge \mu_T(v)]$$

and these two issues will be discussed below.

Determination of $\mu_D(u_0, \ldots, u_{N-1} \mid x_0, Q)$

First, let us confine the class of fuzzy linguistic quantifiers to the monotonic ones (2.111), page 52, i.e. such that

$$\mu_Q(v_3) \geq \mu_Q(v_1) \vee \mu_Q(v_2) \tag{4.83}$$

for any $v_1, v_2, v_3 \in V$ such that $v_3 \in \{v_1 \text{ and } v_2\}$.

Example 4.13 For $N = 3$, V as in (4.76), and $Q =$ "most" given by (4.79), if we assume that $v_1 \in \{P_1\}$, $v_2 \in \{P_2 \text{ and } P_3\}$ and $v_3 \in \{P_1 \text{ and } P_2 \text{ and } P_3\}$, then

$$1 = \mu_{\text{"most"}}(v_3) > [\mu_{\text{"most"}}(v_1)] \vee [\mu_{\text{"most"}}(v_2)] = 0.2 \vee 0.7 = 0.7$$

$\square$

The monotonic property of a fuzzy linguistic quantifier means that the more (at a higher number of control stages) the fuzzy constraints and fuzzy goals are satisfied the better. Clearly, only such quantifiers are practically relevant in our context. Note that the monotonicity meant as above is very similar, though not exactly the same, to that of nondecreasingness defined by (4.67).

Now we introduce the α-level sets of the fuzzy linguistic quantifier Q as [cf. (2.14), page 26]

$$Q_\alpha = \{v \in V : \mu_Q(v) \geq \alpha\}, \qquad \text{for each } \alpha \in (0, 1] \tag{4.84}$$

For any element of Q_α, denoted generically as "P_{k1} and P_{k2} and $\ldots$ and $P_{kp} \in Q_\alpha$," we define its *length* as [cf. (2.109), page 52]

$$g_\alpha = g_\alpha(P_{k1} \text{ and } P_{k2} \text{ and } \ldots \text{ and } P_{kp}) = p \tag{4.85}$$

and the *minimum length* of any $v \in Q_\alpha$ as

$$\overline{g}_\alpha = \min_{v \in Q_\alpha} g_\alpha \tag{4.86}$$

Example 4.14 For $Q =$ "most" given by (4.79) and V as in (4.76), we obtain

$$\begin{aligned}
(\text{"most"})_{0.2} &= \{P_1, P_2, P_3, P_1 \text{ and } P_2, P_1 \text{ and } P_3, P_2 \text{ and } P_3, \\
&\qquad P_1 \text{ and } P_2 \text{ and } P_3\} = V \\
(\text{"most"})_{0.7} &= \{P_1 \text{ and } P_2, P_1 \text{ and } P_3, P_2 \text{ and } P_3, P_1 \text{ and } P_2 \text{ and } P_3\} \\
(\text{"most"})_1 &= \{P_1 \text{ and } P_2 \text{ and } P_3\}
\end{aligned}$$

and for, say, $\alpha = 0.7$, we have

$$\begin{aligned}
g_{0.7}(P_1 \text{ and } P_2) &= g_{0.7}(P_1 \text{ and } P_3) = g_{0.7}(P_2 \text{ and } P_3) = 2 \\
g_{0.7}(P_1 \text{ and } P_2 \text{ and } P_3) &= 3
\end{aligned}$$

so that $\overline{g}_{0.7} = 2$.

$\square$

Let us now consider the set

$$M = \{[\mu_{C^0}(u_0) \wedge \mu_{G^1}(x_1)], \ldots, [\mu_{C^{N-1}}(u_{N-1}) \wedge \mu_{G^N}(x_N)]\} \qquad (4.87)$$

whose elements, for a fixed sequence of controls $u_0, \ldots, u_{N-1}$ and of states $x_0, \ldots, u_{N-1}$, are evidently real numbers between 0 and 1.

If we denote by $m(\overline{g}_\alpha)$ the $\overline{g}_\alpha$-th largest element of M, then we have an important property.

Lemma 4.2 *For any Q_α, $\alpha \in (0, 1]$, there holds*

$$\max_{v \in Q_\alpha} \mu_T(v) = m(\overline{g}_\alpha) \qquad (4.88)$$

Proof. A generic element of Q_α is denoted by P_{k1} and P_{k2} and ... and P_{kp}, $p = \overline{g}_\alpha, \overline{g}_{\alpha+1}, \ldots, N$. Due to (4.83), we need to consider only $p = \overline{g}_\alpha$, because $\mu_T(v)$ given by (4.80) cannot be higher for any $p > \overline{g}_\alpha$. The maximization in (4.88) is therefore to proceed over the v's of length $\overline{g}_\alpha$.

Evidently, due to "$\wedge$" used in (4.80), the highest value of $\mu_T(v)$ is obtained for such a $v \in V$ which contains the first $\overline{g}_\alpha$-largest elements of M. This value is obviously equal to $m(\overline{g}_\alpha)$, i.e. to the $\overline{g}_\alpha$-th largest element of M. $\qquad \square$

We are now in a position to state the main property formulated as the following proposition.

Proposition 4.2 *For any monotonic linguistic quantifier Q defined as a fuzzy set in V, there holds*

$$\mu_D(u_0, \ldots, u_{N-1} \mid x_0, Q) = \max_{v \in V}[\mu_Q(v) \wedge \mu_T(v)] =$$
$$= \max_{\alpha \in (0,1]}[\alpha \wedge \max_{v \in Q_\alpha} \mu_T(v)] = \max_{\alpha \in (0,1]}[\alpha \wedge m(\overline{g}_\alpha)] \qquad (4.89)$$

Proof. The first part, i.e.

$$\max_{v \in V}[\mu_Q(v) \wedge \mu_T(v)] = \max_{\alpha \in (0,1]}[\alpha \wedge \max_{v \in Q_\alpha} \mu_T(v)]$$

is a well-known result stated first by Tanaka, Okuda and Asai (1974) in the context of fuzzy mathematical programming.

And the second part, i.e.

$$\max_{\alpha \in (0,1]}[\alpha \wedge \max_{v \in Q_\alpha} \mu_T(v)] = \max_{\alpha \in (0,1]}[\alpha \wedge m(\overline{g}_\alpha)]$$

results from Lemma 4.2. $\qquad \square$

We have therefore a formula (4.89) which makes it possible to calculate the value of $\mu_D(u_0, \ldots, u_{N-1} \mid x_0, Q)$. For convenience, however, we will use its equivalent form

$$\mu_D(u_0, \ldots, u_{N-1} \mid x_0, Q) =$$
$$= \max_{v \in V}[\mu_Q(v) \wedge \mu_T(v)] = \max_{i=1,\ldots,N}(a_i \wedge b_i) \qquad (4.90)$$

where a_i is the i-th largest element of the set of values M (4.87), and b_i is the grade of membership of elements in Q of length i, i.e. of P_{k1} and P_{k2} and ... and P_{ki}.

Example 4.15 Suppose that we have the same data as in Example 4.7, page 116, that is: $U = \{c_1, c_2\}$, $X = \{s_1, s_2, s_3\}$, $N = 3$, and

$$
\begin{aligned}
C^0 &= 0.5/c_1 + 1/c_2 & G^1 &= 0.1/s_1 + 0.6/s_2 + 1/s_3 \\
C^1 &= 1/c_1 + 0.7/c_2 & G^2 &= 0.61/s_1 + 1/s_2 + 0.5/s_3 \\
C^2 &= 1/c_1 + 0.6/c_2 & G^3 &= 1/s_1 + 0.8/s_2 + 0.3/s_3
\end{aligned}
$$

If we apply $u_0 = c_1$, $u_1 = c_2$ and $u_2 = c_1$ and attain $x_1 = s_2$, $x_2 = s_1$ and $x_3 = s_2$, then we have:

$$
\begin{aligned}
\mu_{C^0}(c_1) \wedge \mu_{G^1}(s_2) &= 0.5 \wedge 0.6 = 0.5 \\
\mu_{C^1}(c_2) \wedge \mu_{G^2}(s_1) &= 0.7 \wedge 0.6 = 0.6 \\
\mu_{C^2}(c_1) \wedge \mu_{G^3}(s_2) &= 1 \wedge 0.8 = 0.8
\end{aligned}
$$

that is, $M = \{0.5, 0.6, 0.8\}$, and $a_1 = 0.8$, $a_2 = 0.6$, $a_3 = 0.55$.

For $Q =$ "most" given by (4.79), we have $b_1 = 0.2$, $b_2 = 0.7$, $b_3 = 1$. Therefore

$$
\max_{v \in V} [\mu_{\text{"most"}}(v) \wedge \mu_T(v)] = \max_{i=1,2,3}(a_i \wedge b_i) =
$$

$$
= (0.8 \wedge 0.2) \vee (0.6 \wedge 0.7) \vee (0.5 \wedge 1) = 0.2 \vee 0.6 \vee 0.5 = 0.6
$$

which is the same result as that obtained in a direct (longer!) way in Example 4.12, page 122.   □

Now, we can proceed to the description of an algorithm for the optimization of control sequence.

Optimization of $u_0, \ldots, u_{N-1}$

We will now consider the second issue, i.e. the determination of an optimal sequence of controls $u_0^*, \ldots, u_{N-1}^*$ such that

$$
\begin{aligned}
\mu_D(u_0^*, \ldots, u_{N-1}^* \mid x_0, Q) &= \\
&= \max_{u_0, \ldots, u_{N-1}} \mu_D(u_0, \ldots, u_{N-1} \mid x_0, Q) = \\
&= \max_{u_0, \ldots, u_{N-1}} \max_{v \in V} [\mu_Q(v) \wedge \mu_T(v)] \qquad (4.91)
\end{aligned}
$$

It can be seen that since the control space U is finite, then the control process considered may be presented as a decision tree of the type shown in Figure 17, page 91. Its nodes correspond to the successive states attained, with the initial state x_0 in the root, and with its arcs there are associated the consecutive controls applied and the values of the respective $\mu_{C^i}(u_t) \wedge \mu_{G^{i+1}}(x_{t+1})$. Each $u_0, \ldots, u_{N-1}$ is therefore equivalent to a path (sequence of subsequent nodes and arcs) from x_0 to x_N, and the solution sought, $u_0^*, \ldots, u_{N-1}^*$, corresponds to such a path that (4.91) holds.

Our next arguments which will finally lead to an algorithm for the determination of optimal controls will deal with such a decision tree.

It is evident that the solution of the problem considered (4.91) may be obtained by full enumeration, i.e. by analyzing all the possible paths and choosing an optimal one [that with the highest values of $\mu_D(\cdot \mid \cdot, \cdot)$].

However, the full enumeration is clearly inefficient and therefore we will develop an *implicit enumeration* algorithm that will make it possible to neglect nonpromising paths and reduce the search space to a relatively small set of promising paths. The following interesting properties form a basis for such an algorithm.

The point of departure is the solution of the conventional problem (that is, for $Q =$ "all"): find a $u_0^*, \ldots, u_{N-1}^*$ such that

$$
\begin{aligned}
\mu_D(u_0^*, \ldots, u_{N-1}^* \mid x_0) &= \\
&= \mu_D(u_0^*, \ldots, u_{N-1}^* \mid x_0, \text{"all"}) = \\
&= \max_{u_0, \ldots, u_{N-1}} \bigwedge_{t=0}^{N-1} [\mu_{C^t}(u_t) \wedge \mu_{G^{t+1}}(x_{t+1})]
\end{aligned}
\tag{4.92}
$$

which may be efficiently solved by dynamic programming (cf. Section 4.1.1) and branch-and-bound (cf. Section 4.1.2), and even using a genetic algorithm (cf. Section 4.1.3) or a special neural network (cf. Section 4.1.4).

We have now some important properties.

Proposition 4.3 *For any monotonic [in the sense of (4.83), page 124] quantifier Q defined in V, and any sequence of controls $u_0, \ldots, u_{N-1}$, we have*

$$
\mu_D(u_0, \ldots, u_{N-1} \mid x_0, Q) \geq \mu_D(u_0, \ldots, u_{N-1} \mid x_0, \text{"all"})
\tag{4.93}
$$

and in consequence

$$
\mu_D(u_0^*, \ldots, u_{N-1}^* \mid x_0, Q) \geq \mu_D(u_0^*, \ldots, u_{N-1}^* \mid x_0, \text{"all"})
\tag{4.94}
$$

Proof. See that P_1 and P_2 and $\ldots$ and $P_N \in V$, so we may write $V = \{\cdot, P_1 \text{ and } P_2 \text{ and } \ldots \text{ and } P_N\}$.

Therefore

$$
\begin{aligned}
\mu_D(u_0, \ldots, u_{N-1} \mid x_0, Q) &= \max_{v \in V}[\mu_Q(v) \wedge \mu_T(v)] = \\
&= \max\{\mu_Q(P_1 \text{ and } P_2 \text{ and } P_3) \wedge \mu_T(P_1 \text{ and } P_2 \text{ and } P_3)], \\
&\quad \max_{v \in V \setminus \{P_1 \text{ and } P_2 \text{ and } P_3\}} [\mu_Q(v) \wedge \mu_T(v)]\} = \\
&= \max\{\mu_D(u_0, \ldots, u_{N-1} \mid x_0, \text{"all"}), \max_{v \in V \setminus \{P_1 \text{ and } P_2 \text{ and } P_3\}} [\mu_Q(v) \wedge \mu_T(v)]\}
\end{aligned}
$$

which obviously implies (4.93). And since (4.93) holds for any sequence of controls $u_0, \ldots, u_{N-1}$, then it holds for an optimal sequence of controls $u_0^*, \ldots, u_{N-1}^*$ in particular. $\square$

Thus the solution of the generalized problem (4.82), page 123, i.e. with a monotonic fuzzy linguistic quantifier Q, is always not worse than the solution of the conventional problem (4.92), i.e. for $Q =$ "all."

We now have a necessary condition for a path, i.e. its corresponding sequence of controls, to be an optimal solution of the generalized problem (4.82).

Proposition 4.4 *If $u_0^*, \ldots, u_{N-1}^*$ is an optimal solution of the generalized problem (4.82) with some monotonic quantifier Q, then in its corresponding path there exists* *as least one arc such that for* u_t^*, $t \in \{0, 1, \ldots, N-1\}$, *corresponding to this path there holds*

$$\mu_{C^t}(u_t^*) \wedge \mu_{G^{t+1}}[f(x_t, u_t^*)] \geq \mu_D(u_0^*, \ldots, u_{N-1}^* \mid x_0, \text{"all"}) \qquad (4.95)$$

Proof. For each u_t^* we evidently have

$$\mu_D(u_0^*, \ldots, u_{N-1}^* \mid x_0, Q) \leq \mu_{C^t}(u_t^*) \wedge \mu_{G^{t+1}}[f(x_t, u_t^*)]$$

and due to Proposition 4.3 there must hold

$$\mu_D(u_0^*, \ldots, u_{N-1}^* \mid x_0, Q) \geq \mu_D(u_0^*, \ldots, u_{N-1}^* \mid x_0, \text{"all"})$$

$\square$

Thus if (4.95) holds, then the path is considered *promising*, i.e. it may correspond to an optimal solution sought.

The properties stated above imply the following implicit enumeration algorithm for solving problem (4.82):

Step 1 Construct a decision tree (of the type shown in Figure 17, page 91), i.e. with x_0 in the root, the consecutive states in the nodes, and the arcs associated with the controls applied and the values of $\mu_{C^t}(u_t) \wedge \mu_{G^{t+1}}(u_{t+1})$.

Step 2 Solve the conventional problem (4.92), i.e. find $u_0^*, \ldots, u_{N-1}^*$ such that

$$\mu_D(u_0^*, \ldots, u_{N-1}^* \mid x_0, \text{"all"}) = \max_{u_0, \ldots, u_{N-1}} \mu_D(u_0, \ldots, u_{N-1} \mid x_0, \text{"all"})$$

Step 3. Find such arcs in the decision tree for which

$$\mu_{C^t}(u_t) \wedge \mu_{G^{t+1}}(x_{t+1}) > \mu_D(u_0^*, \ldots, u_{N-1}^* \mid \text{"all"}) \qquad (4.96)$$

and note that we do not consider the inequality "$\geq$" because we already have some arc(s) for which "$=$" in (4.95) holds in the path corresponding to the solution found in Step 2. We seek a better solution, hence we put "$>$" in (4.96).

Step 4. Determine all the paths from x_0 to x_N containing the arcs found in Step 3.

Step 5. For each path (its corresponding sequence of controls) found in Step 4 determine

$$\mu_D(u_0, \ldots, u_{N-1} \mid x_0, Q) =$$
$$= \max_{v \in V}[\mu_Q(v) \wedge \mu_T(v)] = \max_{i=1, \ldots, N}(a_i \wedge b_i) \qquad (4.97)$$

For an example of how the above algorithm works, we refer the interested reader to the source paper (Kacprzyk, 1983c).

Thus, unlike in the solution when using the algebraic method, while using the substitution method the particular form of the linguistic quantifier has an impact on the solution process and the result obtained which is obviously intuitively appealing.

We will proceed now to the control problems in which both linguistic quantifiers and discounting is involved.

4.1.5.3 Discounting in multistage fuzzy control with fuzzy linguistic quantifiers

As we may remember from Section 3.2, *discounting* is commonly used in control and multistage decision making to reflect an obvious fact that what will happen (as a result of applying controls) at an earlier control stage should be more important than what will happen at later. There are many obvious reasons for this fact as, e.g., the diminishing importance of the later control stages (i.e. of a more distant future) may well account for the fact that in the future our model, developed clearly in advance, may be inappropriate owing to changing conditions, and the more so the later the control stage. Moreover, human nature seems to tend to pay more attention to more immediate results, i.e. at earlier control stages, than to later.

Discounting has been introduced into the multistage control model with linguistic quantifiers (shown in the previous section) by Kacprzyk and Iwański (1987), and we will in principle follow their line of reasoning and analysis.

In the context of multistage fuzzy control, discounting may be equated with the importance of the particular control stages. Importance B is defined as a fuzzy set in the space of control stages $\{0, 1, \ldots, N-1\}$; $\mu_B(t) \in [0, 1]$ is (a degree of) importance of control stage $t = 0, 1, \ldots, N-1$, the higher the more important, from 0 for fully unimportant to 1 for fully important through all intermediate values. As we have already mentioned, in the context of multistage fuzzy control it is only reasonable to assume that

$$t' > t'' \implies \mu_B(t') \le \mu_B(t''), \qquad \text{for each } t', t'' \in \{0, 1, \ldots, N-1\} \qquad (4.98)$$

that is, earlier control stages are assumed to be more (in fact, not less) important.

We assume that the other elements of the problem formulation are the same as in the case of multistage control with a fuzzy linguistic quantifier but without discounting (cf. Section 4.1.5). So, the system under control is deterministic, and its dynamics is described by the state transition equation (4.1), page 85, i.e.

$$x_{t+1} = f(x_t, u_t), \qquad \text{for each } t = 0, 1, \ldots, N-1$$

where $x_t, x_{t+1} \in X\{s_1, \ldots, s_n\}$ are states at control stages t and $t+1$, respectively, and $u_t \in U = \{c_1, \ldots, c_m\}$ is the control at stage t; $t = 0, 1, \ldots, N-1$ where N is fixed and specified in advance.

At each control stage t, a fuzzy constraint $\mu_{C^t}(u_t)$ is imposed on u_t, and the resulting x_{t+1} is subjected to a fuzzy goal $\mu_{G^{t+1}}(x_{t+1})$.

Now, using a linguistic quantifier, to express "how many" control stages are to accounted for) and discounting (to express a varying importance of the particular control stages), we may generally formulate the control problem as "find an optimal sequence of controls that best satisfies the fuzzy constraints and fuzzy goals at Q (e.g., most, almost all, much more than a half, etc.) of the *important* (earlier) control stages."

Therefore, analogously to (4.59) and (4.60), the fuzzy decision of such a problem formulation may be written as

$$
\begin{aligned}
\mu_D(u_0, \ldots, u_{N-1} \mid x_0, Q, B) &= \\
&= \tau(P_1 \text{ and } \ldots \text{ and } P_N \mid Q, B) = \\
&= \left(\bigwedge_{t=0}^{N-1} \mid Q, B \right) \tau(P_{t+1}) = \\
&= \left(\bigwedge_{t=0}^{N-1} \mid Q, B \right) [\mu_{C^t}(u_t) \wedge \mu_{G^{t+1}}(x_{t+1})]
\end{aligned}
\tag{4.99}
$$

and the problem as to find $u_0^*, \ldots, u_{N-1}^*$ such that

$$
\begin{aligned}
\mu_D(u_0^*, \ldots, u_{N-1}^* \mid x_0 \mid Q, B) &= \\
&= \max_{u_0, \ldots, u_{N-1}} \tau(P_1 \text{ and } \ldots \text{ and } P_N \mid Q, B) = \\
&= \max_{u_0, \ldots, u_{N-1}} \left(\bigwedge_{t=0}^{N-1} \mid Q, B \right) \tau(P_{t+1}) = \\
&= \max_{u_0, \ldots u_{N-1}} \left(\bigwedge_{t=0}^{N-1} \mid Q, B \right) [\mu_{C^t}(u_t) \wedge \mu_{G^{t+1}}(x_{t+1})]
\end{aligned}
\tag{4.100}
$$

where Q is a fuzzy linguistic quantifier and B is importance, and as previously:

$$
P_{t+1} : \text{``}C^t \text{ and } G^{t+1} \text{ are satisfied (by } u_t \text{ and the resulting } x_{t+1})\text{''}
\tag{4.101}
$$

and

$$
\tau(P_{t+1}) = \mu_{C^t}(u_t) \wedge \mu_{G^{t+1}}(x_{t+1})
\tag{4.102}
$$

where $t = 0, 1, \ldots, N-1$.

And, as previously, in presumably one of the most natural cases when $Q = $ "most" and $B = $ "earlier", the fuzzy decision (4.99) becomes

$$
\begin{aligned}
\mu_D(u_0, \ldots, u_{N-1} \mid x_0, \text{``most''}, \text{``earlier''}) &= \\
&= \tau(P_1 \text{ and } \ldots \text{ and } P_N \mid \text{``most''}, \text{``earlier''}) = \\
&= \left(\bigwedge_{t=0}^{N-1} \mid \text{``most''} \right) \tau(P_{t+1}) = \\
&= \left(\bigwedge_{t=0}^{N-1} \mid \text{``most''} \right) [\mu_{C^t}(u_t) \wedge \mu_{G^{t+1}}(x_{t+1})]
\end{aligned}
\tag{4.103}
$$

and the problem (4.100) becomes to find a $u_0^*, \ldots, u_{N-1}^*$ such that

$$
\mu_D(u_0^*, \ldots, u_{N-1}^* \mid x_0, \text{``most''}, \text{``earlier''}) =
$$

$$= \max_{u_0,\ldots,u_{N-1}} \tau(P_1 \text{ and } \ldots \text{ and } P_N \mid \text{"most"}, \text{"earlier"}) =$$

$$= \max_{u_0,\ldots,u_{N-1}} \left(\bigwedge_{t=0}^{N-1} \mid \text{"most"}, \text{"earlier"} \right) \tau(P_{t+1}) =$$

$$= \max_{u_0,\ldots u_{N-1}} \left(\bigwedge_{t=0}^{N-1} \mid \text{"most"}, \text{"earlier"} \right) [\mu_{C^i}(u_t) \wedge \mu_{G^{i+1}}(x_{t+1})] \quad (4.104)$$

that is, such that it best satisfies the fuzzy constraints and fuzzy goals at Q (e.g., most) of B (e.g., earlier) control stages.

It is easy to see that the use of the substitution method is clearly much more interesting, so we will now present its use for the formulation and solution of the problem considered.

Analogously as in Section 4.1.5.2, we construct the set

$$V = \{v\} = \{P_{k1} \text{ and } P_{k2} \text{ and } \ldots \text{ and } P_{km}\} = 2^{P_1,\ldots,P_N} \setminus 0$$

and then the fuzzy decision (4.99) becomes

$$\mu_D(u_0,\ldots,u_{N-1} \mid x_0, Q, B) =$$
$$= \tau(P_1 \text{ and } \ldots \text{ and } P_N \mid Q, B) =$$
$$= \left(\bigwedge_{t=0}^{N-1} \mid Q, B \right) \tau(P_{t+1}) =$$
$$= \max_{v=\{P_{k1} \text{ and } \ldots \text{ and } P_{km}\} \in V} \{\mu_Q(v) \wedge$$
$$\wedge \bigwedge_{i=k1-1}^{km-1} [\mu_B(i)(\mu_{C^i}(u_i) \wedge \mu_{G^{i+1}}(x_{i+1}))]\} \quad (4.105)$$

and the problem as to find a $u_0^*,\ldots,u_{N-1}^*$ such that

$$\mu_D(u_0^*,\ldots,u_{N-1}^* \mid x_0 \mid Q, B) =$$
$$= \max_{u_0,\ldots u_{N-1}} \mu_D(u_0,\ldots,u_{N-1} \mid x_0 \mid Q, B) \quad (4.106)$$

where $\mu_D(\cdot \mid Q, B)$ is given by (4.105).

This problem is equivalent [cf. (2.115), page 52] to seeking $u_0^*,\ldots,u_{N-1}^*$ such that

$$\mu_D(u_0^*,\ldots,u_{N-1}^* \mid x_0, Q, B) =$$
$$= \max_{u_0,\ldots,u_{N-1}} \max_{v=\{P_{k1} \text{ and } \ldots \text{ and } P_{km}\} \in V} \{\mu_Q(v) \wedge$$
$$\wedge \bigwedge_{i=k1-1}^{km-1} [\mu_B(i) \implies (\mu_{C^i}(u_i) \wedge \mu_{G^{i+1}}(x_{i+1}))]\} \quad (4.107)$$

where "$\implies$" is an implication operator given by (2.90)–(2.95), page 47.

If we assign the same importance for each control stage, i.e. $B = 1/0 + 1/1 + \cdots + 1/(N-1)$, this problem becomes the same as that without discounting, i.e. (4.82), page 123.

Some properties of the problem and an algorithm for its solution

We will present now a branch-and-bound algorithm for solving the problem considered (4.107), but we should start with some interesting properties.

First, as we have already mentioned, the case without discounting is equivalent to the assumption that $B = 1/0 + 1/1 + \cdots + 1/(N-1)$. Then we have the following property which is an equivalent of Proposition 4.3, page 127.

Proposition 4.5 *For any* $x_0 \in X$, *any monotonic fuzzy quantifier* Q, *and any sequence of controls* $u_0, \ldots, u_{N-1}$, *we have*

$$
\begin{aligned}
\mu_D(u_0, \ldots, u_{N-1} \mid x_0, Q, B) &\geq \\
&\geq \mu_D(u_0, \ldots, u_{N-1} \mid x_0, \text{all}, B = 1/0 + \cdots + 1/(N-1)) = \\
&= \mu_D(u_0, \ldots, u_{N-1} \mid x_0, \text{all}) = \overline{a}
\end{aligned} \tag{4.108}
$$

and in particular

$$
\begin{aligned}
\mu_D(u_0^*, \ldots, u_{N-1}^* \mid x_0, Q, B) &\geq \\
&\geq \mu_D(u_0^*, \ldots, u_{N-1}^* \mid x_0, \text{all}, B = 1/0 + \cdots + 1/(N-1)) = \\
&= \mu_D(u_0^*, \ldots, u_{N-1}^* \mid x_0, \text{all}) = \overline{a}^*
\end{aligned} \tag{4.109}
$$

The proof of this proposition does closely parallel that of Proposition 4.3, page 127, and will not be repeated here. Here $u_0^*, \ldots, u_{N-1}^*$ is an optimal solution of problem (4.58), i.e. for $Q = $ "all," or equivalently, of the source problem (4.3), page 86.

This proposition provides a starting point, $\overline{a}^*$, i.e. the lowest possible value of the fuzzy decision that is worth considering in the case of an algorithm which, as virtually all algorithms, would improve this value step by step, in consecutive iterations. Moreover, its corresponding optimal sequence of controls $u_0^*, \ldots, u_{N-1}^*$ can be easily determined by, say, dynamic programming (cf. Section 4.1.1), branch-and-bound (cf. Section 4.1.2), a genetic algorithm (cf. Section 4.1.3) or a special neural network (cf. Section 4.1.4).

We now have the following counterpart of Proposition 4.4:

Proposition 4.6 *In any optimal sequence of controls* $u_0^*, \ldots, u_{N-1}^*$ *being a solution to problem* (4.107), *there exists at least one optimal control* u_k^*, $k \in \{0, 1, \ldots, N-1\}$, *such that*

$$
\begin{aligned}
\mu_B(k) \Longrightarrow [\mu_{C^k}(u_k^*) \wedge \mu_{G^{k+1}}(f(x_k, u_k^*))] &\geq \\
&\geq \overline{a}^* = \mu_D(u_0^*, \ldots, u_{N-1}^* \mid x_0, \text{all})
\end{aligned} \tag{4.110}
$$

This means, roughly speaking, that by adding a next control, say u_k, we cannot increase the value of $\mu_D(\cdot \mid \cdot, \cdot)$ given by (4.103). This is a consequence of the definitions of "$\Longrightarrow$" and "$\wedge$," and the proof of this proposition closely parallels that of Proposition 4.4, page 128.

Observe that since x_0, and $u_0^*, \ldots, u_{N-1}^*$ is clearly equivalent to a path from x_0 to x_N in the decision tree corresponding to the problem considered, then u_k (or, in fact, x_k and u_k) corresponds to an arc in this decision tree.

Now, it is convenient to introduce the following definition. At control stage $t = k$, a control u_k is said to be *promising* if (4.110) holds, that is

$$\mu_B(k) \Longrightarrow [\mu_{C^k}(u_k) \wedge \mu_{G^{k+1}}(f(x_k, u_k))] \geq a \geq \bar{a}^* \qquad (4.111)$$

where a, as we will see later in this section, is some value that is being successively improved in the course of the solution process.

Evidently, this definition should be meant as that control u_k applied at control stage $t = k$ while in state x_k is promising, i.e. the edge (x_k, u_u) in the decision tree is promising.

The algorithm used by Kacprzyk and Iwański (1987) for solving the problem considered is of the branch-and-bound type. The branching process proceeds by applying the particular controls at the subsequent control stages, i.e. by choosing edges in the decision tree.

The bounding, on the other hand, proceeds as follows. First, assume that we start in state x_0 (the root of the decision tree) and we are currently at control stage $t = k$, having traversed some path, i.e. subsequent nodes and edges from x_0 to x_k:
$x_0, u_0, x_1, u_1, \ldots, x_{k-1}, u_{k-1}, x_k$.

With the above path is associated a vector $Y_k = [y_0, y_1, \ldots, y_{k-1}]$ defined as

$$y_i = \begin{cases} 1 & \text{if edge } (x_i, y_i) \text{ is promising} \\ 0 & \text{otherwise} \end{cases} \qquad (4.112)$$

where $i = 0, 1, \ldots, k - 1$. Evidently, Y_k depends both on the earlier controls $u_0, u_1, \ldots, u_{k-2}$ and on the current control u_{k-1}.

We define now the augmented vector $\bar{Y}_k = [\bar{y}_0, \bar{y}_1, \ldots, \bar{y}_{k-1}, \bar{y}_k, \ldots, \bar{y}_{N-1}]$ such that

$$\bar{y}_i = \begin{cases} y_i & \text{for } i = 0, 1, \ldots, k - 1 \\ 1 & \text{for } i = k, k + 1, \ldots, N - 1 \end{cases} \qquad (4.113)$$

Basically, this means that we take the real promising edges up to the current control stage k, and assume that the rest of the edges in the path, i.e. for the control stages $k + 1, k, \ldots, N - 1$, are promising by definition.

The linguistic quantifier Q is now defined as a fuzzy set in 2^W, where $W = \{w\} = \{t : \bar{y}_t = 1, t \in \{0, 1, \ldots, N - 1\}\}$. Thus, $\mu_Q(w) \in [0, 1]$ gives the degree to which the set of promising edges in $\bar{Y}_k$ satisfies the meaning of Q. Then, denote the fact that Q is defined with respect to the number of 1's in $\bar{Y}_k$, which is equivalent to the length of w, denoted $|w|$ and defined by (2.109), page 52. For instance, if $\bar{Y}_7 = [0, 1, 1, 1, 0, 0, 1, 1]$, then $\mu_{\text{"most"}}(5 \mid \bar{V}_k)$ yields the degree to which 5 out of 8 (i.e. 5 "ones" among 8 elements) satisfy the meaning of "most" as, say, $\mu_{\text{"most"}}(5 \mid \bar{Y}_7) = 0.6$.

It is now easy to see that in view of the definition of a promising path and of Proposition 4.6, if we already have a solution $u_0', \ldots, u_{N-1}'$ for which

$$\mu_D(u_0', \ldots, u_{N-1}' \mid x_0, !, B) = a \geq \bar{a}^* \qquad (4.114)$$

then at each control stage $k = 0, 1, \ldots, N - 1$ we should look for a next control u_k such that

$$\mu_Q(w \mid \bar{Y}_k) > a \qquad (4.115)$$

since only such a u_k may lead to a better solution, i.e. one for which $\mu_D(\cdot \mid \cdot, \cdot) > a$.

Moreover, see that for any monotonic [in the sense of (2.111), page 52] fuzzy quantifier Q we have

$$j \geq i \Longrightarrow \mu_Q(w \mid \overline{Y}_j) \leq \mu_Q(w \mid \overline{Y}_i) \qquad (4.116)$$

which is implied by the definition of $\overline{Y}_k$ (4.113) in that the number of 1's in $\overline{Y}_j$ is not less than the number of 1's in $\overline{Y}_i$, $j \geq i$, and hence, since Q is a monotonic fuzzy quantifier, then (4.116) holds.

The above properties make it possible to construct a branch-and-bound algorithm, similar to that presented in Section 4.1.2). For details on this algorithm we refer the interested reader to the source paper (Kacprzyk and Iwański's, 1987).

We will now prove that this algorithm, which is based on the properties stated above – in the analogous sense as the branch-and-bound algorithm in Section 4.1.2 is based on the properties (4.17)–(4.19), page 92) – does find an optimal solution to the problem considered.

Theorem 4.1 *If* $u_0^*, \ldots, u_{N-1}^*$ *is an optimal solution to the problem considered* (4.100), *i.e.*

$$a^* = \mu_D(u_0^*, \ldots, u_{N-1}^* \mid x_0, Q, B) =$$
$$= \max_{u_0, \ldots, u_{N-1}} \mu_D(u_0, \ldots, u_{N-1} \mid x_0, Q, B)$$

then the branch-and-bound algorithm as mentioned above will attain a^*, *i.e. will find an optimal solution.*

Proof. Let us denote by $\overline{Y}_i^*$, $i = 1, \ldots, N$, the vectors of promising edges determined for the consecutive edges in the decision tree belonging to the path from x_0 to x_N corresponding to an optimal sequence of controls for problem (4.100). In order to traverse the whole path corresponding to an optimal solution, it is necessary that each optimal control u_t^*, $t = 0, \ldots, N-1$, be applied.

Hence, the following inequality should hold:

$$\mu_Q(\cdot \mid \overline{Y}_i^*) > a$$

where $a = \mu_D(u_0, \ldots, u_{N-1} \mid x_0, Q, B)$ for some previous nonoptimal solution $u_0, \ldots, u_{N-1}$.

By the definition of a^* [cf. inequality (4.109), page 132], there holds

$$\mu_D(\cdot \mid \overline{Y}_i^*) \geq a^* > a$$

because $a^* = \mu_Q(\cdot \mid Y_N^*) \wedge \mu_T(\cdot)$, and $\mu_T(\cdot) \in (0, 1]$.

Moreover

$$\mu_Q(\cdot \mid \overline{Y}_i^*) \geq \mu_Q(\cdot \mid Y_N^*), \qquad i = 1, \ldots, N$$

Hence

$$\mu_Q(\cdot \mid \overline{Y}_i^*) > a, \qquad i = 1, \ldots, N$$

that is, the algorithm will attain a better value than the one corresponding to any nonoptimal solution. Therefore, finally, an optimal solution will be found. ☐

Example 4.16 Suppose that the state space is $X = \{s_1, \ldots, s_{10}\}$, the control space is $U = \{c_1, \ldots, c_5\}$, and the dynamics of the system under control is given as the following state transition equation:

$$x_{t+1} = f(x_t, u_t) =$$

	$u_t = c_1$	c_2	c_3	c_4	c_5
$x_t = s_1$	s_1	s_1	s_1	s_2	s_2
s_2	s_1	s_2	s_2	s_2	s_3
s_3	s_1	s_2	s_3	s_3	s_4
s_4	s_3	s_4	s_4	s_4	s_5
s_5	s_1	s_2	s_4	s_5	s_6
s_6	s_4	s_5	s_6	s_6	s_7
s_7	s_7	s_7	s_8	s_8	s_9
s_8	s_5	s_5	s_7	s_8	s_9
s_9	s_8	s_8	s_9	s_{10}	s_{10}
s_{10}	s_8	s_8	s_9	s_{10}	s_{10}

The termination time is $N = 10$, the fuzzy constraints are:

$$C^0 = 1/c_1 + 1/c_2 + 1/c_3 + 0.7/c_4 + 0.3/c_5$$
$$C^1 = 1/c_1 + 1/c_2 + 0.8/c_3 + 0.4/c_4$$
$$C^2 = 0.8/c_1 + 1/c_2 + 0.7/c_3 + 0.3/c_4$$
$$C^3 = 0.6/c_1 + 1/c_2 + 1/c_3 + 0.5/c_4 + 0.1/c_5$$
$$C^4 = 0.8/c_1 + 1/c_2 + 0.6/c_3 + 0.3/c_4 + 0.1/c_5$$
$$C^5 = 1/c_1 + 0.8/c_2 + 0.6/c_3 + 0.4/c_4$$
$$C^6 = 1/c_1 + 1/c_2 + 1/c_3 + 1/c_4 + 1/c_5$$
$$C^7 = 0.6/c_1 + 0.8/c_2 + 1/c_3 + 0.7/c_4 + 0.5/c_5$$
$$C^8 = 1/c_1 + 1/c_2 + 0.9/c_3 + 0.8/c_4 + 0.7/c_5$$
$$C^9 = 1/c_1 + 1/c_2 + 0.8/c_3 + 0.6/c_4 + 0.4/c_5$$

where here, and in the whole book, the singletons in fuzzy sets with the membership grade equal 0 are omitted.

The fuzzy goals are

$$G^1 = 0.3/s_1 + 0.7/s_2 + 1/s_3 + 0.6/s_4 + 0.2/s_5$$
$$G^2 = 0.2/s_1 + 0.5/s_2 + 0.8/s_3 + 1/s_4 + 0.9/s_5 + 0.4/s_6$$
$$G^3 = 0.3/s_2 + 0.7/s_3 + 1/s_4 + 1/s_5 + 0.8/s_6 + 0.3/s_7$$
$$G^4 = 0.1/s_3 + 0.5/s_4 + 0.8/s_5 + 1/s_6 + 1/s_7 + 0.6/s_8 + 0.3/s_9$$
$$G^5 = 0.6/s_3 + 0.7/s_4 + 1/s_5 + 1/s_6 + 1/s_7 + 0.8/s_8 + 0.6/s_9 + 0.2/s_{10}$$
$$G^6 = 0.5/s_4 + 0.7/s_5 + 1/s_6 + 1/s_7 + 1/s_8 + 1/s_9 + 0.8/s_{10}$$
$$G^7 = 0.5/s_5 + 0.8/s_6 + 0.9/s_7 + 1/s_8 + 1/s_9 + 1/s_{10}$$
$$G^8 = 0.3/s_5 + 0.7/s_6 + 0.8/s_7 + 1/s_8 + 1/s_9 + 1/s_{10}$$
$$G^9 = 0.2/s_6 + 0.4/s_7 + 0.7/s_8 + 1/s_9 + 1/s_{10}$$
$$G^{10} = 0.6/s_8 + 0.8/s_9 + 1/s_{10}$$

Importance of the particular control stages is given as

$$B = 1/0 + 1/1 + 0.9/2 + 0.8/3 + 0.7/4 + 0.5/5 + 0.3/6 + 0.2/7 + 0.1/8$$

We assume now that the following fuzzy linguistic quantifiers are available [note that they are defined as $\mu_Q(| v |)$, where $| v |$ is the length of v defined by (2.109), page 52, i.e. as the number of control stages that satisfy the meaning of Q]:

- $Q =$ "most" given as

$$\mu_{\text{"most"}}(|v|) = 1/10 + 1/9 + 0.8/8 +$$
$$+ 0.6/7 + 0.5/6 + 0.3/5 + 0.2/4 + 0.1/3 + 0/2 + 0/1$$

- $Q =$ " $\gg 50\%$ " = "much more than 50%" given as

$$\mu_{\text{"}\gg 50\%\text{"}}(|v|) = 1/10 + 1/9 + 1/8 +$$
$$+ 0.9/7 + 0.7/6 + 0.5/5 + 0.2/4 + 0.1/3 + 0/2 + 0/1$$

- $Q =$ "at least a few" given as

$$\mu_{\text{"atleastafew"}}(|v|) = 1/10 + 1/9 + 1/8 +$$
$$+ 1/7 + 1/6 + 1/5 + 0.8/4 + 0.5/3 + 0.2/2 + 0/1$$

Next, suppose that we may employ the following three types of fuzzy implications [cf. (2.90) – (2.95), page 47]:

$$\begin{aligned}
&\text{Type 1:} \quad a \Longrightarrow b = a^b \\
&\text{Type 2:} \quad a \Longrightarrow b = (1 - b) \vee a \\
&\text{Type 3:} \quad a \Longrightarrow b = 1 \wedge (1 - b + a)
\end{aligned}$$

First, note that the (full) decision tree for this problem has 12,207,030 edges so it is evident that some implicit enumeration scheme has to be employed for the solution.

We start by solving the conventional problem (4.58), i.e. for $Q =$ "all", and obtain the following solution:

$$\begin{aligned}
&u_0^* = c_1 \quad u_1^* = c_1 \quad u_2^* = c_4 \quad u_3^* = c_5 \quad u_4^* = c_1 \\
&u_5^* = c_1 \quad u_6^* = c_1 \quad u_7^* = c_1 \quad u_8^* = c_1 \quad u_9^* = c_1
\end{aligned}$$

with $\mu_D(u_0^*, \ldots, u_9^* \mid s_1, \text{all}) = 0.1$ which becomes our starting point (a); by the way, this solution is obtained by examining 326 edges in the decision tree.

Then we solve the problem (4.107), with the importance B as given above, and for the three fuzzy linguistic quantifiers "most," "much more than 50%," and "at least a few" as defined above. We obtain the results shown in Table 1 in which there are consecutively given the optimal solutions (optimal sequences of controls) for the particular fuzzy linguistic quantifier, and type of implication. For each optimal solution the corresponding value of the membership function of a fuzzy decision is shown, as well as the number of edges in the decision tree which have been examined for finding that optimal solution. □

The results of the example presented above show, first of all, that the new problem formulation obtained by involving both a fuzzy linguistic quantifier and importance (discounting) does yield qualitatively different solutions. Namely, on the one hand, using the conventional problem formulation, i.e. for "all," we end up with an optimal solution whose "goodness," i.e. the value of $\mu_D(\cdot \mid \cdot, \text{all})$, is only equal to 0.1, that is very low and even prohibitive for practical use and implementation of its corresponding optimal sequence of control.

Table 1 Summary of results obtained in Example 4.16

Quantifier	Type of implication	Optimal sequence of of controls $u_0^*, \ldots u_9^*$	$\mu_D(\cdot \mid \cdot, \cdot)$	Number of edges examined
"most"	1	$c_4 c_5 c_3 c_5 c_3 c_5 c_5 c_5 c_3 c_4$	0.6	6,219
	1	$c_4 c_2 c_2 c_5 c_3 c_5 c_5 c_5 c_1 c_1$	0.5	7,440
	2	$c_4 c_5 c_3 c_5 c_3 c_3 c_5 c_5 c_5 c_3$	0.7	2,526
"$\gg 50\%$"	3	$c_4 c_5 c_3 c_5 c_5 c_5 c_5 c_2 c_3 c_4$	0.7	12,679
	1	$c_4 c_5 c_3 c_5 c_5 c_5 c_5 c_2 c_5 c_3$	0.7	12,167
	2	$c_1 c_4 c_5 c_5 c_3 c_3 c_5 c_5 c_5 c_3$	0.7	2,411
"at least a few"	3	$c_1 c_4 c_5 c_5 c_5 c_5 c_5 c_2 c_5 c_3$	0.8	39,309
	2	$c_4 c_5 c_5 c_5 c_5 c_5 c_5 c_3 c_3 c_3$	0.8	87,540
	3	$c_1 c_1 c_4 c_5 c_5 c_5 c_5 c_3 c_5 c_3$	0.8	22,623

On the other hand, the new problem formulation (with a linguistic quantifier and importance) results in a much higher goodness of 0.5–0.8 depending on the choice of parameters (cf. Table 1). This result is clearly much closer to our intuitive perception of the situation in question. Moreover, the fewer control stages are required to be taken into account, i.e. for $Q =$ "$\gg 50\%$", and even more so for $Q =$ "at least a few", the higher the goodness.

The algorithm is also quite efficient judging by a relatively low number of edges in the decision tree that needs to be examined for finding an optimal solution. Notice that we have started from a very poor initial point $a = \mu_D(\cdot \mid \cdot, \text{all}) = 0.1$ which is a very small value as compared to the optimal solutions characterized by the values of 0.5–0.8 obtained. Thus, the optimization algorithm has had to "work hard" to yield such an improvement. A better starting point would speed up the solution process.

4.2 CONTROL OF A STOCHASTIC SYSTEM

In this section we will deal with the control problem of type (4.3), page 86, but with a stochastic system under control. The other elements of the problem will be basically the same, notably the termination time will be fixed and specified in advance. Notice that in this class of problems we will face a joint occurrence of fuzziness and randomness.

The stochastic system under control is assumed here, and throughout this book, to be a Markov chain whose dynamics (state transitions) is governed by a conditional probability function given by (2.135), page 59, i.e.

$$p(x_{t+1} \mid x_t, u_t), \qquad t = 0, 1, \ldots \tag{4.117}$$

which specifies the probability of attaining a state at control stage $t + 1$, $x_{t+1} \in X = \{s_1, \ldots, s_n\}$ from a state at control stage t, $x_t \in X$, under a control at stage t, $u_t \in U = \{c_1, \ldots, c_m\}$.

At each control stage t, control $u_t \in U$ is subjected to a fuzzy constraint $\mu_{C^t}(u_t)$, $t = 0, 1, \ldots, N - 1$, and on the final state $x_N \in X$ a fuzzy goal $\mu_{G^N}(x_N)$ is imposed.

The value of the fuzzy decision $\mu_D(u_0, \ldots, n_{N-1} \mid x_0)$ is now evidently a random variable since the system under control is stochastic. The problem can no longer be stated as to find an optimal sequence of controls maximizing the membership function of the fuzzy decision as in the case of the deterministic system under control in problem (4.3), page 86. As is usually done in the analysis of stochastic problems, the expected value should involved in some way in the problem formulation.

The following basic problem formulations are used:

- Bellman and Zadeh's (1970) formulation: find an optimal sequence of controls $u_0^*, \ldots, u_{N-1}^*$ maximizing the probability of attainment of the fuzzy goal subject to the fuzzy constraints, that is

$$
\begin{aligned}
\mu_D(u_0^*, \ldots, u_{N-1}^* \mid x_0) = \\
= \max_{u_0, \ldots, u_{N-1}} [\mu_{C^0}(u_0) \wedge \ldots \\
\ldots \wedge \mu_{C^{N-1}}(u_{N-1}) \wedge E\mu_{G^N}(x_N)]
\end{aligned}
\tag{4.118}
$$

- Kacprzyk and Staniewski's (1980a) formulation: find an optimal sequence of controls $u_0^*, \ldots, u_{N-1}^*$ maximizing the expected value of the fuzzy decision, that is

$$
\begin{aligned}
\mu_D(u_0^*, \ldots, u_{N-1}^* \mid x_0) = \\
= \max_{u_0, \ldots, u_{N-1}} E\mu_D(u_0, \ldots, u_{N-1} \mid x_0) = \\
= \max_{u_0, \ldots, u_{N-1}} E[\mu_{C^0}(u_0) \wedge \ldots \\
\ldots \wedge \mu_{C^{N-1}}(u_{N-1}) \wedge \mu_{G^N}(x_N)]
\end{aligned}
\tag{4.119}
$$

In both the above formulations we are evidently interested in finding an optimal control strategy $A^* = (a_0^*, \ldots, a_{N-1}^*)$ [cf. (3.56), page 81], where $a_t^* : X \longrightarrow U$, such that $u_t^* = a_0^*(x_t)$, $t = 0, 1, \ldots, N - 1$, is an optimal control policy at control stage t.

Notice that the above two problem formulations are not the same since, quite naturally, the following inequality holds

$$E\mu_D(u_0, \ldots, u_{N-1} \mid x_0) \neq \mu_{C^0}(u_0) \wedge \ldots \wedge \mu_{C^{N-1}}(u_{N-1}) \wedge E\mu_{G^N}(x_N) \tag{4.120}$$

While considering the mathematical expressions for the two problem formulations considered – (4.118) and (4.119) – we should be aware that the equality signs should not be taken literally, i.e. that the left-hand side is equal to the right-hand side, but rather as a symbolic expression – in the convention adopted in the whole book – that the optimal sequence of controls shown in the left-hand side is the one that maximizes the function in the right-hand side.

In this section we will basically assume that the probability of a fuzzy event involved in both the problem formulations (4.118) and (4.119) is meant in Zadeh's (1968b) sense discussed in Section 2.1.8.1, i.e. as a real number in $[0, 1]$, which is the case for virtually all models developed in the literature and employed in practice. The use of the fuzzy probability of a fuzzy event, which is a fuzzy number in $[0, 1]$, assuming Yager's (1979) definition, will be presented in the next part devoted to Bellman and Zadeh's (1970) formulation.

4.2.1 Problem formulation by Bellman and Zadeh

This basic formulation was first proposed in Bellman and Zadeh's (1970) seminal paper on decision making and control in a fuzzy environment.

First, it will be useful to repeat for convenience the problem formulation (4.118): find an optimal sequence of controls $u_0^*, \ldots, u_{N-1}^*$ maximizing the probability of attainment of the fuzzy goal subject to the fuzzy constraints, that is

$$\mu_D(u_0^*, \ldots, u_{N-1}^* \mid x_0) =$$
$$= \max_{u_0, \ldots, u_{N-1}} [\mu_{C^0}(u_0) \wedge \ldots \wedge \mu_{C^{N-1}}(u_{N-1}) \wedge E\mu_{G^N}(x_N)]$$

The probability of attainment of the fuzzy goal, $E\mu_{G^N}(x_N)$, may be specified here, as we may remember from Section 2.1.8, either as a real number from the unit interval, i.e. as a nonfuzzy probability, or as a fuzzy number defined in the unit interval, i.e. as a nonfuzzy probability. In the former case, while specifying $E\mu_{G^N}(x_N)$ we can use the classic Zadeh's (1968b) definition of the (nonfuzzy) probability of a fuzzy event presented in Section 2.1.8.1, while in the latter case we can use, e.g., Yager's (1979) definition of the (fuzzy) probability of a fuzzy event outlined in Section 2.1.8.2.

Virtually all models of multistage control of a stochastic system under fuzziness which have appeared in the literature have employed the classic Zadeh's (1968b) definition, and hence we will use it in most of our next considerations. In our context, the use of a fuzzy probability of a fuzzy event, namely Yager's (1979) definition has only been considered by Kacprzyk (1982a, 1984c). We will present also this approach in some detail.

4.2.1.1 A model using nonfuzzy probability of a fuzzy event

The probability of attainment of the fuzzy goal, $E\mu_{G^N}(x_N)$, is in this case meant in Zadeh's (1968b) sense. Namely, the fuzzy goal is regarded as a fuzzy event in X, and the conditional probability of this event given x_{N-1} and u_{N-1} is expressed by [cf. (2.80), page 44]

$$E\mu_{G^N}(x_N) = E\mu_{G^N}(x_N \mid x_{N-1}, u_{N-1}) =$$
$$= \sum_{x_N \in X} p(x_N \mid x_{N-1}, u_{N-1})\, \mu_{G^N}(x_N) \tag{4.121}$$

It may readily be noticed that the structure of problem (4.118) is essentially the same as that of problem (4.3), page 86, for the deterministic system under control.

Namely, $E\mu_{G^N}(x_N) = E\mu_{G^N}[f(x_{N-1}, u_{N-1})]$, i.e. is a function of x_{N-1} and u_{N-1}. Therefore, the two right-most terms in the right-hand side of equation (4.118) depend on control u_{N-1} and not on the other controls. The second right-most terms depend on u_{N-2}, etc.

The maximization over a sequence of controls $u_0, \ldots, u_{N-1}$ in problem (4.118) can be therefore split into the consecutive maximizations with respect to the particular u_t's, $t = N - 1, N - 2, \ldots, 0$. This leads to

$$\mu_D(u_0^*, \ldots, u_{N-1}^* \mid x_0) =$$
$$= \max_{u_0, \ldots, u_{N-2}} [\mu_{C^0}(u_0) \wedge \ldots \wedge \mu_{C^{N-2}}(u_{N-2}) \wedge$$
$$\wedge \max_{u_{N-1}} (\mu_{C^{N-1}}(u_{N-1}) \wedge E\mu_{G^N}(x_N))] =$$
$$= \max_{u_0, \ldots, u_{N-1}} [\mu_{C^0}(u_0) \wedge \ldots \wedge \mu_{C^{N-3}}(u_{N-3}) \wedge$$
$$\wedge \max_{u_{N-2}} (\mu_{C^{N-2}}(u_{N-2}) \wedge \max_{u_{N-1}} (\mu_{C^{N-1}}(u_{N-1}) \wedge E\mu_{G^N}(x_N)))] \quad (4.122)$$

On repeating this backward iteration, i.e. by consecutively performing the maximization over u_{N-3}, u_{N-4}, $\ldots$, u_0, we arrive at the following set of dynamic programming recurrence equations:

$$\begin{cases} \mu_{G^{N-1}}(x_{N-1}) = \max_{u_{N-1}} [\mu_{C^{N-i}}(u_{N-i}) \wedge E\mu_{G^{N-i+1}}(x_{N-i+1})] \\ E\mu_{G^{N-1+1}}(x_{N-i+1}) = \\ \quad = \sum_{x_{N-i} \in X} p(x_{N-i+1} \mid x_{N-i}, u_{N-i}) \mu_{G^{N-i+1}}(x_{N-i+1}) \\ i = 1, \ldots, N \end{cases} \quad (4.123)$$

where $\mu_{G^{N-i}}(x_{N-i})$ may be viewed as a fuzzy goal at control stage $t = N - i$ induced by the fuzzy goal at $t = N - i + 1$.

The structure of dynamic programming recurrence equations (4.123) is evidently the same as of those for the deterministic system under control, i.e. as (4.7) on page 87.

The successive maximizing values of u_{N-i}, u_{N-i}^*, $i = 1, 2, \ldots, N$, give the optimal sequence of controls to be determined. In fact, we obtain again the optimal control policies a_{N-i}^* such that $u_{N-i}^* = a_{N-i}^*(x_{N-i})$, $i = 1, \ldots, N$, which form the optimal control strategy $A^* = (a_0^*, \ldots, a_{N-1}^*)$ sought. Notice that the optimal policies are here in the sense of (3.56) and (3.57), page 81, i.e. they relate the current optimal control to the current state only.

As an illustration consider a simple example from Bellman and Zadeh (1970).

Example 4.17 Let the state space be $X = \{s_1, s_2, s_3\}$, the control space be $U = \{c_1, c_2\}$, the fuzzy constraints and fuzzy goals be, respectively:

$$C^0 = 0.7/c_1 + 1/c_2$$
$$C^1 = 1/c_1 + 0.6/c_2$$
$$G^2 = 0.3/s_1 + 1/s_2 + 0.8/s_3$$

Finally, let the conditional probabilities, $p(x_{t+1} \mid x_t, u_t)$, representing the dynamics of the stochastic system under control be:

$$p(x_{t+1} \mid x_t, u_t) =$$

$u_t = c_1$ $x_t = s_1$	$x_{t+1} = s_1$	s_2	s_3
	0.8	0.1	0.1
s_2	0	0.1	0.9
s_3	0.8	0.1	0.1

$u_t = c_2$ $x_t = s_1$	$x_{t+1} = s_1$	s_2	s_3
	0.1	0.9	0
s_2	0.8	0.1	0.1
s_3	0.1	0	0.9

In the first iteration of the set of dynamic programming recurrence equations (4.123), for $i = 1$, we obtain first

$$E\mu_{G^2}(x_2) =$$

$u_1 = c_1$	$x_1 = s_1$	s_2	s_3
	0.42	0.82	0.42
c_2	0.93	0.42	0.75

and then $G^1 = 0.6/s_1 + 0.82/s_2 + 0.6/s_3$ which corresponds to the following optimal policy:

$$a_1^*(s_1) = c_2 \quad a_1^*(s_2) = c_1 \quad a_1^*(s_3) = c_3$$

Then, the second iteration of the set of recurrence equations (4.123), for $i = 2$, yields first

$$E\mu_{G^1}(x_1) =$$

$u_0 = c_1$	$x_0 = s_1$	s_2	s_3
	0.62	0.62	0.62
c_2	0.8	0.62	0.6

and then $G^0 = 0.8/s_1 + 0.62/s_2 + 0.62/s_3$ which corresponds to the following optimal policy:

$$a_0^*(s_1) = c_1 \quad a_0^*(s_2) \in \{c_1, c_2\} \quad a_0^*(s_3) = c_1$$

$\square$

Similarly as in the case of a deterministic system under control (cf. Section 4.1.1.1), also in the case of a stochastic system under control, the sets of dynamic programming recurrence equations analogous to (4.123) can be devised for other types of fuzzy decision.

Namely, for instance:

• for the product-type fuzzy decision

$$\begin{cases} \mu_{G^{N-1}}(x_{N-1}) = \max_{u_{N-1}}[\mu_{C^{N-i}}(u_{N-i}) \, E\mu_{G^{N-i+1}}(x_{N-i+1})] \\ E\mu_{G^{N-i+1}}(x_{N-i+1}) = \sum_{x_{N-i} \in X} p(x_{N-i+1} \mid x_{N-i}, u_{N-i}) \times \\ \quad \times \mu_{G^{N-i+1}}(x_{N-i+1}) \\ i = 1, \ldots, N \end{cases}$$

$$(4.124)$$

- for the weighted-sum-type fuzzy decision

$$
\begin{cases}
\mu_{\overline{G}^N}(x_N) = r_N\, \mu_{G^N}(x_N) \\
\mu_{\overline{G}^{N-1}}(x_{N-1}) = \max_{u_{N-1}}[\mu_{C^{N-i}}(u_{N-i})\, E\mu_{\overline{G}^{N-i+1}}(x_{N-i+1})] \\
E\mu_{\overline{G}^{N-1+1}}(x_{N-i+1}) = \sum_{x_{N-i}\in X} p(x_{N-i+1}\mid x_{N-i}, u_{N-i})\times \\
\qquad \times\, \mu_{\overline{G}^{N-i+1}}(x_{N-i+1}) \\
i = 1,\ldots,N; r_0, + \cdots + r_N = 1
\end{cases}
$$

$$(4.125)$$

- for the max-type fuzzy decision

$$
\begin{cases}
\mu_{G^{N-1}}(x_{N-1}) = \max_{u_{N-1}}[\mu_{C^{N-i}}(u_{N-i}) \vee E\mu_{G^{N-i+1}}(x_{N-i+1})] \\
E\mu_{G^{N-1+1}}(x_{N-i+1}) = \sum_{x_{N-i}\in X} p(x_{N-i+1}\mid x_{N-i}, u_{N-i})\times \\
\qquad \times\, \mu_{G^{N-i+1}}(x_{N-i+1}) \\
i = 1,\ldots,N
\end{cases}
$$

$$(4.126)$$

The optimal solution in the case of each of the above sets of recurrence equations is an optimal policy a_{N-i}^* such that $u_{N-i}^* = a_{N-i}^*(x_{N-i})$, $i = 1,\ldots,N$, analogously as in the case of (4.123).

4.2.1.2   A model using fuzzy probability of a fuzzy event

In this section we will present the use of fuzzy probability of a fuzzy event in a multistage control context as proposed by Kacprzyk (1982b, 1984c).

The probability of attainment of the fuzzy goal, $E\mu_{G^N}(x_N)$, is now assumed to be a fuzzy number in $[0,1]$, and meant in Yager's (1979) sense discussed in Section 2.1.8.2, and we will repeat below basic elements of this approach for convenience of the reader.

Suppose therefore that A is a fuzzy set in X whose membership function is $\mu_A : X \longrightarrow [0,1]$, $\mu_A(x) \in [0,1]$. The α-level set (α-cut) of A [cf. (2.14), page 26] is $A_\alpha = \{x \in X : \mu_A(x) \geq \alpha\}$, for each $\alpha \in [0,1]$.

Then, by the representation theorem (Theorem 2.16, page 26), A may be expressed by its α-cuts as follows:

$$A = \bigcup_{\alpha=0}^{1} \alpha A_\alpha \qquad (4.127)$$

where "$\bigcup$" is meant as the set-theoretic union [cf. (2.32), page 30].

If now $Y = \{y\}$ and $W = \{w\}$ are some nonfuzzy sets, $G = 2^Y$ is the family of all nonfuzzy subsets of Y, and A is a fuzzy set in Y, and $f : G \longrightarrow W$, then due to the extension principle discussed in Section 2.1.6 [cf. (2.69), page 40], f may be extended to act on fuzzy sets in Y as follows:

$$f(A) = \bigcup_{\alpha=0}^{1} \alpha/f(A_\alpha) \qquad (4.128)$$

If now A is a fuzzy event in X and $p : G \longrightarrow [0,1]$ is a probability function, then Yager (1979) defines the (fuzzy) probability of a fuzzy event as [cf. (2.81), page 45]

$$P(A) = E_f\mu_A(x) = \bigcup_{\alpha=1}^{1} \alpha p(A_\alpha) \qquad (4.129)$$

i.e. as the (fuzzy) expected value of the fuzzy event's membership function.

Since, $p(A_\alpha) \in [0, 1,]$, then it may be represented [cf. (2.4)] as a fuzzy singleton $\{1/p(A_\alpha)\}$, and hence (4.129) may be rewritten as

$$P(A) = \bigcup_{\alpha=1}^{1} \alpha \cdot \{1/p(A_\alpha)\} \tag{4.130}$$

It is obvious that the fuzzy probability $P(A)$ defined by (4.129) and (4.130) is a fuzzy number in $[0, 1]$.

Example 2.19, page 45, shows how to calculate the fuzzy probability $P(A)$ given by (4.130).

Since now we have a fuzzy probability of a fuzzy event, and the very essence of the control problem considered (4.118) is to maximize the probability of attaining a fuzzy goal, then we need some tools for ordering (comparing) the fuzzy probabilities. This is equivalent to the ordering of fuzzy numbers. In the source Kacprzyk's (1982b, 1984c) works, a simple and efficient approach has been adopted in which a fuzzy number (in our case the fuzzy probability) in $[0, 1]$ is equated with some real numer from $[0, 1]$. Then, the comparison, ordering, etc. is natural. More specifically, Yager's (1981) definition of a real number "subsuming" a fuzzy number has been employed.

Suppose that, as before, A is a fuzzy set in $[0, 1]$ and its α-cut is $A_\alpha = \{x \in [0, 1] : \mu_A(x) \geq \alpha\}$, where $A_\alpha \subseteq [0, 1]$. Let $V = \{v\} \subseteq [0, 1]$. We denote by $M(V)$ the mean value of the elements of V, i.e. if $V = \{v_1, \ldots, v_k\}$, then

$$M(V) = \frac{1}{k} \sum_{i=1}^{k} v_i \tag{4.131}$$

Now, an ordering function $F : \mathcal{L} \longrightarrow [0, 1]$, where $\mathcal{L}$ is the family of all fuzzy sets in $[0, 1]$, is defined as

$$F(A) = \int_0^{\alpha_{max}} M(V) d\alpha \tag{4.132}$$

where α_{max} is the maximum value of $\mu_A(x)$.

For the finite case assumed here, we have

$$F(A) = \sum_{\alpha=0}^{\alpha_{max}} \alpha M(A_\alpha) \tag{4.133}$$

and this function subsumes in fact a fuzzy number in $[0, 1]$ by its "equivalent" real number from the unit interval.

This function $F(A)$ has some interesting properties which are relevant for our considerations as, e.g.:

1. If A is a real number from $[0, 1]$, which may be written as $A = \{a\} = \{1/a\}$, the $\alpha_{max} = 1$, $A_\alpha = \{a\}$, for each $\alpha \in [0, 1]$, and

$$F(A) = \int_0^1 a \, d\alpha = a \tag{4.134}$$

so that $F(A)$ preserves the natural ordering between the real numbers.

2. Since $M(A_\alpha) \leq 1$, for each $\alpha \in [0,1]$, then

$$\int_0^1 M(A_\alpha)d\alpha \leq \int_0^1 1 d\alpha = 1 \qquad (4.135)$$

3. The function $F(A)$ attains its maximum value for $A = \{1/1\}$ which is equal to $F(\{1/1\}) = 1$.
4. If A is a nonfuzzy set, $A \subseteq [0,1]$, then $A_\alpha = A$, for each $\alpha \in [0,1]$, and

$$F(A) = \int_0^1 M(A)d\alpha = M(A) \qquad (4.136)$$

5. If $A = \{b/a\}$, then $\alpha_{max} = b$, $A_\alpha = \{a\}$, for $\alpha \leq b$, and $M(A_\alpha) = a$, for $\alpha \leq b$, and

$$F(A) = \int_0^b a\,d\alpha = ab \qquad (4.137)$$

The above main properties of the ordering function $F(A)$ do indicate that it may be used to meaningfully compare fuzzy sets, and fuzzy sets and real numbers.

Example 4.18 Suppose that we have a fuzzy set $A = 0.1/0.1 + 0.4/0.3 + 0.6/0.5 + 0.8/0.7 + 1/1$. Then

$$\begin{array}{ll}
A_{0.1} = \{0.1, 0.3, 0.5, 0.7, 1\} & M(A_{0.1}) = 0.52 \\
A_{0.4} = \{0.3, 0.5, 0.7, 1\} & M(A_{0.4}) = 0.625 \\
A_{0.6} = \{0.5, 0.7, 1\} & M(A_{0.6}) = 0.733 \\
A_{0.8} = \{0.7, 1\} & M(A_{0.8}) = 0.85 \\
A_1 = \{1\} & M(A_1) = 1
\end{array}$$

Hence

$$F(A) =$$
$$= \int_0^{0.1} 0.52 d\alpha + \int_{0.4}^{0.6} 0.733 d\alpha + \int_{0.6}^{0.8} 0.85 d\alpha + \int_{0.8}^1 1\alpha =$$
$$= 0.52 \cdot 0.1 + 0.3 \cdot 0.525 + 0.2 \cdot 0.733 + 0.2 \cdot 1 = 0.7561$$

$\square$

We are now in a position to proceed to the formulation of the multistage control problem with a stochastic system, and using the fuzzy probability of a fuzzy event defined above.

The stochastic system under control is again a Markov chain characterized by a conditional probability $p(x_{t+1} \mid x_t, u_t)$, where $x_t, x_{t+1} \in X = \{x\} = \{s_1, \ldots, s_n\}$ are the states at control stages t and $t+1$, and $u_t \in U = \{u\} = \{c_1, \ldots, c_m\}$ is the control at t [cf. (4.117), page 138]. At each t, u_t is subjected to a fuzzy constraint $\mu_{C^t}(u_t)$, $t = 0, 1, \ldots, N-1$, and on x_N a fuzzy goal $\mu_{G^N}(x_N)$ is imposed.

The fuzzy decision is therefore [cf. (4.118), page 138]

$$\mu_D(u_0, \ldots, u_{N-1} \mid x_0) =$$
$$= \mu_{C^0}(u_0) \wedge \mu_{C^1}(u_1) \wedge \ldots \wedge \mu_{C^{N-1}}(u_{N-1}) \underline{\Delta} E_f \mu_{G^N}(x_N) \qquad (4.138)$$

where, regarding the fuzzy goal G^N as a fuzzy event in X, we have [cf. (4.130)]

$$E_f \mu_{G^N}(x_N) = E_f[\mu_{G^N}(x_N) \mid x_{N-1}, u_{N-1}] = \bigcup_{\alpha=0}^{1} \alpha\{1/p(G_\alpha^N)\} \qquad (4.139)$$

and notice that since for any realization of the sequence of controls $u_0, \ldots, u_{N-1}$ the value of each $\mu_{C^t}(u_t)$, $t = 0, \ldots, N-1$, is a real number from $[0,1]$, while the value of $E_f \mu_{G^N}(x_N)$ is a fuzzy number in $[0,1]$, these two entities are not directly comparable so that "$\underline{\wedge}$" instead of "$\wedge$" is used in (4.138).

The problem is therefore to find an optimal sequence of controls $u_0^*, \ldots, u_{N-1}^*$ such that

$$\mu_D(u_0^*, \ldots, u_{N-1}^* \mid x_0) =$$
$$= \max_{u_0, \ldots, u_{N-1}} [\mu_{C^0}(u_0) \wedge \mu_{C^1}(u_1) \wedge \ldots \wedge \mu_{C^{N-1}}(u_{N-1}) \underline{\wedge} E_f \mu_{G^N}(x_N)] \, (4.140)$$

This problem cannot be directly solved due to the above-mentioned incompatibility between the real numbers, i.e. the values of $\mu_{C^t}(u_t)$, $t = 0, \ldots, N-1$ and the fuzzy number $E_f \mu_{G^N}(x_N)$, but if we employ the ordering function $F(.)$ defined by (4.132) [or, in fact, (4.133)], which replaces the fuzzy $E_f \mu_{G^N}(x_N)$ by its "equivalent" real $F[E_f \mu_{G^N}(x_N)]$, then this incompatibility disappears and the problem (4.140) becomes

$$\mu_D(u_0^*, \ldots, u_{N-1}^* \mid x_0) =$$
$$= \max_{u_0, \ldots, u_{N-1}} [\mu_{C^0}(u_0) \wedge \mu_{C^1}(u_1) \wedge \ldots$$
$$\ldots \wedge \mu_{C^{N-1}}(u_{N-1}) \wedge F(E_f \mu_{G^N}(x_N))] \qquad (4.141)$$

It may readily be noticed that the structure of this problem (4.141) is essentially the same as that of problem (4.118) in the case of using Zadeh's (1968) (nonfuzzy) probability of a fuzzy event since now, again, $E_f \mu_{G^N}(x_N) = E_f[\mu_{G^N}(f(x_{N-1}, u_{N-1}))]$, i.e., is a function of x_{N-1} and u_{N-1} only. Therefore, the two right-most terms in the right-hand side of equation (4.141) depend on control u_{N-1} and not on the other controls. The second right-most terms depend on u_{N-2}, etc.

Hence

$$\max_{u_0, \ldots, u_{N-1}} [\mu_{C^0}(u_0) \wedge \ldots$$
$$\ldots \wedge \mu_{C^{N-1}}(u_{N-1}) \wedge F(E_f \mu_{G^N}(x_N))] =$$
$$= \max_{u_0}[\mu_{C^0}(u_0) \wedge \ldots \wedge \max_{u_{N-2}}(\mu_{C^{N-2}}(u_{N-2}) \wedge$$
$$\wedge \max_{u_{N-2}}(\mu_{C^{N-1}}(u_{N-1}) \wedge F(E_f \mu_{G^N}(x_N)) \ldots] \qquad (4.142)$$

The maximization over a sequence of controls $u_0, \ldots, u_{N-1}$ in problem (4.141) can be therefore split, as may be seen from (4.142), into the consecutive maximizations with respect to the particular u_t's, $t = N-1, N-2, \ldots, 0$. This leads to the following set of fuzzy dynamic programming recurrence equations which are analogous to (4.123), page 140:

$$\begin{cases} \mu_{G^{N-i}}(x_{N-1}) = \max_{u_{N-1}}[\mu_{C^{N-i}}(u_{N-i}) \wedge E_f \mu_{G^{N-i+1}}(x_{N-i+1})] \\ E_f \mu_{G^{N-i+1}}(x_{N-i+1}) = \bigcup_{\alpha=0}^{1} \alpha\{1/p(G_\alpha^{N-i+1})\} \\ i = 1, \ldots, N \end{cases} \qquad (4.143)$$

where $\mu_{G^{N-i}}(x_{N-i})$ may be viewed as a fuzzy goal at control stage $t = N - i$ induced by the fuzzy goal at $t = N - i + 1$.

The solution of (4.143) yields the successive maximizing values of u_{N-i}, u^*_{N-i}, $i = 1, 2, \ldots, N$ sought. In fact, we obtain again the optimal control policies a^*_{N-i} such that $u^*_{N-i} = a^*_{N-i}(x_{N-i})$, $i = 1, \ldots, N$, which form the optimal control strategy $A^* = (a^*_0, \ldots, a^*_{N-1})$.

Now we will solve the simple example given in the source paper (Bellman and Zadeh, 1970), and presented as Example 4.17, page 140, in which the nonfuzzy probability of a fuzzy event has been used.

Example 4.19 Suppose that the state space is $X = \{s_1, s_2, s_3\}$, the control space is $U = \{c_1, c_2\}$, the fuzzy constraints and fuzzy goals are, respectively:

$$C^0 = 0.7/c_1 + 1/c_2$$
$$C^1 = 1/c_1 + 0.6/c_2$$
$$G^2 = 0.3/s_1 + 1/s_2 + 0.8/s_3$$

and the conditional probabilities, $p(x_{t+1} \mid x_t, u_t)$, representing the dynamics of the stochastic system under control are:

$p(x_{t+1} \mid x_t, u_t) =$

$u_t = c_1$		$x_{t+1} = s_1$	s_2	s_3
$x_t = s_1$		0.8	0.1	0.1
	s_2	0	0.1	0.9
	s_3	0.8	0.1	0.1

$u_t = c_2$		$x_{t+1} = s_1$	s_2	s_3
$x_t = s_1$		0.1	0.9	0
	s_2	0.8	0.1	0.1
	s_3	0.1	0	0.9

In the first iteration of the fuzzy dynamic programming recurrence equation (4.143), for the control stage $t = 1$, we obtain

$E_f \mu_{G^2}(x_2) =$

	$u_1 = c_1$	c_2
$x_1 = s_1$	$1/0.1 + 0.8/0.2 + 0.3/1$	$1/0.9 + 0.3/1$
s_2	$1/0.1 + 0.8/1$	$1/0.1 + 0.8/0.2 + 0.3/1$
s_3	$1/0.1 + 0.8/0.2 + 0.3/1$	$1/0 + 0.8/0.9 + 0.3/1$

and due to (4.132)

$F[E_f \mu_{G^2}(x_2)] =$		$x_1 = s_1$	s_2	s_3
	$u_1 = c_1$	0.225	0.46	0.225
	c_2	0.795	0.225	0.415

Then, we obtain

$\mu_{C^1}(u_1) \wedge F[E_f \mu_{G^2}(x_2)] =$		$x_1 = s_1$	s_2	s_3
	$u_1 = c_1$	0.225	0.46	0.225
	c_2	0.6	0.225	0.415

and
$$G^1 = 0.6/s_1 + 0.46/s_2 + 0.415/s_3$$

and the following optimal control policy at $t = 1$:

$$a_1^*(s_1) = c_2 \quad a_1^*(s_2) = c_1 \quad a_1^*(s_3) = c_2$$

In the second iteration of (4.143), i.e. for the control stage $t = 0$, we obtain

$E_f\mu_{G^1}(x_1) =$

	$u_0 = c_1$	c_2
$x_0 = s_1$	$0.46/0.2 + 0.6/0.8 + 0.415/1$	$0.6/0.1 + 0.46/1$
s_2	$0.6/0 + 0.46/0.1$	$0.415/1 + 0.46/0.2 + 0.6/0.8 + 0.415/1$
s_3	$0.46/0.2 + 0.6/0.8 + 0.415/1$	$0.6/0.1 + 0.415/1$

and due to (4.132)

$$F[E_f\mu_{G^1}(x_1)] = \quad \begin{array}{c|ccc} & x_1 = s_1 & s_2 & s_3 \\ \hline u_1 = c_1 & 0.449 & 0.161 & 0.449 \\ c_2 & 0.254 & 0.449 & 0.408 \end{array}$$

Then, we obtain

$$\mu_{C^0}(u_0) \wedge F[E_f\mu_{G^1}(x_1)] = \quad \begin{array}{c|ccc} & x_0 = s_1 & s_2 & s_3 \\ \hline u_0 = c_1 & 0.449 & 0.161 & 0.449 \\ c_2 & 0.254 & 0.449 & 0.408 \end{array}$$

and
$$G^0 = 0.449/s_1 + 0.449/s_2 + 0.449/s_3$$

and the following optimal control policy at $t = 0$:

$$a_0^*(s_1) = c_1 \quad a_0^*(s_2) = c_2 \quad a_0^*(s_3) = c.$$

$\square$

The solution in the case of the fuzzy probability of a fuzzy event is clearly different than that for the nonfuzzy probability (cf. Example 4.17), page 140, but quite close.

This concludes our brief analysis of the multistage control problem with a stochastic system under fuzzy constraints and goals over a fixed and specified planning horizon when the fuzzy probability of a fuzzy event is employed.

Finally, it may be seen that all the defuzzification procedures presented in Section 2.1.9 and other ordering functions presented in Section 2.1.7 in the discussion of ordering and comparing fuzzy numbers do serve the same purpose as Yager's (1979) ordering function (4.132) [or (4.133), page 143] , and they may also be used analogously in the problem considered (4.141).

4.2.2   Problem formulation by Kacprzyk and Staniewski

This formulation, which appeared later that the original Bellman and Zadeh's (1970), was proposed by Kacprzyk and Staniewski (1980a). For convenience of the reader, let us repeat this problem statement (4.119), page 138: find an optimal sequence of controls $u_0^*, \ldots, u_{N-1}^*$ maximizing the expected value of membership function of the fuzzy decision, that is

$$\mu_D(u_0^*, \ldots, u_{N-1}^* \mid x_0) =$$
$$= \max_{u_0, \ldots, u_{N-1}} E[\mu_{C^0}(u_0) \wedge \ldots \wedge \mu_{C^{N-1}}(u_{N-1}) \wedge \mu_{G^N}(x_N)]$$

and, as previously, we will be in fact interested in finding optimal control policies.

First, due to an evident property of "$\wedge$," the following inequality holds:

$$E\mu_D(u_0, \ldots, u_{N-1} \mid x_0) \neq$$
$$\neq E[\mu_{C^0}(u_0) \wedge E(\mu_{C^1}(u_1) \wedge \ldots \wedge E\mu_{G^N}(x_N)) \ldots] \qquad (4.144)$$

so that the dynamic programming recurrence equations of type (4.123), page 140, cannot be used, and a different approach should be employed, and it will be presented below.

First, the probability of attaining the final state x_N from the initial state x_0 through the sequence of controls $u_0, \ldots, u_{N-1}$ is evidently equal to

$$p(x_1 \mid x_0, u_0) \cdot p(x_2 \mid x_1, u_1) \cdot \ldots \cdot p(x_N \mid x_{N-1}, u_{N-1})$$

and the expected value of the random variable $\mu_D(u_0, \ldots, u_{N-1} \mid x_0)$ is equal to

$$E\mu_D(u_0, \ldots, u_{N-1} \mid x_0) =$$
$$= \sum_{(x_0, \ldots, x_N) \in X^{N+1}} \mu_D(u_0, \ldots, u_{N-1} \mid x_0) \times$$
$$\times p(x_1 \mid x_0, u_0) \cdot p(x_2 \mid x_1, u_1) \cdot \ldots \cdot p(x_N \mid x_{N-1}, u_{N-1}) \qquad (4.145)$$

We introduce now a sequence of functions $\hat{h}_i, h_j, i = 0, 1, \ldots, N, j = 1, \ldots, N-1$, defined as follows:

$$\begin{cases} \hat{h}_N(x_N, u_0, \ldots, u_{N-1}) = \mu_D(u_0, \ldots, u_{N-1} \mid x_0) \\ \cdots\cdots\cdots\cdots\cdots\cdots\cdots\cdots\cdots\cdots\cdots\cdots\cdots\cdots \\ \hat{h}_k(x_k, u_0, \ldots, u_{k-1}) = \max_{u_k} h_k(x_k, u_0, \ldots, u_k) \\ h_{k-1}(x_{k-1}, u_0, \ldots, u_{k-1}) = \\ \quad = \sum_{x_k \in X} \hat{h}_k(x_k, u_0, \ldots, u_{k-1}) \cdot p(x_k \mid x_{k-1}, u_{k-1}) \\ \cdots\cdots\cdots\cdots\cdots\cdots\cdots\cdots\cdots\cdots\cdots\cdots\cdots\cdots \\ \hat{h}_0(x_0) = \max_{u_0} h_0(x_0, u_0) \end{cases} \qquad (4.146)$$

The functions $h_{k-1}(.)$ are meant as follows. If the consecutive controls applied at $t = 0, \ldots, k-1$ are $u_0, \ldots, u_{k-1}$, respectively, and the successive states attained are $x_1, \ldots, x_k$, then $h_{k-1}(.)$ is the expected value of the fuzzy decision, $E\mu_D(. \mid x_0)$, assuming an optimal continuation of control from control stage $t = k$ to the end of the control process ($t = N$), i.e. $u_k^*, \ldots, u_{N-1}^*$.

The meaning of $\hat{h}_k$ is then, roughly speaking, the highest possible expected value of the fuzzy decision that may be attained if we are at the current control stage.

Let us now observe that since the state space X and control space U are finite, then there obviously exist some functions $w_k : X \times \underbrace{U \times \cdots \times U}_{k} \longrightarrow U$, $k = 0, 1, \ldots, N-1$, such that

$$\hat{h}_k(x_k, u_0, \ldots, u_{k-1}) = h_k[x_k, u_0, \ldots, u_{k-1}, w_k(x_k, u_0, \ldots, u_{k-1})] \qquad (4.147)$$

If we now introduce a sequence of functions $g_k^* : \underbrace{X \times \cdots \times X}_{k} \longrightarrow U$, $k = 0, 1, \ldots, N-1$, such that

$$g_k^*(x_0, \ldots, x_k) = w_k[x_k, g_0^*(x_0), \ldots, g_{k-1}^*(x_0, \ldots, x_{k-1})] \qquad (4.148)$$

where $g_0^*(x_0) = w_0(x_0)$, then we have the following important property (Kacprzyk and Staniewski, 1980a):

Proposition 4.1 *If the functions h_k, $\hat{h}_k$, w_k and g_k^* are given by (4.146), (4.147) and (4.148), respectively, then the functions g_k^* are the optimal policies sought, that is*

$$u_k^* = g_k^*(x_0, \ldots, x_k), \qquad k = 0, \ldots, N-1 \qquad (4.149)$$

and the optimal strategy to be determined is $A^* = (g_0^*, \ldots, g_{N-1}^*)$.
Moreover

$$h_0(x_0) = \max_{u_0, \ldots, u_{N-1}} E\mu_D(u_0, \ldots, u_{N-1} \mid x_0) \qquad (4.150)$$

where, evidently, the policy is meant nere in the sense of (3.61), or even (3.59), i.e. it relates the optimal control not only to the current state but also to a "summary" of the past trajectory, as we will see in the following.

Proof. Consider first some arbitrary strategy $S = (e_0, \ldots, e_{N-1})$ such that $u_k = e_k(x_0, \ldots, x_k)$, $k = 0, \ldots, N-1$. Denoting by E_S the mathematical expectation given strategy S, we obtain for each $0 \le k \le N-1$

$$
\begin{aligned}
E_S \hat{h}_k(x_k, u_0, \ldots, u_{k-1}) &= \\
&= E_S[E_S(\hat{h}_k(x_k, u_0, \ldots, u_{k-1}) \mid x_{k-1}] = \\
&= E_S \sum_{x_k \in X} \hat{h}_k(x_k, u_0, \ldots, u_{k-1}) p(x_k \mid x_{k-1}, u_{k-1}) = \\
&= E_S h_{k-1}(x_{k-1}, u_0, \ldots, u_{k-1}) \le E_S \hat{h}_{k-1}(x_{k-1}, u_0, \ldots, u_{k-2}) \quad (4.151)
\end{aligned}
$$

The last inequality in 4.151) results from an evident relation $h_k(.) \le \hat{h}_k(.)$ and the fact that control u_k is a function of the consecutive states, $x_0, \ldots, x_k$, $k = 0, \ldots, N-1$, as already mentioned.

On repeating the above reasoning, we obtain

$$E_S \hat{h}_N(x_N, u_0, \ldots, u_{N-1}) \le E_S \hat{h}_0(x_0) \qquad (4.152)$$

Therefore, if we have $u_k^* = g_k^*(x_0, \ldots, x_k)$ and $A^* = (g_0^*, \ldots, g_{N-1}^*)$, then for each $0 \le k \le N-1$ there holds

$$
\begin{aligned}
E_{A} \cdot \hat{h}_k(x_k, u_0, \ldots, u_{k-1}) &= \\
&= E_{A} \cdot [E_{A} \cdot (\hat{h}_k(x_k, u_0, \ldots, u_{k-1}) \mid x_{k-1})] = \\
&= E_{A} \cdot \sum_{x_k \in X} \hat{h}_k(x_k, u_0, \ldots, u_{k-1}) p(x_k \mid x_{k-1}, u_{k-1}) = \\
&= E_{A} \cdot h_{k-1}(x_{k-1}, u_0, \ldots, u_{k-1}) = \\
&= E_{A} \cdot h_{k-1}(x_{k-1}, u_0, \ldots, u_{k-2}, g_{k-1}^*(x_0, \ldots, x_{k-1})) = \\
&= E_{A} \cdot \hat{h}_{k-1}(x_{k-1}, u_0, \ldots, u_{k-2}) \qquad (4.153)
\end{aligned}
$$

and, finally,

$$
E_{A} \cdot \hat{h}_N(x_N, u_0, \ldots, u_{N-1}) = E_{A} \cdot \hat{h}_0(x_0) \qquad (4.154)
$$

Thus, since for each arbitrary strategy S the relation (4.152) holds, and for A^* the equality (4.154) is satisfied, then A^* is naturally an optimal strategy. $\qquad\square$

The above properties immediately lead to an algorithm for the determination of an optimal strategy sought in the problem formulation considered, i.e. satisfying (4.144), page 148. Such an algorithm boils down to the consecutive applications of the set of formulas (4.146) with the controls represented by the optimal policies given by (4.148).

The algorithm is as follows:

Step 1. Input the problem description, i.e. $X = \{s_1, \ldots, s_n\}$, $U = \{c_1, \ldots, c_m\}$, $x_0 \in X$, N, and $p(x_{t+1} \mid x_t, u_t)$, for all $x_t, x_{t+1} \in X$, and $u_t \in U$.

Step 2. Set $k := N$.

Step 3. Calculate h_{k-1} and $\hat{h}_{k-1}$ via (4.146). Express u_k^* by g_k^* using (4.148), with w_k given by (4.147).

Step 4. If $k = 0$, then STOP. Otherwise, set $k := k - 1$ and proceed to Step 3.

For illustration we will solve now using the above algorithm the same example as for the Bellman and Zadeh's (1970) problem formulation (Example 4.19 on page 146). A comparison of the results obtained will therefore be possible.

Example 4.20 For convenience of the reader let us briefly repeat the problem formulation: the state space is $X = \{s_1, s_2, s_3\}$, the control space is $U = \{c_1, c_2\}$, the fuzzy constraints and fuzzy goals are, respectively:

$$
\begin{aligned}
C^0 &= 0.7/c_1 + 1/c_2 \\
C^1 &= 1/c_1 + 0.6/c_2 \\
G^2 &= 0.3/s_1 + 1/s_2 + 0.8/s_3
\end{aligned}
$$

and the conditional probabilities, $p(x_{t+1} \mid x_t, u_t)$, representing the dynamics of the stochastic system under control (4.117) are:

$$p(x_{t+1} \mid x_t, u_t) =$$

$u_t = c_1$		$x_{t+1} = s_1$	s_2	s_3
	$x_t = s_1$	0.8	0.1	0.1
	s_2	0	0.1	0.9
	s_3	0.8	0.1	0.1

$u_t = c_2$		$x_{t+1} = s_1$	s_2	s_3
	$x_t = s_1$	0.1	0.9	0
	s_2	0.8	0.1	0.1
	s_3	0.1	0	0.9

First, the fuzzy decision may clearly be written as

$$\mu_D(u_0, u_1 \mid x_0) =$$
$$= \mu_{C^0}(u_0) \wedge \mu_{C^1}(u_1) \wedge \mu_{G^2}(x_2) = v_0(u_0) \wedge \mu_{C^1}(u_1) \wedge \mu_{G^2}(x_2)$$

where $v_0(u_0) \in \{1, 0.7\}$.

The consecutive steps of the algorithm are now as follows:

1. Introduce the problem description as given above.

2. $k = 2$.

3. We obtain

$$h_1(x_1, u_0, u_1) = \sum_{i=1}^{3} h_2(s_i, u_0, u_1) p(s_1 \mid x_1, u_1) =$$

$$= \sum_{i=1}^{3} [v_0(u_0) \wedge \mu_{C^1}(u_1) \wedge \mu_{G^2}(x_2)] p(s_i \mid x_1, u_1)$$

which may be represented in the following tabular form:

$$h_1(x_1, u_0, u_1) =$$

		$v_0(u_0) = 1$		
		$x_{t+1} = s_1$	s_2	s_3
$u_1 = c_1$		0.42	<u>0.82</u>	0.42
	c_2	<u>0.57</u>	0.36	<u>0.57</u>

		$v_0(u_0) = 0.7$		
		$x_{t+1} = s_1$	s_2	s_3
$u_1 = c_1$		0.38	<u>0.7</u>	0.38
	c_2	<u>0.57</u>	0.36	<u>0.57</u>

with the underlined values corresponding to the maximal values of h_1, and for which the respective u_1's yield the following optimal policy:

$$a_1^*(1, s_1) = c_2 \quad a_1^*(1, s_2) = c_1 \quad a_1^*(1, s_3) = c_2$$
$$a_1^*(0.7, s_1) = c_2 \quad a_1^*(0.7, s_2) = c_1 \quad a_1^*(0.7, s_3) = c_2$$

4. We set $k = 1$, and go to Step 3.

3. We obtain

$$h_0(x_0, u_0) = \sum_{i=1}^{3} h_1(s_i, u_0) p(s_i \mid x_0, u_0)$$

which can be represented in a tabular form as

$$h_1(x_1, u_0, u_1) =$$

	$x_{t+1} = s_1$	s_2	s_3
$u_1 = c_1$	0.583	0.583	0.583
c_2	0.795	0.595	0.57

with the maximal values of h_0 shown underlined, and for which the respective u_0's yield the following optimal policy

$$a_0^*(s_1) = c_2 \quad a_0^*(s_3) = c_2 \quad a_0^*(s_3) = c_1$$

4. We obtain $k = 0$, i.e. STOP.

$\square$

Note that the optimal policies obtained here are in the sense of (3.60), page 82, i.e. they relate the current optimal control not only to the current state but also to a "summary" of the past trajectory which is in our case represented by the value of $v_0(u_0) \in [0, 1]$. Evidently, the optimal policy at $t = 0$, $a_0^*(x_0)$ is of the type (3.56), page 81, i.e. it relates the optimal control to the current state only as at $t = 0$ there is no past trajectory.

The optimal policy found in Example 4.20 is slightly different – if we neglect the fact that it is meant in a different sense, i.e. (3.60), page 82 – than the one found in Example 4.17, page 140, using Bellman and Zadeh's (1970) formulation in which it is defined in the sense of (3.56), page 81.

Finally, note that Kacprzyk and Staniewski's (1980a) formulation can also be solved for many other types of the fuzzy decision mentioned in Section 4.1.1.1. This will not be discussed here due to lack of space.

Among some other, not numerous, alternative approaches to this (class of) problems of multistage control of a stochastic system in a fuzzy environment, with a fixed and specified termination time, we can recommend the interested reader to consult, e.g., Jacobson (1976), Odanaka (1987) and Yoshida (1994).

4.3 CONTROL OF A FUZZY SYSTEM

In this section we will consider the case of a fuzzy system under control whose dynamics is given as a state transition equation (2.140), page 61, i.e.

$$X_{t+1} = F(X_t, U_t), \qquad t = 0, 1, \ldots \qquad (4.155)$$

where $X_t, X_{t+1} \in \mathcal{X}$ are fuzzy states at control stage t and $t + 1$, respectively, and $U_t \in \mathcal{U}$ is a fuzzy control at control stage t, $t = 0, 1, \ldots, N - 1$; $\mathcal{U} = \{C_1, \ldots, C_l\}$ is the set of fuzzy controls, and $\mathcal{X} = \{S_1, \ldots, S_q\}$ is the set of fuzzy states.

Observe that we assume here for simplicity that the fuzzy system under control is time-invariant (cf. Section 2.4) though it is easy to see that the assumption of a time-varying system under control does not change the essence of our discussion and its generality. Moreover, we assume for the time being that the control is fuzzy, U_t.

Before considering the case of a fuzzy system under control it should first be noticed that in both the previously discussed cases of a deterministic and stochastic system under control, the consecutive controls applied, $u_0, \ldots, u_{N-1} \in U$, and the states attained, $x_1, \ldots, x_N \in X$, were nonfuzzy, hence we could directly determine their grade of membership in the fuzzy constraints, $\mu_{C^0}(u_0), \ldots, \mu_{C^{N-1}}(u_{N-1})$, and in the fuzzy goals, $\mu_{G^1}(x_1), \ldots, \mu_{G^N}(x_N)$, respectively (cf. Section 3).

The situation changes, however, in the case of a fuzzy system as the control applied and states attained are fuzzy. Thus, their grade of membership in the fuzzy constraints, $\mu_{C^0}(u_0), \ldots, \mu_{C^{N-1}}(u_{N-1})$, and in the fuzzy goal $\mu_{G^N}(x_N)$ cannot be directly determined, and some manipulation ("trickery") is needed which will be employed below.

Suppose therefore that at each t, the fuzzy control applied $U_t \in \mathcal{U}$ is subjected to a fuzzy constraint $\mu_{C^t}(u_t)$, and on the resulting fuzzy state $X_{t+1} \in \mathcal{X}$ a fuzzy goal $\mu_{G^{t+1}}(x_{t+1})$ is imposed, $t = 0, 1, \ldots, N-1$.

To account for the fuzziness of the controls and states, the above manipulation is basically some redefinition of the fuzzy constraints and fuzzy goals, e.g., as follows:

$$\mu_{\overline{C}^t}(U_t) = 1 - diss(C^t, U_t), \qquad t = 0, 1, \ldots, N-1 \qquad (4.156)$$

and

$$\mu_{\overline{G}^{t+1}}(U_{t+1}) = 1 - diss(G^{t+1}, U_{t+1}), \qquad t = 0, 1, \ldots, N-1 \qquad (4.157)$$

where $diss : [0,1] \times [0,1] \longrightarrow [0,1]$ is some measure of dissemblance.

Traditionally, this measure is assumed to be a normalized distance between fuzzy sets, i.e.

$$d : \mathcal{X} \times \mathcal{X} \longrightarrow [0,1] \qquad (4.158)$$

where $\mathcal{X}$ is a family of all fuzzy sets defined in X.

For the normalized distance, we evidently have that if $U_t = C^t$, then $d(U_t, C^t) = 0$, and the more U_t differs from C_t the closer $d(U_t, C^t)$ is to 1. These relations are clearly opposite to what we require for the distance, so that the following straightforward expression

$$\mu_{\overline{C}^t}(U_t) = 1 - d(U_t, C^t), \qquad t = 0, 1, \ldots, N-1 \qquad (4.159)$$

will obviously serve the purpose of measuring the closeness (similarity) of U_t's and C^t's, and may be used instead of $\mu_{C^t}(u_t)$'s in the control problem formulation.

And similarly for the fuzzy goals: employing the same line of reasoning, we obtain

$$\mu_{\overline{G}^{t+1}}(X_{t+1}) = 1 - d(X_{t+1}, G^{t+1}), \qquad t = 0, 1, \ldots, N-1 \qquad (4.160)$$

will obviously serve the purpose of measuring the closeness (similarity) of the X_{t+1}'s and G^{t+1}'s, and may be used instead of the $\mu_{G^{t+1}}(x_{t+1})$'s.

For the normalized distances we have an abundance to choose from, with the simplest and most traditional options, working very well in applications, being:

- the normalized linear (Hamming) distance (2.26), page 28, i.e.

$$d_l(X_N, G^N) = \frac{1}{N} \sum_{i=1}^{N} \mid \mu_{X_N}(s_i) - \mu_{G^N}(s_i) \mid \qquad (4.161)$$

- the normalized quadratic (Euclidean) distance (2.27), page 28, i.e.

$$d_q(X_N, G^N) = \sqrt{\frac{1}{N} \sum_{i=1}^{N} [\mu_{X_N}(s_i) - \mu_{G^N}(s_i)]^2} \qquad (4.162)$$

As to other choices, the use of a degree of equality of two fuzzy sets proposed by Kacprzyk and Staniewski (1982) is also a plausible choice (cf. Section 8.7), i.e.

$$\mu_{\overline{G}^N}(X_N) = e(X_N, G^N) \qquad (4.163)$$

where $e(X_N, G^N) \in [0, 1]$ is a degree of equality between X_N and G^N given, e.g., by (2.7)–(2.10), page 24. In this case the degree of equality does evidently fulfill our requirements as to the measure of closeness (similarity).

We may also use other indices or measures of (dis)similiarity, and one of them – Kaufmann and Gupta's (1985) dissimilarity index – will be discussed in Section 4.3.4.

Then, generally, the fuzzy decision is

$$\begin{aligned} \mu_D(U_0, \dots, U_{N-1} \mid X_0) &= \\ &= \mu_{\overline{C}^0}(U_0) \wedge \mu_{\overline{G}^1}(X_1) \wedge \dots \wedge \mu_{\overline{C}^{N-1}}(U_{N-1}) \wedge \mu_{\overline{G}^N}(X_N) \end{aligned} \qquad (4.164)$$

and we seek an optimal sequence of fuzzy controls $U_0^*, \dots, U_{N-1}^*$ such that

$$\mu_D(U_0^*, \dots, U_{N-1}^* \mid X_0) = \max_{U_0, \dots, U_{N-1}} \mu_D(U_0, \dots, U_{N-1} \mid X_0) \qquad (4.165)$$

Now we will first show the application of more traditional dynamic programming and branch-and-bound approaches to solve the resulting control problems, and then the use of newer ones, a non-standard technique of interpolative reasoning and a genetic algorithm.

4.3.1 Solution by dynamic programming

The application of dynamic programming to solving the problem of optimal control of a fuzzy system under control, under fuzziness and with a fixed and specified termination time is due to Baldwin and Pilsworth (1982).

The fuzzy system under control is assumed as previously mentioned, to be described by a fuzzy state transition equation (4.155), i.e.

$$X_{t+1} = F(X_t, U_t), \qquad t = 0, 1, \dots$$

where $X_t, X_{t+1} \in \mathcal{X}$ are fuzzy states at control stages t and $t + 1$, respectively, and $U \in \mathcal{U}$ is a fuzzy control at control stage t.

At each control stage t, the fuzzy control U_t is subjected to a fuzzy constraint $\mu_{C^t}(u_t)$, and on the resulting fuzzy state X_{t+1} a fuzzy goal $\mu_{G^{t+1}}(x_{t+1})$ is imposed,

$t = 0, 1, \ldots, N - 1$. Thus, we assume here that the fuzzy goals are imposed on all the control stages, not only on the final one as before (which is evidently a particular case of the more general approach currently assumed).

Both the control at stage t, U_t, and the state at $t + 1$, X_{t+1}, are now fuzzy, and hence their grades of membership in the fuzzy constraints C^t and G^{t+1} cannot be directly determined as the values of $\mu_{C^t}(u_t)$ and $\mu_{G^{t+1}}(x_{t+1})$, respectively.

Therefore, for each control stage t we construct a fuzzy relation R defined in $U \times X$ such that

$$R^t = C^t \times G^{t+1}, \qquad t = 0, 1, \ldots, N - 1 \qquad (4.166)$$

where "$\times$" is the Cartesian product of two fuzzy sets in the sense of (2.56), page 37; or in terms of the membership functions

$$\mu_{R^t}(u_t, x_t) =$$
$$= \mu_{C^t}(u_t) \wedge \mu_{G^{t+1}}(x_{t+1}), \qquad \text{for each } u_t \in U, x_{t+1} \in X \qquad (4.167)$$

which represents the degree of how well (to which degree, between 0 and 1) the fuzzy constraint C^t and fuzzy goal G^{t+1} are satisfied.

In turn the degree to which a particular fuzzy control U_t and the resulting fuzzy state X_{t+1} satisfy C^t and G^{t+1}, respectively, is now defined as

$$\overline{T}(U_t, R^t, x_{t+1}) = U_t \circ R^t \circ X_{t+1}, \qquad t = 0, 1, \ldots, N - 1 \qquad (4.168)$$

where "$\circ$" is the max–min composition given as (2.53), page 36; or, in terms of membership functions

$$T(U_t, R^t, X_{t+1}) = \max_{x_{t+1} \in X} [\max_{u_t \in U} (\mu_{U_t}(u_t) \wedge \mu_{R^t}(u_t, x_t)) \wedge \mu_{X_{t+1}}(x_{t+1})] =$$
$$= \max_{x_{t+1} \in X} [\max_{u_t} (\mu_{U_t}(u_t) \wedge \mu_{C^t}(u_t) \wedge \mu_{G^{t+1}}(x_{t+1}) \wedge \mu_{X_{t+1}}(x_{t+1}))] =$$
$$= \max_{u_t \in U} [\mu_{U_t}(u_t) \wedge \mu_{C^t}(u_t)] \wedge \max_{x_{t+1} \in X} [\mu_{X_{t+1}}(x_{t+1}) \wedge \mu_{G^{t+1}}(x_{t+1})] \qquad (4.169)$$

Example 4.21 Let $X = \{s_1, s_2, s_3\}$ and $U = \{c_1, c_2\}$. If now

$$C^t = 0.5/c_1 + 1/c_2$$
$$G^{t+1} = 1/s_1 + 0.7/s_2 + 0.3/s_3$$

then

$$\mu_{R^t}(u_t, x_t) = \quad \begin{array}{c|ccc} & x_{t+1} = s_1 & s_2 & s_3 \\ \hline u_t = c_1 & 0.3 & 0.5 & 0.3 \\ c_2 & 1 & 0.7 & 0.3 \end{array}$$

Now, if $U_t = 1/c_1 + 0.7/c_2$ and $X_{t+1} = 0.5/s_1 + 0.8/s_2 + 1/s_3$, then

$$T(U_t, R^t, X_{t+1}) =$$
$$= \max_{u_t \in \{c_1, c_2\}} [\mu_{U_t}(u_t) \wedge \mu_{C^t}(u_t)] \wedge \max_{x_{t+1} \in \{s_1, s_2, s_3\}} [\mu_{X_{t+1}}(x_{t+1}) \wedge \mu_{G^{t+1}}(x_{t+1})] =$$
$$= [(1 \wedge 0.5) \vee (1 \wedge 0.7)] \wedge [(0.5 \wedge 1) \vee (0.8 \wedge 0.7)] \vee (1 \wedge 0.3)) =$$
$$= (0.5 \wedge 0.7) \wedge (0.5 \wedge 0.7 \wedge 0.3) = 0.7$$

$\square$

The fuzzy decision is now given as

$$\mu_D(U_0,\ldots,U_{N-1}\mid X_0) =$$
$$= T(U_0,R^0,X_1)\wedge\ldots\wedge T(U_{N-1},R^{N-1},X_N) \qquad (4.170)$$

which yields the degree to which a particular sequence of fuzzy controls applied, $U_0,\ldots,U_{N-1}$, and the resulting fuzzy states obtained, $X_1,\ldots,X_N$, satisfy the fuzzy constraints and fuzzy goals at the successive control stages.

The problem is now to determine an optimal sequence of fuzzy controls $U_0^*,\ldots,U_{N-1}^*$ such that

$$\mu_D(U_0^*,\ldots,U_{N-1}^*\mid X_0) =$$
$$= \max_{U_0,\ldots,U_{N-1}} \mu_D(U_0,\ldots,U_{N-1}\mid X_0) =$$
$$= \max_{U_0,\ldots,U_{N-1}} [T(U_0,R^0,X_1)\wedge\ldots\wedge T(U_{N-1},R^{N-1},X_N)] \qquad (4.171)$$

From now on let us assume for simplicity that the fuzzy constraints are imposed at control stages $t = 0,1,\ldots,N-1$ but the fuzzy goal is imposed only at control stage $t = N$. Problem (4.171) becomes therefore: find an optimal sequence of fuzzy controls $U_0^*,\ldots,U_{N-1}^*$ such that

$$\mu_D(u_0^*,\ldots,U_{N-1}^*\mid X_0) =$$
$$= \max_{U_0,\ldots,U_{N-1}} \max_{u_0\in U}[\mu_{U_0}(u_0)\wedge\mu_{C^0}(u_0)\wedge\ldots\wedge \max_{U_{N-1}\in U}(\mu_{U_{N-1}}(u_{N-1})\wedge$$
$$\wedge\mu_{C^{N-1}}(u_{N-1})\wedge \max_{x_N}(\mu_{X_N}(x_N)\wedge\mu_{G^N}(x_N))\ldots] \qquad (4.172)$$

where the final state X_N is derived from the initial state X_0 through the sequence of fuzzy controls $U_0,\ldots,U_{N-1}$ through the state transition equation (4.155), page 152.

It is now easy to see that the structure of (4.172) is essentially the same as that of (4.3), page 86, i.e. the two right-most terms depend only on the fuzzy control U_{N-1} and not on the other controls, the next right-most term depends only on U_{N-2}, etc.

Problem (4.172) may be therefore written as

$$\mu_D(U_0^*,\ldots,U_{N-1}^*\mid X_0) =$$
$$= \max_{U_0\in\mathcal{U}}[((\max_{u_0\in U}(\mu_{U_0}(u_0)\wedge\mu_{C^0}(u_0)\wedge\ldots$$
$$\ldots\wedge \max_{U_{N-2}\in\mathcal{U}}((\max_{u_0\in U}(\mu_{U_{N-2}}(u_{N-2})\wedge\mu_{C^{N-2}}(u_{N-2}))\wedge$$
$$\wedge \max_{U_{N-1}\in\mathcal{U}}((\max_{u_{N-1}\in U}(\mu_{U_{N-1}}(u_{N-1})\wedge\mu_{C^{N-1}}(u_{N-1}))\wedge$$
$$\wedge \max_{X_N\in\mathcal{X}}(\max_{x_N\in X}(\mu_{X_N}(x_N)\wedge\mu_{G^N}(x_N))))\ldots] \qquad (4.173)$$

which leads to the following set of dynamic programming recurrence equations:

$$\begin{cases}
\mu_{\overline{G}^N}(X_N) = \max_{x_N\in X}[\mu_{X_N}(x_N)\wedge\mu_{G^N}(x_N)] \\
\mu_{\overline{G}^{N-i}}(X_{N-i}) = \max_{U_{N-i}\in\mathcal{U}}[\max_{u_{N-i},\in U}(\mu_{U_{N-i}}(u_{N-i})\wedge \\
\qquad\wedge\mu_{C^{N-i}}(u_{N-i}))\wedge\mu_{\overline{G}^{N-i+1}}(x_{N-i+1})] \\
\mu_{X_{N-i+1}}(x_{N-i+1}) = \max_{x_{N-i}\in X}[\max_{u_{N-i}\in U}(\mu_{U_{N-i}}(u_{N-i})\wedge \\
\qquad\wedge\mu_{X_{N-i+1}}(x_{N-i+1}\mid x_{N-i},u_{N-i}))\wedge\mu_{X_{N-i}}(x_{N-i})] \\
i = 1,\ldots N
\end{cases} \qquad (4.174)$$

In principle, the above set of dynamic programming recurrence equations may be solved. However, a serious difficulty may be seen just at first glance. First, $\mu_{\overline{G}^{N-i}}(X_{N-i})$ is to specified for each possible fuzzy state $X_{N-i} \in \mathcal{X}$. Second, the maximization, $\mu_{U_{N-i}}(.)$ is to proceed over all (well, maybe not all but a large subset of all) the fuzzy controls $U_{N-i} \in \mathcal{U}$. Evidently, the number of all the possible fuzzy controls U_{N-i} and fuzzy states X_{N-i} may be very high: infinite in the general case but at least very high in our context as we consider fuzzy sets defined in finite universes of discourse. This clearly makes the solution of (4.174) practically impossible.

A first reflexion is that the number of possible fuzzy controls and fuzzy states should be considerably reduced to make the set (4.174) solvable in a reasonable time. This is virtually the essence of Baldwin and Pilsworth's (1982) approach [it is also virtually the same as Kacprzyk and Staniewski's (1982) approach to be discussed in Section 8.7]. This reduction is there called a *fuzzy interpolation*.

First, let us consider $\mu_{\overline{G}^{N-i+1}}(X_{N-i+1})$, $i = 1, \ldots, N$. To overcome its troublesome dependence on a very high number of fuzzy states X_{N-i+1}, a relatively small number, say r, of the so-called *reference fuzzy states* $\overline{X}^1_{N-i+1}, \ldots, \overline{X}^r_{N-i+1}$ is introduced. Then, using the fuzzy conditional statements of type (2.62), page 38, $\mu_{G^{N-i+1}}(X_{N-i+1})$ is approximately specified as

$$
\begin{cases}
\text{IF } (X_{N-i+1} = \overline{X}^1_{N-i+1}) \text{ THEN } [\mu_{\overline{G}^{N-i+1}}(X_{N-i+1}) = \mu_{\overline{G}^{N-i+1}}(X^1_{N-i+1})] \\
\text{ELSE} \ldots \text{ELSE} \\
\text{IF } (X_{N-i+1} = \overline{X}^r_{N-i+1}) \text{ THEN } [\mu_{\overline{G}^{N-i+1}}(X_{N-i+1}) = \mu_{\overline{G}^{N-i+1}}(X^r_{N-i+1})]
\end{cases}
$$
$$(4.175)$$

which evidently corresponds to a fuzzy relation R^{N-i+1} defined in $X \times [0,1]$ whose membership function is $\mu_{R^{N-i+1}}(x_{N-i+1}, w_{N-i+1})$, $x_{N-i+1} \in X$ and $w_{N-i+1} \in [0,1]$.

Now, if we have a fuzzy state X_{N-i+1}, not necessarily a reference one, then it yields via the above fuzzy relation R^{N-i+1} the following induced fuzzy goal at control stage $N - i + 1$:

$$
\begin{aligned}
\mu_{\overline{G}^{N-i+1}}(X_{N-i+1}) &= \\
&= \max_{x_{N-i+1} \in X} [\mu_{X_{N-i+1}}(x_{N-i+1}) \wedge \mu_{R^{N-i+1}}(x_{N-i+1}, w_{N-i+1})]
\end{aligned}
\quad (4.176)
$$

which evidently results from the use of the max–min composition of X_{N-i+1} and R^{N-i+1} via (2.53), page 36.

Thus, for a given fuzzy control U_{N-i} and fuzzy state X_{N-i+1} the expression to be maximized with respect to U_{N-i} in (4.174) is

$$
\begin{aligned}
b(U_{N-i}, X_{N-i}) = \max_{u_{N-i} \in U} [\mu_{U_{N-i}}(u_{N-i}) \wedge \\
\wedge \max_{x_{N-i+1} \in X} (\mu_{X_{N-i+1}}(x_{N-i+1}) \wedge \mu_{R^{N-i+1}}(x_{N-i+1}, w_{N-i+1}))]
\end{aligned}
\quad (4.177)
$$

As we have already mentioned, the number of possible U_{N-i}'s, with respect to which the maximization in (4.177) is to proceed, may be very high. Thus, similarly as in the case of fuzzy states, we introduce a relatively small number, say p, of the so-called *reference fuzzy controls* $\overline{U}^1_{N-i}, \ldots, \overline{U}^p_{N-i}$. Then, for a given X_{N-i}, the following fuzzy conditional statements are constructed:

$$\begin{cases} \text{IF } (U_{N-i} = \overline{U}^1_{N-i}) \text{ THEN } [b(X_{N-i}, U_{N-i}) = b(X_{N-i}, \overline{U}^1_{N-i})] \\ \text{ELSE} \dots \text{ELSE} \\ \text{IF } (U_{N-i} = \overline{U}^p_{N-i}) \text{ THEN } [b(X_{N-i}, U_{N-i}) = b(X_{N-i}, \overline{U}^p_{N-i})] \end{cases} \quad (4.178)$$

which is equivalent to a fuzzy relation R'_{N-i} defined in $U \times [0,1]$ whose membership function is given in a simplified notation by $\mu_{R'}(u_{N-i}, v_{N-i} \mid X_{N-i})$, $u_{N-i} \in U$, $v_{N-i} \in [0,1]$. For simplicity, this fuzzy relation will be denoted by $R'(X_{N-i})$.

Now we need to find a method to determine such a U^*_{N-i} that maximizes $b(X_{N-i}, U_{N-i})$ with respect to U_{N-i}, for a given X_{N-i}.

First, a fuzzy set labelled "k–large" is defined as

$$\mu_{k--\text{large}}(x) = \begin{cases} 0 & \text{for } x \leq k \\ z > 0 & \text{for } x > k \end{cases} \quad (4.179)$$

such that if $x' > x'' > k$, then $\mu_{k--\text{large}}(x') > \mu_{k--\text{large}}(x'')$, and $\mu_{k--\text{large}}(1) = 1$. The value of k should be carefully chosen such that $\mu_{k'--\text{large}}(.)$ is preferred to $\mu_{k''--\text{large}}(.)$ if $k' > k''$.

The fuzzy sets "k–large" are now assumed to be test fuzzy sets. Using the fuzzy relation $R'(X_{N-i})$ obtained via (4.178), we determine the following fuzzy sets induced by "k–large:"

$$U(k) = (k\text{–large}) \circ R'(X_{N-i}), \qquad \text{for } k = 1, 0.9, 0.8, \dots \quad (4.180)$$

or in terms of membership functions

$$\mu_{U(k)}(u_{N-i}) = \max_{x_{N-i} \in X} [\mu_{R'}(u_{N-i}, x_{N-i} \mid X_{N-i}) \wedge$$

$$\wedge \mu_{k--\text{large}}(x_{N-i})], \qquad \text{for each } u_{N-i} \in U \quad (4.181)$$

We accept now as an optimal fuzzy control U^*_{N-i} such a $U(k)$ which corresponds to the highest value of k, i.e.

$$\max_{u_{N-i} \in U} \mu_{U(k)}(u_{N-i}) \geq m \quad (4.182)$$

where m is some parameter to be carefully chosen.

Note that all the above is performed for a particular X_{N-i} so that U^*_{N-i} determined using (4.182) is in fact a function of X_{N-i}, i.e. is a control policy. The above procedure is repeated for each reference fuzzy state $\overline{X}^1_{N-i}, \dots, \overline{X}^r_{N-i}$ yielding $U^*_{N-i}(\overline{X}^1_{N-i}), \dots, U^*_{N-i}(\overline{X}^r_{N-i})$, respectively, which may be represented as the following fuzzy conditional statement:

$$\begin{cases} \text{IF } (X_{N-i} = \overline{X}^1_{N-i}) \text{ THEN } [U^*_{N-i}(X_{N-i}) = U^*_{N-i}(\overline{X}^1_{N-i})] \\ \text{ELSE} \dots \text{ELSE} \\ \text{IF } (X_{N-i} = \overline{X}^r_{N-i}) \text{ THEN } [U^*_{N-i}(X_{N-i}) = U^*_{N-i}(\overline{X}^1_{N-r})] \end{cases} \quad (4.183)$$

which is equivalent to a fuzzy relation $R^{N-i}_{U^*}$ defined in $X \times U$ whose membership function is $\mu_{R^{N-i}_{U^*}}(x_{N-i}, u_{N-i})$; this fuzzy relation represents the *optimal fuzzy control policy*.

If we are therefore currently at control stage $t = N - i$ in a state X_{N-i}, then the *optimal fuzzy control* given by this optimal fuzzy control policy is

$$U_{N-i}^* = X_{N-i} \circ R_{U^*}^{N-i}, \qquad i = 1, \ldots, N \qquad (4.184)$$

or in terms of membership functions

$$\mu_{U_{N-i}^*}(u_{N-i}) = \max_{x_{N-i} \in X} [\mu_{X_{N-i}}(x_{N-i}) \wedge$$

$$\wedge \mu_{R_{U^*}^{N-i}}(x_{N-i}, u_{N-i})], \qquad \text{for each } u_{N-i} \in U \qquad (4.185)$$

Therefore, through the use of reference fuzzy states and reference fuzzy controls the solution of the problem has become in principle possible.

Example 4.22 Suppose that: $X = \{s_1, s_2, s_3\}$, $U = \{c_1, c_2\}$, $N = 3$, the fuzzy constraints are:

$$C^0 = 1/c_1 + 0.5/c_2$$
$$C^1 = 0.8/c_1 + 0.7/c_2$$
$$C^2 = 1/c_1 + 1/c_2$$

and the fuzzy goal is

$$G^3 = 1/s_1 + 0.4/s_2 + 0.1/s_3$$

Let the fuzzy system under control be given by

$$\mu_{X_{t+1}}(x_{t+1} \mid x_t, u_t) =$$

$u_t = c_1$	$x_t = s_1$	$x_{t+1} = s_1$	s_2	s_3
	$x_t = s_1$	1	0.7	0.3
	s_2	0.7	1	0.7
	s_3	0.3	0.7	1

$u_t = c_2$	$x_t = s_1$	$x_{t+1} = s_1$	s_2	s_3
	$x_t = s_1$	1	0.7	0.3
	s_2	1	0.7	0.3
	s_3	1	0.7	0.3

First, we introduce the following three reference fuzzy states:

$$\overline{X}_1 = 1/s_1 + 0.4/s_2 + 0.1/s_3$$
$$\overline{X}_2 = 0.4/s_1 + 1/s_2 + 0.4/s_3$$
$$\overline{X}_3 = 0.1/s_1 + 0.4/s_2 + 1/s_3$$

and the following two reference fuzzy controls:

$$\overline{U}_1 = 1/c_1 + 0.2/c_2$$
$$\overline{U}_2 = 0.2/c_1 + 1/c_2$$

Now, we present $\mu_{X_{t+1}}(x_{t+1} \mid x_t, u_t)$ in the form involving only the reference fuzzy states and controls, that is

$X_{t+1} =$		$U_t = \overline{U}^1$	$\overline{U}^2$
	$X_t = \overline{X}^1$	$1/s_1 + 0.7/s_2 + 0.4/s_3$	$1/s_1 + 0.7/s_2 + 0.3/s_3$
	$\overline{X}^2$	$0.7/s_1 + 1/s_2 + 0.7/s_3$	$1/s_1 + 0.7/s_2 + 0.3/s_3$
	$\overline{X}^3$	$0.4/s_1 + 0.7/s_2 + 1/s_3$	$1/s_1 + 0.7/s_2 + 0.3/s_3$

We solve (4.174) for $i = 0$, i.e. for $N - i = 3$, and obtain

$$\mu_{\overline{G}^3}(\overline{X}_1) = \max_{x_3 \in \{s_1, s_2, s_3\}} [\mu_{\overline{X}_1}(x_3) \wedge \mu_{G^3}(x_3)] =$$

$$= (1 \wedge 1) \vee (0.4 \wedge 0.4) \vee (0.1 \wedge 0.1) = 1$$

and analogously we obtain

$$\mu_{\overline{G}^3}(\overline{X}_2) = 0.4 \quad \mu_{\overline{G}^3}(\overline{X}_3) = 0.4$$

The above is equivalent to

$$\left\{ \begin{array}{l} \text{IF } \overline{X}_1 \text{ THEN } \mu_{\overline{G}^3}(\overline{X}_1) = 1 \\ \text{ELSE} \\ \text{IF } \overline{X}_2 \text{ THEN } \mu_{\overline{G}^3}(\overline{X}_2) = 0.4 \\ \text{ELSE} \\ \text{IF } \overline{X}_3 \text{ THEN } \mu_{\overline{G}^3}(\overline{X}_3) = 0.4 \end{array} \right.$$

which, due to (4.175), corresponds to the following fuzzy relation R^3 in $X \times [0, 1]$:

$$\mu_{R^3}(x_3, w_3) =$$

		$x_3 = s_1$	s_2	s_3
	$w_3 = 0.4$	0.4	0.4	1
	1	1	1	0.4

Now, for $i = 2$, i.e. $N - i = 2$, we obtain via (4.176) the following fuzzy goal $\mu_{\overline{G}^3}(X_3)$ induced by X_3 and R^3:

$$\mu_{\overline{G}^3}(X_3) = \max_{x_3}[\mu_{X_3}(x_3) \wedge \mu_{R^3}(x_3, w_3)] =$$

		$U_2 = \overline{U}^1$	$\overline{U}^2$
$=$	$X_2 = \overline{X}^1$	$0.4/0.7 + 1/1$	$0.4/0.7 + 1/1$
	$\overline{X}^2$	$0.4/0.7 + 1/0.7$	$0.4/0.7 + 1/1$
	$\overline{X}^3$	$1/0.4 + 0.4/1$	$0.4/0.7 + 1/1$

and via (4.177) we obtain

$$b(X_2, U_2) =$$

		$U_2 = \overline{U}^1$	$\overline{U}^2$
$= \max_{U_2 \in \{\overline{U}_1, \overline{U}_2\}}$	$X_2 = \overline{X}^1$	$1 \wedge \{0.4/0.7 + 1/1\}$	$1 \wedge \{0.4/0.7 + 1/1\}$
	$\overline{X}^2$	$1 \wedge \{0.4/0.7 + 1/0.7\}$	$1 \wedge \{0.4/0.7 + 1/1\}$
	$\overline{X}^3$	$1 \wedge \{1/0.4 + 0.4/1\}$	$1 \wedge \{0.4/0.7 + 1/1\}$

which is to be meant as

• for $X_2 = \overline{X}^1$

$$\text{IF } \overline{U}^1 \text{ THEN } b(\overline{X}^1, \overline{U}^1) \text{ ELSE IF } \overline{U}^2 \text{ THEN } b(\overline{X}^1, \overline{U}^2)$$

- for $X_2 = \overline{X}^2$

  IF $\overline{U}^1$ THEN $b(\overline{X}^2, \overline{U}^1)$ ELSE IF $\overline{U}^2$ THEN $b(\overline{X}^2, \overline{U}^2)$

- for $X_2 = \overline{X}^3$

  IF $\overline{U}^1$ THEN $b(\overline{X}^3, \overline{U}^1)$ ELSE IF $\overline{U}^2$ THEN $b(\overline{X}^3, \overline{U}^2)$

which in turn gives, via (4.178), the following fuzzy relation $R'_2(X_1)$ defined in $U \times [0, 1]$:

$$\mu_{R'_2}(u_2, w_2) = \quad \begin{array}{c|cc} & w_2 = 0.4 & 1 \\ \hline u_2 = c_1 & 1 & 0.7 \\ c_2 & 0.7 & 1 \end{array}$$

Now, if we suppose the test function "k–large" $= 1/1$, then due to (4.180), for $m = 1$ we find the optimal control at $t = 2$ to be

$$U_2^* = 0.7/c_1 + 1/c_2$$

and, in a similar way, we can determine optimal controls for $X_2 = \overline{X}^2$ and $X_2 = \overline{X}^3$.

The same is repeated for the next control stages, i.e., for $t = 1$ and $t = 0$ which, however, will not be shown here for lack of space. We will give below only the consecutive optimal fuzzy policies obtained [given as fuzzy relation – cf. (2.62, page 38]:

$$R_{u^*}^0 = \quad \begin{array}{c|cc} & u_0 = c_1 & c_2 \\ \hline x_0 = s_1 & 1 & 0.2 \\ s_2 & 1 & 0.2 \\ s_3 & 1 & 0.2 \end{array}$$

$$R_{u^*}^1 = \quad \begin{array}{c|cc} & u_1 = c_1 & c_2 \\ \hline x_1 = s_1 & 1 & 0.2 \\ s_2 & 1 & 0.2 \\ s_3 & 1 & 0.2 \end{array}$$

$$R_{u^*}^2 = \quad \begin{array}{c|cc} & u_2 = c_1 & c_2 \\ \hline x_2 = s_1 & 1 & 1 \\ s_2 & 0.7 & 1 \\ s_3 & 0.4 & 1 \end{array}$$

If, for instance, the initial fuzzy state at control stage $t = 0$ is $X_0 = 0.2/s_2 + 1/s_3$, then – using consecutively the above optimal fuzzy policies – we obtain for the consecutive control stages $t = 1, 2, 3$ the following fuzzy states X_t [via (4.155), page 152] and the optimal fuzzy controls U_t^* [via (4.184), page 159]:

	X_t	U_t^*
$t = 0$	$0.2/s_2 + 1/s_3$	$1/c_1 + 0.2/c_2$
1	$0.3/s_1 + 0.7/s_2 + 1/s_3$	$1/c_1 + 0.2/c_2$
2	$0.7/s_1 + 0.7/s_2 + 1/s_3$	$0.7/c_1 + 1/c_2$
3	$1/s_1 + 0.7/s_2 + 0.7/s_3$	

□

This concludes our short exposition of Baldwin and Pilsworth's (1982) dynamic-programming-based approach. Notice, first, that the complexity of the problem considered implied that this approach be based on some simplification, namely an approximation by using reference fuzzy controls and reference fuzzy states – a similar approach by Kacprzyk and Staniewski (1982) will be discussed in Section 8.7.

Now we will proceed to a conceptually and computationally simpler branch-and-bound approach, which has appeared even earlier.

4.3.2   Solution by branch-and-bound

A branch-and-bound approach for solving the problem of controlling a fuzzy system under fuzzy constraints and fuzzy goals and with a fixed and specified termination time was proposed by Kacprzyk (1979). Suppose that the fuzzy system under control is given by a fuzzy state transition equation (4.155), page 152, i.e.

$$X_{t+1} = F(X_t, u_t), \qquad t = 0, 1 \ldots \qquad (4.186)$$

where X_t, X_{t+1} are fuzzy states at control stages t and $t + 1$, respectively, defined in $X = \{s_1, \ldots, s_n\}$, i.e. $X_t, X_{t+1} \in \mathcal{X}$, and $u_t \in U = \{c_1, \ldots, c_m\}$ is a nonfuzzy control at control stage t. Notice that we assume here a fuzzy system under control but a nonfuzzy control; this is done for simplicity since, as we will see soon, the controls will correspond to branches (edges) in a decision tree. We should, however, bear in mind that one can also assume fuzzy controls, $U_t \in \mathcal{U}$, and then use some predefined reference fuzzy controls to correspond to the branches going out of nodes as in Section 4.3.1. This will not be considered here – the method presented works analogously in that case.

As mentioned in Section 2.4.3 [cf. (2.146), page 62], the fuzzy system under control (4.186) may be equated with a conditioned fuzzy set whose membership function is $\mu_{X_{t+1}}(x_{t+1} \mid x_t, u_t)$, and its state transition equation (4.186) is equivalent to

$$\mu_{X_{t+1}}(x_{t+1}) =$$
$$= \max_{x_t \in X} [\mu_{X_t}(x_t) \wedge \mu_{X_{t+1}}(x_{t+1} \mid x_t, u_t)], \qquad \text{for each } x_{t+1} \in X \quad (4.187)$$

At each control stage t, the control $u_t \in U$ is subjected to a fuzzy constraint $\mu_{C^t}(u_t)$, and on the final fuzzy state attained a fuzzy goal $\mu_{G^N}(x_N)$ is imposed. The initial fuzzy state is $X_0 \in \mathcal{X}$.

The final fuzzy state $X_N \in \mathcal{X}$ evidently cannot be introduced directly into the fuzzy goal $\mu_{G^N}(x_N)$ (this membership function gives a grade of membership of a nonfuzzy state only). A "trickery" is therefore needed to be able to apply a branch-and-bound scheme proposed by Kacprzyk (1978a) for the control of a nonfuzzy system in a fuzzy environment. Its idea is based on the fact that the very essence of the control problem, as far as the fuzzy goal is concerned, is to attain a final fuzzy state X_N [or, equivalently, $\mu_{X_N}(x_N)$] being as close as possible to the fuzzy goal G^N (or, equivalently, $\mu_{G^N}(x_N)$). A measure of closeness (similarity) between X_N and G^N does therefore serve the same purpose for the fuzzy state in this respect as the fuzzy constraint for the nonfuzzy state, provided it takes on its values in $[0, 1]$: with 1 standing for full closeness (similarity, i.e. identity, and 0 for full dissimilarity (difference), with intermediate values for cases in between.

In the original Kacprzyk's (1979) approach a normalized distance measure (4.158), page 153, between two fuzzy sets was used, i.e. $d : \mathcal{X} \times \mathcal{X} \longrightarrow [0,1]$, where $\mathcal{X}$ is a family of all fuzzy sets defined in X. Since, evidently, if $X_N = G^N$, then $d(X_M, G^N) = 0$, and the more X_N differs from G^N the closer $d(X_N, G^N)$ is to 1, then we should use

$$\mu_{\overline{G}^N}(X_N) = 1 - d(X_N, G^N) \tag{4.188}$$

to measure the closeness (similarity) of X_N and G^N, and to use it instead of $\mu_{G^N}(x_N)$ is the control problem formulation.

In Kacprzyk's (1979) source paper, the most traditional definitions of:

- the normalized linear (Hamming) distance (2.26), page 28, i.e.

$$d_l(X_N, G^N) = \frac{1}{N} \sum_{i=1}^N | \mu_{X_N}(s_i) - \mu_{G^N}(s_i) | \tag{4.189}$$

- the normalized quadratic (Euclidean) distance (2.27), page 28, i.e.

$$d_q(X_N, G^N) = \sqrt{\frac{1}{N} \sum_{i=1}^N [\mu_{X_N}(s_i) - \mu_{G^N}(s_i)]^2} \tag{4.190}$$

have been employed.

Moreover, the use of a degree of equality of two fuzzy sets proposed by Kacprzyk and Staniewski (1982) might also be a plausible choice (cf. Section 8.7), that is

$$\mu_{\overline{G}^N}(X_N) = e(X_N, G^N) \tag{4.191}$$

where $e(X_N, G^N) \in [0,1]$ is a degree of equality between X_N and G^N given, e.g., by (2.7)–(2.10), page 24. In this case the degree of equality (4.163) does evidently fulfill our requirements as to the measure of closeness (similarity).

Now, whatever definition of $\mu_{\overline{G}^N}(X_N)$ we may assume, the problem to be solved is to find an optimal sequence of (nonfuzzy) controls $u_0^*, \ldots, u_{N-1}^*$ such that [cf. (4.165), page 154]

$$\mu_D(u_0^*, \ldots, u_{N-1}^* \mid X_0) =$$
$$= \max_{u_0, \ldots, u_{N-1}} [\mu_{C^0}(u_0) \wedge \ldots \wedge \mu_{C^{N-1}}(u_{N-1}) \wedge \mu_{\overline{G}^N}(X_N)] \tag{4.192}$$

It is clear that this problem satisfies all the conditions of type (4.17)–(4.19) in Section 4.1.2 which form a basis of the branch-and-bound procedure. The block diagram of the algorithm is evidently the same as in Figure 18 on page 94, for the case of a nonfuzzy system under control, though the following replacement should be made: X_N (i.e. a fuzzy final state) instead of x_N (i.e. a nonfuzzy final state), and $\mu_{\overline{G}^N}(X_N)$ instead of $\mu_{G^N}(x_N)$. All the tables used are the same but their format is different in that their rows are as follows:

1	2	3	...	$k+3$	...	$N+3$	$N+4$	...	$N+2+n$
v_k	k	u_0	...	u_k	...	$\mu_{X_k}(s_1)$	$\mu_{X_k}(s_2)$	...	$\mu_{X_k}(s_n)$

For illustration let us solve the following simple example.

Example 4.23 Suppose that the state space is $X = \{s_1, \ldots, s_5\}$, the control space is $U = \{c_1, c_2, c_3\}$, the termination time is $N = 3$, the fuzzy constraints are

$$C^0 = 0.4/c_1 + 0.7/c_2 + 1/c_3$$
$$C^1 = 0.5/c_1 + 1/c_2 + 0.7/c_3$$
$$C^2 = 1/c_1 + 0.8/c_2 + 0.6/c_3$$

and the final fuzzy goal is

$$G^3 = 0.4/s_1 + 0.7/s_2 + 1/s_3 + 0.7/s_4 + 0.4/s_5$$

The fuzzy system under control is assumed to be given by

$\mu_{X_{t+1}}(x_{t+1} \mid x_t, u_t) =$

$u_t = c_1$	$x_t = s_1$	$x_{t+1} = s_1$	s_2	s_3	s_4	s_5
	s_1	1	0.1	0.9	0.1	0.2
	s_2	0.8	0.5	0.7	0.3	0.5
	s_3	0.7	0.9	0.5	0.5	0.7
	s_4	0.5	0.7	0.7	0.3	0.4
	s_5	0.2	0.3	0.9	0.7	0.3

$u_t = c_2$	$x_t = s_1$	$x_{t+1} = s_1$	s_2	s_3	s_4	s_5
	s_1	0.3	0.9	1	0.4	0.6
	s_2	0.5	0.7	0.5	0.2	0.3
	s_3	0.8	0.5	0.3	0.5	0.2
	s_4	0.9	0.7	0.7	0.9	0.5
	s_5	0.7	0.9	0.7	1	0.7

$u_t = c_3$	$x_t = s_1$	$x_{t+1} = s_1$	s_2	s_3	s_4	s_5
	s_1	0.5	0.7	0.7	1	0.7
	s_2	0.7	0.8	0.1	0.5	0.9
	s_3	0.8	0.1	0.2	0.3	1
	s_4	0.9	0.2	0.3	0.5	0.8
	s_5	1	0.5	0.4	0.7	0.4

and the initial fuzzy state is $X_0 = 0.1/s_1 + 0.2/s_2 + 0.5/s_3 + 0.7/s_4 + 1/s_5$. The normalized linear distance (4.189) is employed to represent $\mu_{\overline{G}^N}(X_N)$.

The solution process is shown in Figure 22, page 165. The notation $X_{tac\ldots d}$ means the fuzzy state attained at the t-th control stage through the sequence of controls $c_1, c_b, \ldots, c_d$.

The optimal decisions are

$$u_0^* = c_3 \quad u_1^* = c_2 \quad u_2^* = c_2$$

with $\mu_D(c_3, c_2, c_2 \mid X_0) = 0.8$. $\square$

We have assumed here that the control is nonfuzzy. This has made is possible to directly construct the corresponding decision tree (cf. Figure 22). The case of a fuzzy

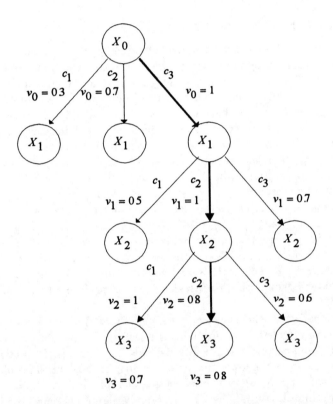

Figure 22   Decision tree with the optimal solution

control can be dealt with here too. Basically, one has to assume some (finite) number of *reference fuzzy controls*. All the controls are then approximated by these reference fuzzy controls, and the branches of the decision tree are associated with the particular reference fuzzy control. This manipulation makes it possible to use the branch-and-bound algorithm presented above. However, one should bear in mind that we obtain here optimal reference fuzzy controls, which are not the same as the "real" fuzzy controls. So, to implement these reference fuzzy controls obtained we need to employ some procedure to infer proper "real" fuzzy controls. The interpolative reasoning based scheme presented in Section 4.3.3 may be used here.

4.3.3 Solution by interpolative reasoning

In Section 4.3, devoted to the multistage control of a fuzzy system, we have presented two basic solution techniques: dynamic programming (Section 4.3.1) and branch-and-bound (Section 4.3.2). The former technique, dynamic programming, is unfortunately very complicated and difficult to implement, if at all implementable. The latter one, branch-and-bound, is much simpler, both from the conceptual and numerical points of view. However, since both of these solution techniques have a combinatorial character, they are very sensitive to the dimensionality of the problem. In many non-trivial practical cases this may be a serious obstacle, hence some other approaches may be needed, and the next two sections will deal with two of them, one based on a fuzzy-logic-based interpolative reasoning scheme, and one employing a genetic algorithm.

To show the essence of an interpolative reasoning scheme which was proposed in a series of Kacprzyk's (1993a, b, d, e, 1994c, f) papers, see that while solving the control problem (4.171), page 156, using Baldwin and Pilsworth's (1982) dynamic programming approach shown in Section 4.3.1, we have to resort to (a few) reference fuzzy states and controls to reduce the search space. This is also the case, though maybe to a lesser extent, while using Kacprzyk's (1979) branch-and-bound algorithm shown in Section 4.3.2.

A lower dimensionality of the (auxiliary, i.e. not "real"!) control problem was obtained in such a way by not involving all possible fuzzy states and controls, but only (much lower) numbers of reference fuzzy states and controls. Needless to say that the lower these numbers the lower the dimensionality of the control problem, and hence the easier its solution. A "sparse" data set is preferable from this point of view.

However, for meaningful results, the compositional rule of inference used in the implementation of an (auxiliary!) optimal control policy [cf. (4.176), page 157] requires sufficiently high numbers of "overlapping" reference fuzzy states and controls, i.e. a "dense" data set is needed from this point of view.

The essence of Kacprzyk's (1993a, b, d, e; 1994c, f) interpolative reasoning approach is an attempt to find an efficient compromise between these two above contradicting requirements. Basically, a very small number of "non-overlapping" reference fuzzy states and controls is assumed, and in their terms an auxiliary (much simpler!) control problem is formulated.

Its solution yields an auxiliary optimal control policy relating optimal reference fuzzy controls to reference fuzzy states. Such a policy is equated with a fuzzy relation which is then used to determine an auxiliary optimal control (not necessarily a reference one) for a particular fuzzy state (not necessarily a reference one). Then, the (auxiliary) optimal solution (control) obtained should be in some way adjusted to become a "real" optimal fuzzy control.

Suppose therefore that we solve problem (4.171), page 156, and – as we may remember – its solution by using the set of dynamic programming recurrence equations (4.174), page 156, is practically impossible due to an excessive dimensionality. A natural idea is that the number of possible fuzzy controls and fuzzy states should be considerably reduced to make the set (4.174) solvable in a reasonable time; this is virtually the essence of Baldwin and Pilsworth's (1982), and also Kacprzyk and Staniewski's (1982) approaches, and is called a *fuzzy interpolation*.

So, first, a relatively small number, r, of reference fuzzy states $\overline{X}^1_{N-i+1}, \ldots, \overline{X}^r_{N-i+1}$ is introduced. Then, we assume a relatively small number, p, of reference fuzzy controls $\overline{U}^1_{N-i}, \ldots, \overline{U}^p_{N-i}$.

The control problem is then expressed in terms of reference fuzzy states and controls only, and its solution, i.e. an (auxiliary!) optimal control policy is expressed by the following fuzzy conditional statement [cf. (4.183), page 158]

$$
\left\{
\begin{array}{l}
\text{IF } (X_{N-i} = \overline{X}^1_{N-i}) \text{ THEN } [U^*_{N-i}(X_{N-i}) = U^*_{N-i}(\overline{X}^1_{N-i})] \\
\text{ELSE} \ldots \text{ELSE} \\
\text{IF } (X_{N-i} = \overline{X}^r_{N-i}) \text{ THEN } [U^*_{N-i}(X_{N-i}) = U^*_{N-i}(\overline{X}^r_{N-i})]
\end{array}
\right.
\tag{4.193}
$$

which is equivalent to a fuzzy relation $R_{U^*}^{N-i}$ defined in $X \times U$ whose membership function is $\mu_{R_{U^*}^{N-i}}(x_{N-i}, u_{N-i})$.

This optimal policy should now be implemented. Suppose therefore that we we are currently at control stage $t = N - i$ in a fuzzy state X_{N-i}. Then, the optimal fuzzy control given by this optimal fuzzy control policy is [cf. (4.184), page 159]

$$
U^*_{N-i} = X_{N-i} \circ R_{U^*}^{N-i}, \qquad \text{for } i = 1, \ldots, N
\tag{4.194}
$$

or in terms of membership functions

$$
\mu_{U^*_{N-i}}(u_{N-i}) = \max_{x_{N-i} \in X} [\mu_{X_{N-i}}(x_{N-i}) \wedge
$$
$$
\wedge \mu_{R_{U^*}^{N-i}}(x_{N-i}, u_{N-i})], \qquad \text{for each } u_{N-i} \in U
\tag{4.195}
$$

Let us now use t instead of $N - i$, for simplicity, and let X_t be a fuzzy number between the two reference fuzzy states $\overline{S}_i$ and $\overline{S}_{i+1}$. We seek therefore an optimal fuzzy control U^*_t corresponding to X_t via the above optimal policy a^*_t specified via the set of fuzzy IF–THEN statements (4.193). Notice that since X_t is not a reference fuzzy state, then the corresponding fuzzy control U^*_t need not be a reference fuzzy optimal control either.

The problem is now the determination of U^*_t meant – assuming for simplicity its representation by a triangular fuzzy number [cf. Section 2.1.7] – as the determination of its mean value and width. It is reasonable to assume that U^*_t be similar (close) to one of these optimal $\overline{C}_{ti}$ and $\overline{C}_{t(i+1)}$ corresponding to $\overline{S}_i$ and $\overline{S}_{i+1}$.

The first aspect is the determination of the mean value of the fuzzy optimal control sought. We apply here Kóczy and Hirota's (1992) idea the essence of which may be expressed as

$$
\frac{d(\overline{S}_i, X_t)}{d(X_t, \overline{S}_{i+1})} = \frac{d(\overline{C}_i, U^*_t)}{d(U^*_t, \overline{C}_{t(i+1)})}
\tag{4.196}
$$

where $d(.,.)$ is a normalized distance between the two fuzzy numbers, e.g., the normalized Hamming (4.161), page 154, or the normalized Euclidean (4.162), page 154. distance.

Notice that the very essence of (4.196) reflects an interpolation, and that is why the approach presented above is termed interpolative reasoning. Clearly, "reasoning" here is not meant in a logical sense.

The sense of (4.196) is that the relative position of U_t^* with respect to its closest reference counterparts should be the same as that concerning X_t and its reference counterparts.

The second aspect is the width of U_t^*. The reasoning is that the lower the number of reference fuzzy states and controls, i.e. the more sparse the data set, the less precise is the available information. Hence, the fuzzier (of a larger width) U_t^* should be. For instance, we can use a formula

$$\overline{w}(U_t^*) = \frac{1}{5}[\overline{w}(\overline{S}_i) + \overline{w}(\overline{S}_{i+1}) + \overline{w}(\overline{X}_t) + \overline{w}(\overline{C}_{ti}) + \overline{w}(\overline{C}_{t(i+1)})] \qquad (4.197)$$

where $\overline{w}(.)$ is a relative width (related to the universe of discourse of the fuzzy states and controls), and the simplest arithmetic mean (4.197) can be replaced by another formula expressing the above rationale as, e.g., a weighted mean, a fuzzy linguistic-quantifier-based aggregation of the type shown in Section 2.3, etc.

Moreover, it may often be expedient to include in (4.197) some term(s) accounting for the fuzziness of the problem in the fuzzy constraints, fuzzy goals, fuzzy system under control, fuzzy termination time, etc. As a first, somehow *ad hoc* attempt, we can use normal degrees of fuzziness of fuzzy sets and calculate the mean fuzziness of fuzzy constraints, fuzzy goals, etc. For the degree of fuzziness of the fuzzy system under control (given as IF–THEN rules), we can use Kacprzyk's (1994b) approach to the determination of a degree of specificity of IF–THEN rules.

Thus, the relative width of U_t^* defined initially by (4.197), which involves the fuzziness of the fuzzy states and fuzzy controls only may be further modified to involve the fuzziness of fuzzy constraints, fuzzy goals and fuzzy system under control.

In general, this approach works well, and helps attain more realistic results.

4.3.4 Solution by a genetic algorithm

It is easy to see that the case of a fuzzy system under control discussed in this section is difficult, and although effective means are available, exemplified by dynamic programming, branch-and-bound, and interpolative reasoning disscussed in the previous subsections, their practical use may be problematic in many non-trivial practical problems as, first, they may be not efficient enough, and second, they may be difficult to implement. Therefore, a genetic algorithm may come to the rescue as in the case of a deterministic system discussed in Section 4.1.3.

Here we will present a genetic algorithm for solving the multistage fuzzy control problem with a fixed and specified termination time and a fuzzy system under control. This was proposed by Kacprzyk (1995a, b), and our discussion will basically follow those works.

First, we use here the same basic framework of a genetic algorithm as presented earlier in Section 4.1.3. That is, by an *individual* we mean a particular solution, i.e. the particular values of the fuzzy controls at the consecutive control stages, $U_0, \ldots, U_{N-1}$. Observe that we assume a general case with fuzzy controls.

An individual is evaluated by the fuzzy decision (4.171), page 156, which is here the *evaluation function*. This function expresses the "fitness" (goodness) of the particular solutions.

A set of potential solutions is termed a *population*. We assume that the algorithm will operate within a population of a fixed size. So, we initially assume some number of potential solutions (the initial population) which has been, e.g., randomly generated. Then, some members of the population, which serve the role of parents, will undergo *reproduction* through *crossover* and *mutation* to produce offsprings (children), i.e. some new solutions. The best ones (the fittest) will "survive," i.e. will be used while repeating this process. Finally, at the end of such a process one may expect to find a very good (if not optimal) solution.

The structure of a genetic algorithm may be portrayed as follows:

begin
    $t := 0$
    set an initial population $P(t)$
    evaluate strings in $P(t)$
    while termination condition is not fulfilled do:
    begin
        $t := t + 1$
        select a current population $P(t)$ from $P(t-1)$
        perform reproduction on elements of $P(t)$
        calculate the evaluation function for each element of $P(t)$
    end
end

and the basic problems are:

- how to represent a potential solution,
- how to create (generate) an initial population,
- how to define an evaluation function,
- how to perform the reproduction (crossover and mutation), and
- how to choose some parameters as, e.g., the population size, probabilities of applying some operations, the termination condition, etc.

Before the description of the genetic algorithm for the problem considered, we will briefly restate the problem formulation for convenience of the reader (cf. Section 4.3).

The dynamics of a fuzzy system under control is given by a state transition equation [cf. (4.155), page 152], i.e.

$$X_{t+1} = F(X_t, U_t), \qquad t = 0, 1, \ldots \qquad (4.198)$$

where $X_t, X_{t+1} \in \mathcal{X}$ are fuzzy states at control stage t and $t+1$, and $U_t \in \mathcal{X}$ is a fuzzy control at t, $t = 0, 1, \ldots$. This equation is equivalent to a conditioned fuzzy set $\mu_{X_{t+1}}(x_{t+1} \mid x_t, u_t)$ or a fuzzy relation in $X \times X \times U$ [cf. Section 2.4.3].

At each t, the fuzzy control applied U_t is subjected to a fuzzy constraint $\mu_{C^t}(u_t)$, and on the resulting fuzzy state X_{t+1} a fuzzy goal $\mu_{G^{t+1}}(x_{t+1})$ is imposed, $t = 0, 1, \ldots, N - 1$.

Both the fuzzy controls, U_t's, and fuzzy states, X_{t+1}'s, are now fuzzy, hence – as we may remember from Section 4.3 – their grades of membership in the fuzzy constraints and goals cannot be directly determined as the values of $\mu_{C^t}(u_t)$ and $\mu_{G^{t+1}}(x_{t+1})$, and some "trickery" is needed. In the source papers (Kacprzyk's, 1995a, b), on which our discussion is based, it was generally used as a measure of dissimilarity (4.156), page 153, i.e.

$$\mu_{\overline{C}^t}(U_t) = 1 - diss(C^t, U_t), t = 0, 1, \ldots, N - 1 \cdot$$

and, to be more specific, Kaufmann and Gupta's (1985) dissimilarity index (4.202), to be shown below.

Thus, generally, the fuzzy decision is then [cf. (4.164), page 154], i.e.

$$\mu_D(U_0, \ldots, U_{N-1} \mid X_0) =$$
$$= \mu_{\overline{C}^0}(U_0) \wedge \mu_{\overline{G}^1}(X_1) \wedge \ldots \wedge \mu_{\overline{C}^{N-1}}(U_{N-1}) \wedge \mu_{\overline{G}^N}(X_N) \qquad (4.199)$$

and the problem is (4.165), page 154, i.e. to find $U_0^*, \ldots, U_{N-1}^*$ such that

$$\mu_D(U_0^*, \ldots, U_{N-1}^* \mid X_0) = \max_{U_0, \ldots, U_{N-1}} \mu_D(U_0, \ldots, U_{N-1} \mid X_0) \qquad (4.200)$$

As an alternative to the three techniques employed for solving this problem, i.e. dynamic programming (Section 4.3.1), branch-and-bound (Section 4.3.2), and interpolative reasoning (Section 4.3.3), Kacprzyk (1995a, b) proposed the use of a genetic algorithm. The algorithm is standard in the sense that it follows the basic structure previously presented.

Due to the specifics of the problem with a fuzzy system, the algorithm's basic elements are to be understood as follows:

- the problem is represented by strings of fuzzy controls $U_0, \ldots, U_{N-1}$ so that we use real coding; in fact, we use triangular fuzzy numbers to represent fuzzy controls (moreover, some reference fuzzy controls, $\overline{U}_0, \ldots, \overline{U}_{N-1}$ are also used);
- the evaluation (objective) function is (4.199), i.e.

$$\mu_D(U_0, \ldots, U_{N-1} \mid X_0) = \mu_{\overline{C}^0}(U_0) \wedge \mu_{G^1}(X_1) \wedge \ldots$$
$$\ldots \wedge \mu_{\overline{C}^{N-1}}(U_{N-1}) \wedge \mu_{\overline{G}^N}(X_N) \qquad (4.201)$$

and for its calculation [of $\mu_{\overline{C}^t}(U_t)$ and $\mu_{\overline{G}^{t+1}}(X_{t+1})$] we use the dissemblance index by Kaufmann and Gupta (1985) defined, for triangular fuzzy numbers assumed, as: if A and B are triangular fuzzy numbers, then the *degree of dissemblance* of A and B is

$$diss(A, B) = \int_{\alpha=0}^{1} \frac{1}{2} (\mid \underline{a}^\alpha - \underline{b}^\alpha \mid + \mid \overline{a}^\alpha - \overline{b}^\alpha \mid) \, d\alpha \qquad (4.202)$$

where $[\underline{a}^\alpha, \overline{a}^\alpha]$ and $[\underline{b}^\alpha, \overline{b}^\alpha]$ are the α-cuts (intervals) of A and B, $\forall \alpha \in (0, 1][cf.(2.14)]$; therefore, if $f_t(U_t, C^t, X_{t+1}, G^{t+1}) = [1 - diss(U_t, C^t)] \wedge [1 - diss(X_{t+1}, G^{t+1})]$, $t = 0, 1, \ldots N - 1$, then the evaluation function (4.201) becomes

$$f(U_0, X_1, \ldots, U_{N-1}, X_N) =$$
$$= \mu_D(U_0, \ldots, U_{N-1} \mid X_0) = f_0(U_0, C^0, X_1, G^1) \wedge \ldots$$
$$\ldots \wedge f_{N-1}(U_{N-1}, C^{N-1}, X_N, G^N) \qquad (4.203)$$

- standard random selections of elements from the consecutive populations, standard concepts of crossover and mutation (evidently, applied to real coded strings), and a standard termination condition, mainly a predefined number of iterations, or iteration-to-iteration improvement lower than a threshhold) are used;

Further, we assume that:

- fuzzy controls are fuzzy sets in $[0, 1]$ defined as triangular fuzzy numbers in $[0,1]$, i.e. as the triples (a, b, c), $0 \leq a \leq b \leq x \leq 1$; the left and right spreads (widths) are assumed to be equal to 5% each, for simplicity, hence only the mean value (b) is practically generated; moreover, 10 reference fuzzy controls are introduced;
- fuzzy states are defined as fuzzy sets in the state space $X = \{s_1, \ldots, s_{10}\}$;
- fuzzy constraints are defined as trapezoid fuzzy numbers in $[0, 1]$;
- fuzzy goals are defined as fuzzy sets in $\{s_1, \ldots, s_{10}\}$;
- the dynamics of the fuzzy system under control (4.198), page 169, i.e. the state transition equation, is defined as a set of fuzzy relations $R_{\overline{U}}$ in $S \times S$, for each of the reference fuzzy control (see that here we need reference fuzzy controls as otherwise we would need infinitely many fuzzy relations, for each possible fuzzy control); so, to choose an appropriate table to determine the state transition, first we find a reference fuzzy control that is the closest [in the sense of the dissemblance index used (4.202)] to the current control, and then we take its corresponding fuzzy relation, and employ the compositional rule of inference [cf. (2.66) and (2.67), page 39] to find the resulting fuzzy state X_{t+1}.

The genetic algorithm employed works therefore as follows:

begin

    $t := 0$

    set the initial population $P(t)$

        which consists of randomly generated

        strings of triangular fuzzy controls

        (i.e. of randomly generated mean

        values from $[0, 1]$, with 5% left and right spreads);

    for each $U_0, \ldots, U_{N-1}$ in each string in the population $P(t)$:

        find the resulting X_{t+1} (by finding first the closest reference

            fuzzy control to choose an appropriate relation

            which is followed by using

            the compositional rule of inference),

        and use the evaluation function (4.203)

            to evaluate each string in $P(t)$;

    while $t < maximum\ number\ of\ iterations$ do

    begin

        $t := t + 1$

        assign the probabilities to each

            string in $P(t - 1)$ which are proportional

            to the value of the evaluation function for each string;

randomly (using those probabilities)
        generate the new population $P(t)$;
    perform crossover and mutation on the strings in $P(t)$;
    calculate the evaluation function (4.203) for each string in $P(t)$.
      end
  end

We will illustrate the algorithm on the following simple example.

Example 4.24 Suppose that the number of control stages is $N = 10$, the state space is $X = \{s_1, \ldots, s_{10}\}$, the controls are triangular fuzzy numbers in $[0, 1]$, and there are 10 "equally-spaced" (with the mean values at 0.1, ..., 0.9, 1) reference fuzzy controls defined as the trapezoid fuzzy numbers in $[0, 1]$ as follows:

$$\begin{aligned}
\overline{C}_1 &= (0.0, 0.1, 0.1, 0.2) & \overline{C}_2 &= (0.1, 0.2, 0.2, 0.3) \\
\overline{C}_3 &= (0.2, 0.3, 0.3, 0.4) & \overline{C}_4 &= (0.3, 0.4, 0.4, 0.5) \\
\overline{C}_5 &= (0.4, 0.5, 0.5, 0.6) & \overline{C}_6 &= (0.5, 0.6, 0.6, 0.7) \\
\overline{C}_7 &= (0.6, 0.7, 0.7, 0.8) & \overline{C}_8 &= (0.7, 0.8, 0.8, 0.9) \\
\overline{C}_9 &= (0.8, 0.9, 0.9, 1.0) & \overline{C}_{10} &= (0.9, 1.0, 1.0, 1.0)
\end{aligned}$$

The initial fuzzy state is

$$X_0 = 1.0/s_1 + 0.7/s_2 + 0.4/s_3 + 0.1/s_4$$

The fuzzy constraints at the particular control stages are also given as the following trapezoid fuzzy numbers:

$$\begin{aligned}
\overline{C}^0 &= (0.0, 0.0, 0.5, 0.8) & \overline{C}^1 &= (0.0, 0.0, 0.5, 0.8) \\
\overline{C}^2 &= (0.0, 0.0, 0.5, 0.8) & \overline{C}^3 &= (0.0, 0.0, 0.5, 0.8) \\
\overline{C}^4 &= (0.0, 0.0, 0.5, 0.8) & \overline{C}^5 &= (0.0, 0.0, 0.5, 0.8) \\
\overline{C}^6 &= (0.0, 0.0, 0.5, 0.8) & \overline{C}^7 &= (0.0, 0.0, 0.5, 0.8) \\
\overline{C}^8 &= (0.0, 0.0, 0.5, 0.8) & \overline{C}^9 &= (0.0, 0.0, 0.5, 0.8)
\end{aligned}$$

The fuzzy goals at the particular control stages are:

$$\overline{G}^1 = 0.1/s_1 + 0.2/s_2 + 0.3/s_3 + 0.6/s_4+ \\ + 1.0/s_5 + 0.6/s_6 + 0.3/s_7 + 0.2/s_8 + 0.1/s_9 + 0.0/s_{10}$$
$$\overline{G}^2 = 0.1/s_1 + 0.2/s_2 + 0.3/s_3 + 0.6/s_4+ \\ + 1.0/s_5 + 0.6/s_6 + 0.3/s_7 + 0.2/s_8 + 0.1/s_9 + 0.0/s_{10}$$
$$\overline{G}^3 = 0.1/s_1 + 0.2/s_2 + 0.3/s_3 + 0.6/s_4+ \\ + 1.0/s_5 + 0.6/s_6 + 0.3/s_7 + 0.2/s_8 + 0.1/s_9 + 0.0/s_{10}$$
$$\overline{G}^4 = 0.1/s_1 + 0.2/s_2 + 0.3/s_3 + 0.6/s_4 + 1.0/s_5 + 0.6/s_6+ \\ + 0.3/s_7 + 0.2/s_8 + 0.1/s_9 + 0.0/s_{10}$$
$$\overline{G}^5 = 0.1/s_1 + 0.2/s_2 + 0.3/s_3 + 0.6/s_4+ \\ + 1.0/s_5 + 0.6/s_6 + 0.3/s_7 + 0.2/s_8 + 0.1/s_9 + 0.0/s_{10}$$
$$\overline{G}^6 = 0.1/s_1 + 0.2/s_2 + 0.3/s_3 + 0.6/s_4+ \\ + 1.0/s_5 + 0.6/s_6 + 0.3/s_7 + 0.2/s_8 + 0.1/s_9 + 0.0/s_{10}$$
$$\overline{G}^7 = 0.1/s_1 + 0.2/s_2 + 0.3/s_3 + 0.6/s_4+ \\ + 1.0/s_5 + 0.6/s_6 + 0.3/s_7 + 0.2/s_8 + 0.1/s_9 + 0.0/s_{10}$$
$$\overline{G}^8 = 0.1/s_1 + 0.2/s_2 + 0.3/s_3 + 0.6/s_4+ \\ + 1.0/s_5 + 0.6/s_6 + 0.3/s_7 + 0.2/s_8 + 0.1/s_9 + 0.0/s_{10}$$
$$\overline{G}^9 = 0.1/s_1 + 0.2/s_2 + 0.3/s_3 + 0.6/s_4+ \\ + 1.0/s_5 + 0.6/s_6 + 0.3/s_7 + 0.2/s_8 + 0.1/s_9 + 0.0/s_{10}$$
$$\overline{G}^{10} = 0.1/s_1 + 0.2/s_2 + 0.3/s_3 + 0.6/s_4+ \\ + 1.0/s_5 + 0.6/s_6 + 0.3/s_7 + 0.2/s_8 + 0.1/s_9 + 0.0/s_{10}$$

The fuzzy state transitions (4.198), page 169, are specified as conditioned fuzzy sets for each particular reference fuzzy control, $\overline{C}_1, \ldots, \overline{C}_{10}$. Due to lack of space we will only present below the state transtion equations for the first and last reference fuzzy control, i.e. $\overline{C}_1$ and $\overline{C}_{10}$, and these are:

- for $\overline{C}_1$

	$x_{t+1} = s_1$	s_2	s_3	s_4	s_5	s_6	s_7	s_8	s_9	s_{10}
$x_t = s_1$	0.0	1.0	0.9	0.8	0.7	0.6	0.5	0.4	0.3	0.1
s_2	0.0	1.0	0.9	0.8	0.7	0.6	0.5	0.4	0.3	0.1
s_3	0.0	0.0	0.9	0.8	0.7	0.6	0.5	0.4	0.3	0.1
s_4	0.0	0.0	0.9	0.8	0.7	0.6	0.5	0.4	0.3	0.1
s_5	0.0	1.0	0.9	0.8	0.7	0.6	0.5	0.4	0.3	0.1
s_6	0.0	1.0	0.9	0.8	0.7	0.6	0.5	0.4	0.3	0.1
s_7	0.0	1.0	0.9	0.8	0.7	0.6	0.5	0.4	0.3	0.1
s_8	0.0	1.0	0.9	0.8	0.7	0.6	0.5	0.4	0.3	0.1
s_9	0.0	1.0	0.9	0.8	0.7	0.6	0.5	0.4	0.3	0.1
s_{10}	0.0	1.0	0.9	0.8	0.7	0.6	0.5	0.4	0.3	0.1

- ...

- for $\overline{C}_{10}$

$x_{t+1} = s_1$	s_2	s_3	s_4	s_5	s_6	s_7	s_8	s_9	s_{10}
$x_t = s_1$	0.0	0.0	0.0	0.0	0.0	0.0	0.0	0.0	0.0	1.0
s_2	0.0	0.0	0.0	0.0	0.0	0.0	0.0	0.0	0.0	1.0
s_3	0.0	0.0	0.0	0.0	0.0	0.0	0.0	0.0	0.0	1.0
s_4	0.0	0.0	0.0	0.0	0.0	0.0	0.0	0.0	0.0	1.0
s_5	0.0	0.0	0.0	0.0	0.0	0.0	0.0	0.0	0.0	1.0
s_6	0.0	0.0	0.0	0.0	0.0	0.0	0.0	0.0	0.0	1.0
s_7	0.0	0.0	0.0	0.0	0.0	0.0	0.0	0.0	0.0	1.0
s_8	0.0	0.0	0.0	0.0	0.0	0.0	0.0	0.0	0.0	1.0
s_9	0.0	0.0	0.0	0.0	0.0	0.0	0.0	0.0	0.0	1.0
s_{10}	0.0	0.0	0.0	0.0	0.0	0.0	0.0	0.0	0.0	1.0

Suppose that the main parameters are:

- the population size is 50,
- the maximum number of iterations (termination condition) is 1000,
- the crossover rate is 0.6, and
- the mutation rate is 0.001.

The ten best results obtained may be summarized as follows:

- the optimal fuzzy controls (triangular fuzzy numbers $U_t = (a, b, c)$) at the particular control stages $t = 0, 1, \ldots, 10$, i.e. the best result obtained, are:

$$U_0^* = (0.4885, 0.5142, 0.5399) \qquad U_1^* = (0.5031, 0.5296, 0.5561)$$
$$U_2^* = (0.4236, 0.4459, 0.4682) \qquad U_3^* = (0.4842, 0.5097, 0.5352)$$
$$U_4^* = (0.4651, 0.4895, 0.5140) \qquad U_5^* = (0.4916, 0.5175, 0.5434)$$
$$U_6^* = (0.3218, 0.3387, 0.3556) \qquad U_7^* = (0.5225, 0.5500, 0.5775)$$
$$U_8^* = (0.3451, 0.3633, 0.3815) \qquad U_9^* = (0.2615, 0.2752, 0.2890)$$

and the value of the fuzzy decision (4.203) is

$$\mu_D(U_0^*, \ldots, U_9^* \mid X_0) = 0.681881$$

- the second best result is

$$U_0 = (0.4885, 0.5142, 0.5399) \qquad U_1 = (0.5031, 0.5296, 0.5561)$$
$$U_2 = (0.4236, 0.4459, 0.4682) \qquad U_3 = (0.4842, 0.5097, 0.5352)$$
$$U_4 = (0.4651, 0.4895, 0.5140) \qquad U_5 = (0.4916, 0.5175, 0.5434)$$
$$U_6 = (0.3218, 0.3387, 0.3556) \qquad U_7 = (0.5225, 0.5500, 0.5775)$$
$$U_8 = (0.3451, 0.3633, 0.3815) \qquad U_9 = (0.2615, 0.2752, 0.2890)$$

and the value of the fuzzy decision (4.203) is

$$\mu_D(U_0^*, \ldots, U_9^* \mid X_0) = 0.681881$$

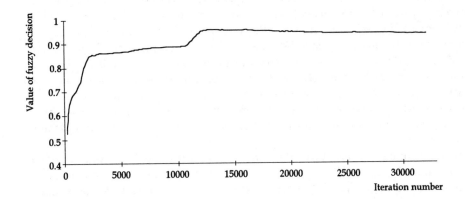

Figure 23   The value of the fuzzy decision obtained in the course of iterations
in Example 4.24

- while the tenth best result is:

$$U_0^* = (0.2510, 0.2642, 0.2774) \qquad U_1^* = (0.4758, 0.5008, 0.5259)$$
$$U_2^* = (0.4855, 0.5111, 0.5366) \qquad U_3^* = (0.5432, 0.5718, 0.6004)$$
$$U_4^* = (0.4780, 0.5032, 0.5284) \qquad U_5^* = (0.5182, 0.5455, 0.5728)$$
$$U_6^* = (0.5100, 0.5368, 0.5637) \qquad U_7^* = (0.3316, 0.3491, 0.3665)$$
$$U_8^* = (0.4639, 0.4883, 0.5127) \qquad U_9^* = (0.3816, 0.4016, 0.4217)$$

and the value of the fuzzy decision (4.203) is

$$\mu_D(U_0^*, \ldots, U_9^* \mid X_0) = 0.679795$$

In Figure 23 we show the best values of the fuzzy decision (4.203) obtained in the course of iterations. It may readily be seen that the optimal solution has been attained quite early, before the 1,000 iterations assumed. In general, also for many different problems solved, the algorithm has proven to be efficient.

The result obtained is quite close to the one obtained using the branch-and-bound and dynamic programming (with interpolative reasoning!) techniques, but the solution is clearly simpler. □

To summarize our considerations, it seems that a genetic algorithm may be a very promising, efficient and easily implementable means for solving the class of control problems with a fuzzy system under control.

This concludes our discussion of the case of a fuzzy system under control with a fixed and specified termination time. Both the approaches presented, one based on dynamic programming (Section 4.3.1) and one based on branch-and-bound (Section 4.3.2), use some "trickery" whose essence is the introduction of some (finite, relatively low) number of reference fuzzy states and controls leading to a manageable size of the control problem.

As to difficulties that may be encountered in the control of fuzzy systems, one should point out the following. In the course of fuzzy state transitions, the fuzzy states consecutively attained become usually more and more fuzzy, i.e. their membership functions become "flatter" and "flatter." For longer termination times this may be a serious obstacle, because they may become too fuzzy to be meaningful. A solution consists of some "sharpening" of the respective fuzzy sets, e.g., by raising them to some powers due to (2.51), page 34. An efficient approach is also to approximate the fuzzy states successively attained by some properly chosen reference fuzzy sets.

As to some other works related to the control of a fuzzy system in a fuzzy environment, which are not so numerous, one should mention Britov and Reznik (1981).

This also concludes the analysis of the basic case with the fixed and specified termination time. Now we will proceed to the next important class, that is, with the termination time specified implicitly by entering for the first time some given termination set of states.

5

Control Processes with an Implicitly Specified Termination Time

The previous chapter has been devoted to the fundamental and most straightforward case of a fixed and specified in advance termination time. In practice there are, however, many control processes in which the termination time is unknown in advance, or is even irrelevant, as we may be mostly interested in attaining some final state, no matter how long this might take. Thus the process terminates when the state attains for the first time some specific value or enters some specific subset of the state space. In such a case the termination time is *implicitly* defined by specifying that value or set.

This chapter will be concerned with such a class of control problems. For clarity and simplicity we will, however, focus our attention on the basic case of a deterministic system under control whose dynamics is assumed to be described by the state transition equation (2.130), page 59, i.e.

$$x_{t+1} = f(x_t, u_t), \qquad t = 0, 1, \ldots \tag{5.1}$$

where $x_t, x_{t+1} \in X = \{s_1, \ldots, s_p, s_{p+1}, \ldots, s_n\}$ are the states at control stages (times) t and $t+1$, respectively, and $u_t \in U = \{c_1, \ldots, c_m\}$ is the control at time t, $t = 0, 1, \ldots$

The process terminates when the state enters for the first time a (specified in advance) termination set of states $W = \{s_{p+1}, s_{p+2}, \ldots, s_n\}, W \subset X$. Let us denote this time, which is evidently unknown in advance, by $\overline{N}$; evidently, only $\overline{N} < \infty$ is interesting for our purposes. The initial state is $x_0 \in X \setminus W = \{s_1, \ldots, s_p\}$. Moreover, if $x_t \in W$, then $f(x_t, u_t) = x_t$, for any $u_t \in U$, by assumption.

The fuzzy goal to be attained should be defined as a fuzzy set in the termination set of states, i.e. in $W \subset X$. The membership function of the fuzzy goal is therefore $\mu_{G^{\overline{N}}}(x_{\overline{N}})$, and $\mu_{G^{\overline{N}}}(x_{\overline{N}}) = 0$ for each $x_{\overline{N}} \notin W$.

At each control stage t, the control u_t is subjected to a fuzzy constraint. However, since the termination time is now unknown in advance, then the time-dependent form of the fuzzy constraint used in the previous chapter, i.e. $\mu_{C^t}(u_t)$, is evidently not applicable, and one should rather use a state-dependent form, i.e. $\mu_C(u_t \mid x_t)$, in which the grade of membership of a particular current control in the fuzzy constraint is a function of the current state. Such a form of a fuzzy constraint is often adequate as, for instance, in a company what we can invest depends on what we have produced.

First, the fuzzy decision is defined as conventionally [cf. (3.51), page 80], i.e.

$$\mu_D(u_0, \ldots, u_{\overline{N}-1} \mid x_0) =$$
$$= \mu_C(u_0 \mid x_0) \wedge \ldots \wedge \mu_C(u_{\overline{N}-1} \mid x_{\overline{N}-1}) \wedge \mu_{G^{\overline{N}}}(x_{\overline{N}}) \qquad (5.2)$$

The problem is to determine an optimal sequence of controls $u_0^*, \ldots, u_{\overline{N}-1}^*$ such that

$$\mu_D(u_0^*, \ldots, u_{\overline{N}-1}^* \mid x_0) =$$
$$= \max_{u_0, \ldots, u_{\overline{N}-1}} [\mu_C(u_0 \mid x_0) \wedge \ldots \wedge \mu_C(u_{\overline{N}-1} \mid x_{\overline{N}-1}) \wedge \mu_{G^{\overline{N}}}(x_{\overline{N}})] \qquad (5.3)$$

where $x_0, \ldots, x_{\overline{N}-1} \in X \setminus W$, and $x_{\overline{N}} \in W$. Evidently, as we have already indicated many times, we would rather be interested in finding an optimal control strategy, in fact an optimal stationary control strategy (cf. Section 3.2).

The termination set of states $W \subseteq X$ need not be reachable from $x_0 \in X \setminus W$, and $\mu_D(u_0, \ldots, u_{N-1} \mid x_0) = 0$ for any $\overline{N} < \infty$ means that the termination set of states W is not reachable from $x_0 \in X \setminus W$, i.e. there exists no finite sequence of controls that can move the system under control from the initial state $x_0 \in X \setminus W$ to any state in the termination set W, i.e. to any $x_{\overline{N}} \in W$.

The solution of the problem sketched above may proceed by using the following three basic approaches:

- an iterative approach by Bellman and Zadeh (1970),
- a graph-theoretic approach by Komolov *et al.* (1979),
- a branch-and-bound approach by Kacprzyk (1983b),

which will be subsequently presented below.

5.1   AN ITERATIVE APPROACH

This approach has been proposed in the seminal Bellman and Zadeh's (1970) paper. It is based on the use of some *functional equation* that relates the fuzzy decision from some specified (current) control stage (time) on, to the fuzzy decision from the next control stage on.

First, the fuzzy decision is defined as (5.2), i.e.

$$\mu_D(u_0, \ldots, u_{\overline{N}-1} \mid x_0) =$$
$$= \mu_C(u_0 \mid x_0) \wedge \ldots \wedge \mu_C(u_{\overline{N}-1} \mid x_{\overline{N}-1}) \wedge \mu_{G^{\overline{N}}}(x_{\overline{N}})$$

We introduce the concept of a *fuzzy decision from control stage t on* (i.e. until the end of the control process) as

$$\mu_D(u_t, u_{t+1}, \ldots, u_{\overline{N}-1} \mid x_t) =$$
$$= \mu_C(u_t \mid x_t) \wedge \mu_C(u_{t+1} \mid u_{t+1}) \wedge \ldots \wedge \mu_C(u_{\overline{N}-1} \mid x_{\overline{N}-1}) \wedge \mu_{G^{\overline{N}}}(x_{\overline{N}}) \quad (5.4)$$

The time-invariance of the fuzzy system under control (5.1) and the time-independence of the fuzzy constraints [which are $\mu_C(. \mid .)$, i.e. the same for each

t] and fuzzy goal [which is $\mu_{G\overline{N}}(.)$, i.e. specified just for $t = \overline{N}$] imply evidently that

$$\mu_D(u_t, u_{t+1}, \ldots, u_{\overline{N}-1} \mid x_t) =$$
$$= \mu_C(u_t \mid x_t) \wedge \mu_C(u_{t+1} \mid x_{t+1}) \wedge \ldots$$
$$\ldots \wedge \mu_C(u_{\overline{N}-1} \mid x_{\overline{N}-1}) \wedge \mu_{G\overline{N}}(x_{\overline{N}}) \qquad (5.5)$$

and

$$\mu_D(u_{t+1}, u_{t+2}, \ldots, u_{\overline{N}-1} \mid x_t) =$$
$$= \mu_C(u_{t+1} \mid x_{t+1}) \wedge \mu_C(u_{t+2} \mid x_{t+2}) \wedge \ldots$$
$$\ldots \wedge \mu_C(u_{\overline{N}-1} \mid x_{\overline{N}-1}) \wedge \mu_{G\overline{N}}(x_{\overline{N}}) \qquad (5.6)$$

These yield

$$\mu_D(u_t, u_{t+1}, \ldots, u_{\overline{N}-1} \mid x_t) =$$
$$= \mu_C(u_t \mid x_t) \wedge \mu_C(u_{t+1} \mid x_{t+1}) \wedge \mu_D(u_{t+1}, u_{t+2}, \ldots, u_{\overline{N}-1} \mid x_{t+1}) =$$
$$= \mu_C(u_t \mid x_t) \wedge \mu_C(u_{t+1} \mid x_{t+1}) \wedge \mu_D[u_{t+1}, u_{t+2}, \ldots, u_{\overline{N}-1} \mid f(x_t, u_t)] \quad (5.7)$$

where $t = 0, 1, \ldots, \overline{N} - 1$ in all the above equations.

This is the functional equation for the problem considered (5.3). Since the termination time is unknown in advance, then it is obviously more realistic and convenient to express the controls by a *stationary policy function* or, briefly, a *stationary policy* $a : X \setminus W \longrightarrow U$ such that

$$u_t = a(x_t), \qquad t = 0, 1, \ldots \qquad (5.8)$$

where $x_t \in X \setminus W$, $u_t \in U$. This stationary policy is evidently in the sense of (3.63) or, equivalently, (3.64), page 82, in that it relates the controls to the states only. The stationary strategy is here $a_{\overline{N}} = \underbrace{(a, \ldots, a)}_{\overline{N}}$.

The functional equation (5.7) may now be written in terms of stationary policies and a stationary strategy, e.g., for $t = 0$, as

$$\mu_D(a_{\overline{N}} \mid x_0) = \mu_C[a(x_0) \mid x_0] \wedge \mu_D[a_{\overline{N}} \mid f(x_0, a(x_0))] \qquad (5.9)$$

which should evidently be read as the following set of n equations, for $x_0 = s_1, \ldots, s_n$:

$$\begin{cases} \mu_D(a_{\overline{N}} \mid s_1) = \mu_C[a(s_1) \mid s_1] \wedge \mu_D[a_{\overline{N}} \mid f(s_1, a(s_1))] \\ \cdots\cdots\cdots\cdots\cdots\cdots\cdots\cdots\cdots\cdots\cdots\cdots\cdots\cdots\cdots \\ \mu_D(a_{\overline{N}} \mid s_n) = \mu_C[a(s_n) \mid s_n] \wedge \mu_D[a_{\overline{N}} \mid f(s_n, a(s_n))] \end{cases} \qquad (5.10)$$

and it can be proven that this set of equations has a unique solution (Bellman and Zadeh, 1970).

Let us now proceed to the functional equation for the fuzzy decision corresponding to an optimal strategy. First we need to reformulate the problem (5.3) in terms of a (stationary) strategy, Namely, an optimal strategy $a^*_{\overline{N}} = (a^*, \ldots, a^*)$ is sought such that

$$\mu_D(a^*_{\overline{N}} \mid x_0) = \max_{a_{\overline{N}}} \mu_D(a_{\overline{N}} \mid x_0) \qquad (5.11)$$

with the following ordering ordering "$\succeq$" among the strategies

$$a'_{\overline{N}} \succeq a''_{\overline{N}} \Longleftrightarrow \mu_D(a'_{\overline{N}} \mid x_0) \geq \mu_D(a''_{\overline{N}} \mid x_0), \qquad \text{for each } x_0 \in X \setminus W \qquad (5.12)$$

that is, strategy $a'_{\overline{N}}$ is not worse than strategy $a''_{\overline{N}}$ if and only if $\mu_D(a'_{\overline{N}} \mid x_0) \geq \mu_D(a''_{\overline{N}} \mid x_0)$. Evidently, $a^*_{\overline{N}}$ is *optimal* if and only if it is not worse than any other possible strategy, i.e. $a^*_{\overline{N}} \succeq a_{\overline{N}}$, for each $a_{\overline{N}}$.

Now, denote by $M(a)$ an $n \times n$ 0–1 matrix whose (i,j)-th element is equal to 1 if and only if $s_j = f(s_i, a(s_i))$, i.e. if state s_j is an immediate successor of state s_i under policy a, and is equal to 0 otherwise.

Then

$$\mu_D(a^*_N \mid x_0) = \max_{a_{\overline{N}}} [\mu_C(a(x_0) \mid x_0) \wedge$$

$$\wedge M(a) \mu_D(a^*_N \mid f(x_0, a(x_0)))], \qquad \text{for each } x_0 \in X \setminus W \qquad (5.13)$$

and since there are $r = (\text{card}\, U) \cdot [\text{card}\,(X \setminus W)]$ distinct policies, i.e. $a^1, \ldots, a^r$, then (5.13) may be written as

$$\mu_D(a^*_N \mid x_0) =$$
$$= [\mu_C(a^1(x_0) \mid x_0) \wedge M(a^1) \mu_D(a^*_N \mid f(x_0, a^1(x_0)))] \vee \ldots$$
$$\ldots \vee [\mu_C(a^r(x_0) \mid x_0) \wedge M(a^r) \mu_D(a^*_N \mid f(x_0, a^r(x_0)))] \qquad (5.14)$$

Due to the distributivity of the operations "$\wedge$" and "$\vee$," by factoring similar terms, we may rewrite (5.14) as

$$\mu_D(a^*_{\overline{N}} \mid s_i) = \max_{c_j} [\mu_C(c_j \mid s_i) \wedge \mu_D(a_{\overline{N}} \mid f(s_i, c_j))] \qquad (5.15)$$

$i = 1, \ldots, n; j = 1, \ldots, m$, where: $c_j = a(s_i)$ and, by definition, $f(s_i, c_j) = s_i$, for each $c_j \in U$, if $s_i \in W$; moreover, $\mu_C(c_j \mid s_i) = 1$, for $s_i \in W$, and $\mu_D(a^*_{\overline{N}} \mid s_i) = \mu_{G^{\overline{N}}}(s_i)$, for each $s_i \in W$.

Therefore, the $\mu_D(a^*_{\overline{N}} \mid s_i)$'s, for $s_i \in X \setminus W$, are unknown, and the $\mu_C(c_j \mid s_i)$'s and $\mu_D(a^*_{\overline{N}})$'s, for $s_i \in W$, are constants.

The problem (5.15) may be therefore represented in matrix form as

$$w = (B \wedge w) \vee z \qquad (5.16)$$

where $w^T = [w_i]^T = [w_1, \ldots, w_p]$, $w_i = \mu_D(a^*_{\overline{N}} \mid s_i)$, is the vector of unknowns, $B = [b_{il}]$ is an $n \times n$ matrix given by

$$b_{il} = \begin{cases} 0 & \text{if } s_j \text{ is not an immediate} \\ & \text{successor of } s_i \\ \max_{c_q : f(s_i, c_q) = s_i} \mu_C(c_q \mid s_i) & \text{otherwise} \end{cases} \qquad (5.17)$$

and $z^T = [z_i]^T = [z_1, \ldots, z_p]$ is a p-element vector such that

$$z_i = \max_{c_j} [\mu_C(c_j \mid s_i) \wedge \mu_G(f(s_i, c_j))] \qquad (5.18)$$

where in both (5.17) and (5.18) $i, l = 1, \ldots, p$.

The notation (5.16) is to be meant as follows: $B \wedge w = [(Bw)_1, \ldots, (Bw)_p]^T$, and

$$(Bw)_i = \max_{l \in \{1, \ldots, p\}} (b_{il} \wedge w_i), \qquad i = 1, \ldots, p \qquad (5.19)$$

and

$$w_i = (Bw)_i \vee z_i = ((Bw)_i \vee z_i)_i, \qquad i = 1, \ldots, p \qquad (5.20)$$

As shown by Bellman and Zadeh (1970), the set of equations (5.16) may be solved iteratively through the following scheme:

$$w^{(t+1)} = (B \wedge w^{(t)}) \vee z, \qquad t = 0, 1, \ldots \qquad (5.21)$$

where $w^{(t)}$ is the vector w obtained in the t-th iteration, and $w^{(0)} = [0, \ldots, 0]^T$, by assumption. The sequence $w^{(0)}, w^{(1)}, \ldots$ is monotone nondecreasing and bounded from above by $w = [1, \ldots, 1]^T$, hence it converges to a solution of (5.16). Moreover, this solution is obtained in no more than p iterations.

Thus, by solving (5.16) we obtain the $\mu_D(a^*_{\overline{N}} \mid s_i)$'s, $s_i \in X \setminus W$, with an optimal stationary policy as derived from (5.15).

For illustration let us solve Bellman and Zadeh's (1970) example.

Example 5.1 Suppose that $X = \{s_1, \ldots, s_5\}$, $U = \{c_1, c_2\}$, $W = \{s_4, s_5\}$, the state transition equations are given by

$$x_{t+1} = \quad$$

		$x_t = s_1$	s_2	s_3	s_4	s_5
	$u_t = c_1$	s_4	s_3	s_5	s_4	s_5
	c_2	s_2	s_2	s_1	s_4	s_5

the fuzzy constraints are

$$C(s_1) = 0.6/c_1 + 1/c_2$$
$$C(s_2) = 0.8/c_1 + 1/c_2$$
$$C(s_3) = 1/c_1 + 0.7/c_2$$

and the fuzzy goal is

$$G^3 = 1/s_4 + 0.7/s_5$$

The equation (5.10) becomes

$$\begin{cases} \mu_D(a^*_{\overline{N}} \mid s_1) = (0.6 \wedge \mu_D(a^*_{\overline{N}} \mid s_4)) \vee (1 \wedge \mu_D(a_{\overline{N}} \mid s_2)) \\ \mu_D(a^*_{\overline{N}} \mid s_2) = (0.8 \wedge \mu_D(a^*_{\overline{N}} \mid s_3)) \vee (1 \wedge \mu_D(a_{\overline{N}} \mid s_2)) \\ \mu_D(a^*_{\overline{N}} \mid s_2) = (1 \wedge \mu_D(a^*_{\overline{N}} \mid s_5)) \vee (0.7 \wedge \mu_D(a_{\overline{N}} \mid s_1)) \\ \mu_D(a^*_{\overline{N}} \mid s_4) = \mu_{G^3}(s_5) = 1 \\ \mu_D(a^*_{\overline{N}} \mid s_5) = \mu_{G^s}(s_5) = 0.8 \end{cases}$$

which is equal, in matrix form (5.16), to

$$\begin{bmatrix} \mu_D(a^*_{\overline{N}} \mid s_1) \\ \mu_D(a^*_{\overline{N}} \mid s_2) \\ \mu_D(a^*_{\overline{N}} \mid s_3) \end{bmatrix} = \left(\begin{bmatrix} 0 & 1 & 0 \\ 0 & 1 & 0.8 \\ 0.7 & 0 & 0 \end{bmatrix} \wedge \begin{bmatrix} \mu_D(a^*_{\overline{N}} \mid s_1) \\ \mu_D(a^*_{\overline{N}} \mid s_2) \\ \mu_D(a^*_{\overline{N}} \mid s_3) \end{bmatrix} \right) \vee \begin{bmatrix} 0.6 \\ 0 \\ 0.8 \end{bmatrix}$$

We solve this equation iteratively using (5.21), assuming $w^{(0)} = [0, 0, \ldots, 0]^T$, and successively obtain

$$w^{(1)} = [0.6, 0, 0.8]^T$$
$$w^{(2)} = [0.6, 0.8, 0.8]^T$$
$$w^{(3)} = [0.8, 0.8, 0.8]^T$$
$$w^{(4)} = [0.8, 0.8, 0.8]^T$$

The solution obtained is therefore

$$\mu_D(a^*_{\overline{N}} \mid s_1) = 0.8 \qquad \mu_D(a^*_{\overline{N}} \mid s_2) = 0.8 \qquad \mu_D(a^*_{\overline{N}} \mid s_3) = 0.8$$

and by introducing these values into (5.15) we readily obtain the optimal stationary strategy to be $a^*_{\overline{N}} = \underbrace{(a^*, \ldots, a^*)}_{\overline{N}}$ such that

$$a^*(s_1) = c_2 \qquad a^*(s_2) \in \{c_1, c_2\} \qquad a^*(s_3) = c_1$$

while, by obvious reasons, $a^*(s_4)$ and $a^*(s_5)$ are irrelevant because s_4 and s_5 belong to the termination set of states, $s_4, s_5 \in W$. $\qquad \square$

Some elements of this approach will also be used in another approach to be presented below.

5.2   A GRAPH-THEORETIC APPROACH

The second approach to the solution of the problem considered has been proposed by Komolov *et al.* (1979). It is based on a graph-theoretic analysis. Some of its elements have already been used in the formerly described Bellman and Zadeh's (1970) approach though mostly for illustrative purposes. In this approach they will be explicitly employed.

First, for convenience, let us repeat the formulation of the problem considered (5.3), page 178: an optimal stationary strategy $a^*_{\overline{N}} = \underbrace{(a^*, \ldots, a^*)}_{\overline{N}}$ is sought such that

$$\mu_D(a^*_{\overline{N}} \mid x_0) = \max_{a_{\overline{N}}} \mu_D(a_{\overline{N}} \mid x_0) =$$

$$= \max_{a_{\overline{N}}} [\mu_C(a(x_0) \mid x_0) \wedge \ldots \wedge \mu_C(a(x_{\overline{N}-1}) \mid x_{\overline{N}-1}) \wedge \mu_{G_{\overline{N}}}(x_{\overline{N}})]$$

where the consecutively states attained, $x_1, \ldots, x_{\overline{N}} \in X$, are given by the state transition equation (5.1), page 177, i.e.

$$x_{t+1} = f(x_t, u_t), \qquad t = 0, 1, \ldots$$

and $x_{\overline{N}} \in W$ while $x_0, \ldots, x_{\overline{N}-1} \in X \setminus W$.

With the above state transition equation of the system under control (5.1) we associate an oriented *state transition graph* $XV = (X, V)$ such that the set of nodes, representing possible states is $X = \{s_1, \ldots, s_p, s_{p+1}, \ldots, s_n\}$, and $V \subseteq X \times X$ is the set of ordered pairs $(s_i, s_j) \in V$ which represent an oriented arcs from node s_i to node s_j. The nodes s_i and s_j are connected with arc (s_i, s_j) if and only if there exists a control moving the system from s_i to s_j.

To each arc $(s_i, s_j) \in V$ we assign its *weight*

$$\mu_{ij} = \max_{c_r:f(s_i,c_r)=s_j} \mu_C(c_r \mid s_j) \qquad (5.22)$$

Now, we introduce an additional node s_{n+1} and $n-p$ arcs $(s_q, s_{n+1}), q = p+1, \ldots, n$, where p is such that $W = \{s_{p+1}, \ldots, s_n\}$. The weights of these arcs are

$$\mu_{q,n+1} = \mu_{G\overline{N}}(s_q), \qquad q = p+1, p+2, \ldots, n \qquad (5.23)$$

To each stationary strategy $a_{\overline{N}} = \underbrace{(a, \ldots, a)}_{\overline{N}}$ moving the system from the initial state x_0 to some $s_q \in W$ there corresponds in the transition graph some path (a sequence of consecutive arcs) b_a from x_0 to s_{n+1}. The arc (s_i, s_j) belongs evidently to this path b_a if policy a moves the system from s_i to s_j. By assumption, (s_q, s_{n+1}) belongs to b_a as well, for each $q = p+1, p+2, \ldots, n$. And, vice versa, to each path from x_0 to s_{n+1} there corresponds some stationary policy.

Now, let $b_{il,ik} = (s_{i1}, s_{i2}) \ldots (s_{i(k-1)}, s_{ik})$ be a path from s_{i1} to s_{ik}. Its weights are defined as

$$\mu(b_{i1,ik}) = \mu_{i1,i2} \wedge \ldots \wedge \mu_{i(k-1),ik} \qquad (5.24)$$

It is clear that the problem to be solved, (5.3), is equivalent to the determination of a path from x_0 to s_{n+1} with the maximum weight (5.24). We should evidently consider only simple paths, i.e. those which pass at most once through each node, since an optimal stationary strategy is sought.

Let us denote now by y_i the weight of a maximal (in the sense of its weight) simple path from s_i to s_{ik}. Then we have

$$y_i = \begin{cases} \max_{s_j \in v(s_i)}(\mu_{ij} \wedge y_j) & \text{for } i = 1, \ldots, n \\ 1 & \text{for } i = n+1 \text{ (by definition)} \end{cases} \qquad (5.25)$$

where $v(s_i)$ is the set of nodes incident from node s_i, i.e. those which represent states that are immediate successors of s_i. Notice that the above equation is to be considered for $i = 1, \ldots, p$ only, because the other states belong to the termination set of states.

It is now easy to see that the equation (5.25) may be written in matrix form as

$$y = (H \wedge y) \vee e \qquad (5.26)$$

where: $y^T = [y_1, \ldots, y_n]$, with y_i's given by (5.25), $e^T = [e_1, \ldots, e_n]$, with e_i's given by

$$e_i = \max_{k \in \{p+1, \ldots, n\}}(\mu_{ik} \wedge \mu_{k,n+1}), \qquad i = 1, \ldots, n \qquad (5.27)$$

with $H = [\mu_{ij}]$, and $H \wedge y = [(Hy)_1, \ldots, (Hy)_n]$, with $(Hy)_i$'s given by

$$(Hy)_i = \max_{k \in \{1, \ldots, p\}}(\mu_{ik} \wedge y_k), \qquad i = 1, \ldots, n \qquad (5.28)$$

As shown by Komolov *et al.* (1979), equation (5.26) has the solution expressed by

$$y = H^* \wedge e \tag{5.29}$$

where

$$H^* = H^0 \vee \ldots \vee H^{\overline{N}-1} \tag{5.30}$$

and H^0 is the identity matrix.

To find H^* we may either use an iterative procedure of Bellman and Zadeh (1970), presented in Section 5.1, or some algebraic method (cf. Backhouse and Carré, 1975). The former one is evidently better in the case when we intend to solve the problem only once, while the latter one is preferable for a repetitive solution with various fuzzy goals. In case of the latter it is necessary to calculate H^* only once, and then solve equation (5.29) for various values of e. The determination of H^* is, however, beyond the scope of our discussion, and will not be considered here. We refer the reader to Backhouse and Carré's (1975) work, and to Komolov *et al.* (1979) for some refinements. In simple cases H^* can also be calculated by employing its definition (5.30).

Finally see that in the solution $y^T = [y_1, \ldots, y_n]$ given by (5.29) only the first p elements, $y_1, \ldots, y_p$, are obviously relevant.

Example 5.2 We will solve for illustration the same example as in Section 5.1 (Example 5.1, page 181). For convenience of the reader, we will repeat the example here.

Let $X = \{s_1, \ldots, s_5\}$, $U = \{c_1, c_2\}$, $W = \{s_4, s_5\}$, the state transition equations being given by

$$x_{t+1} = \quad
\begin{array}{c|ccccc}
 & x_t = s_1 & s_2 & s_3 & s_4 & s_5 \\
\hline
u_t = c_1 & s_4 & s_3 & s_5 & s_4 & s_5 \\
c_2 & s_2 & s_2 & s_1 & s_4 & s_5
\end{array}$$

the fuzzy constraints are

$$C(s_1) = 0.6/c_1 + 1/c_2$$
$$C(s_2) = 0.8/c_1 + 1/c_2$$
$$C(s_3) = 1/c_1 + 0.7/c_2$$

and the fuzzy goal is

$$G^3 = 1/s_4 + 0.7/s_5$$

The state transition graph corresponding to the state transition equation of the system under control is shown in Figure 24. The nodes represent the particular "real" states $s_1, \ldots, s_5 \in X$, and the auxiliary state s_6 [cf. (5.23), page 183]. With the edges there are associated the respective controls ($\in \{c_1, c_2\}$) and the corresponding values of $\mu_C(u_t \mid x_t)$, except, evidently, for the edges from s_4 and s_5 to s_6.

First, applying (5.22) we obtain

$$\mu_{ij} = \quad
\begin{array}{c|ccccc}
 & j = 1 & 2 & 3 & 4 & 5 \\
\hline
i = 1 & 0 & 1 & 0 & 0.6 & 0 \\
2 & 0 & 1 & 0.8 & 0 & 0 \\
3 & 0.7 & 0 & 0 & 0 & 1 \\
4 & 0 & 0 & 0 & 0 & 0 \\
5 & 0 & 0 & 0 & 0 & 0
\end{array}$$

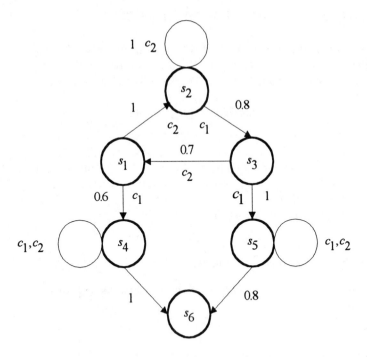

Figure 24   The state transition graph for Example 5.2

and, due to (5.23), $\mu_{4,6} = 1$ and $\mu_{5,6} = 0.8$. Moreover, due to (5.27), $e^T = [0.6, 0, 0.8, 0, 0]$.

Now, by using (5.29), we obtain

$$y = H^* \wedge e = (H^0 \vee H^1 \vee H^2 \vee H^3) \wedge e =$$

$$= \begin{bmatrix} 1 & 1 & 0.8 & 0.6 & 0.8 \\ 0 & 1 & 0.8 & 0 & 0 \\ 0.7 & 0 & 0 & 0 & 1 \\ 0 & 0 & 0 & 0 & 0 \\ 0 & 0 & 0 & 0 & 0 \end{bmatrix} \wedge \begin{bmatrix} 0.6 \\ 0 \\ 0.8 \\ 0 \\ 0 \end{bmatrix} = \begin{bmatrix} 0.8 \\ 0.8 \\ 0.8 \\ 0 \\ 0 \end{bmatrix}$$

Therefore

$$\mu_D(a_{\overline{N}}^* \mid s_1) = 0.8 \quad \mu_D(a_{\overline{N}}^* \mid s_2) = 0.8 \quad \mu_D(a_{\overline{N}}^* \mid s_3) = 0.8$$

which readily implies that optimal stationary policy sought is $a_{\overline{N}}^* = \underbrace{(a^*, \ldots, a^*)}_{\overline{N}}$ such

that

$$a^*(s_1) = c_2 \qquad a^*(s_2) \in \{c_1, c_2\} \qquad a^*(s_3) = c_1$$

while $a^*(s_4)$ and $a^*(s_5)$ are evidently irrelevant since they correspond to the termination states.   □

5.3   A BRANCH-AND-BOUND APPROACH

This approach was proposed in Kacprzyk (1978a, 1978b); see also Kacprzyk (1983b). To present its idea, it is expedient to repeat the problem formulation.

The deterministic system under control is described by the state transition equation (5.1), page 177, i.e.

$$x_{t+1} = f(x_t, u_t), \qquad t = 0, 1 \ldots$$

where $x_t, x_{t+1} \in X = \{s_1, \ldots, s_p, s_{p+1}, \ldots, s_n\}$ and $u_t \in U = \{c_1, \ldots, c_m\}$. The termination set of states is $W = \{s_{p+1}, \ldots, s_n\} \subset X$, and $x_0 \in X \setminus W$.

The control at control stage t, u_t, is subjected to a state-dependent fuzzy constraint $\mu_C(u_t \mid x_t)$, and the fuzzy goal is $\mu_{G^{\overline{N}}}(x_{\overline{N}})$; $\overline{N}$ is a (finite) termination time which is evidently not specified in advance.

First, the fuzzy decision is defined as (5.2), page 178, i.e.

$$\mu_D(u_0, \ldots, u_{\overline{N}-1} \mid x_0) =$$
$$= \mu_C(u_0 \mid x_0) \wedge \ldots \wedge \mu_C(u_{\overline{N}-1} \mid x_{\overline{N}-1}) \wedge \mu_{G^{\overline{N}}}(x_{\overline{N}})$$

The problem is [cf. (5.3), page 178] to find an optimal sequence of controls, $u_0^*, \ldots, u_{\overline{N}-1}^*$, such that

$$\mu_D(u_0^*, \ldots, u_{\overline{N}}^* \mid x_0) =$$
$$= \max_{u_0, \ldots, u_{\overline{N}}} [\mu_C(u_0 \mid x_0) \wedge \ldots \wedge \mu_C(u_{\overline{N}} \mid x_{\overline{N}}) \wedge \mu_{G^{\overline{N}}}(x_{\overline{N}})]$$

where $\overline{N}$ is such that $x_{\overline{N}} \in W$ while $x_0, \ldots, x_{\overline{N}-1} \in X \setminus W$.

It is now easy to see that the fuzzy decision for the problem considered, (5.31), i.e.

$$\mu_D(u_0, \ldots, u_{\overline{N}} \mid x_0) = \mu_C(u_0 \mid x_0) \wedge \ldots \wedge \mu_C(u_{\overline{N}-1} \mid x_{\overline{N}-1}) \wedge \mu_{G^{\overline{N}}}(x_{\overline{N}})$$

satisfies the conditions analogous to (4.17), page 92, for the case of the fixed and specified termination time that make the use of the branch-and-bound procedure possible (cf. Section 4.1.2).

Namely, if

$$\left\{ \begin{array}{l} v_0 = \mu_C(u_0 \mid x_0) \\ \ldots\ldots\ldots\ldots\ldots\ldots\ldots\ldots\ldots\ldots\ldots\ldots\ldots\ldots\ldots\ldots\ldots\ldots\ldots\ldots\ldots\ldots\ldots \\ v_k = \mu_C(u_0 \mid x_0) \wedge \ldots \wedge \mu_C(u_k \mid x_k) = v_{k-1} \wedge \mu_C(u_{k-1} \mid x_{k-1}) \\ \ldots\ldots\ldots\ldots\ldots\ldots\ldots\ldots\ldots\ldots\ldots\ldots\ldots\ldots\ldots\ldots\ldots\ldots\ldots\ldots\ldots\ldots\ldots \\ v_{\overline{N}-1} = \mu_C(u_0 \mid x_0) \wedge \ldots \wedge \mu_C(u_{\overline{N}-1} \mid x_{\overline{N}-1}) = \\ \qquad = v_{\overline{N}-2} \wedge \mu_C(u_{\overline{N}-1} \mid x_{\overline{N}-1}) \\ v_{\overline{N}} = \mu_C(u_0 \mid x_0) \wedge \ldots \wedge \mu_C(u_{\overline{N}-1} \mid x_{\overline{N}-1}) \wedge \mu_{G^{\overline{N}}}(x_{\overline{N}}) = \\ \qquad = \mu_D(u_0, \ldots, u_{\overline{N}} \mid x_0) \end{array} \right. \qquad (5.31)$$

then we evidently have

$$v_k \geq v_w, \qquad \text{for each } x < w \leq \overline{N} - 1 \qquad (5.32)$$

and, in particular,

$$v_k \geq v_{\overline{N}} = \mu_D(u_0, \ldots, u_{\overline{N}} \mid x_0) \qquad (5.33)$$

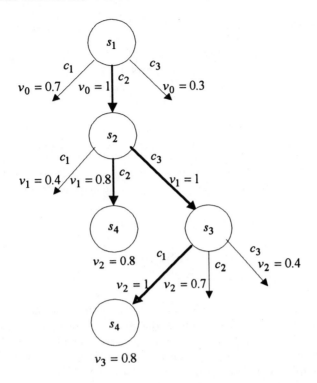

Figure 25 Decision tree with the optimal solutions for Example 5.3

The block diagram of the branch-and-bound procedure for the problem considered is evidently analogous to that for solving the case of a fixed and specified termination time presented in Section 4.1.2 (Figure 18 on page 94). The only relevant difference is that now the termination condition labelled "Block 1" in Figure 18 should now be "$A(\overline{N}+3) \in W?$" instead of "$A(2) = N?$." On the other hand, all the tables and their sizes and formats remain the same, and are given in detail in Section 4.1.2.

Note that by using the branch-and-bound procedure presented we actually obtain as a solution an optimal sequence of control at the consecutive control stages, as opposed to an optimal stationary strategy obtained by using the two previous methods considered in Sections 5.1 and 5.2. This makes no difference for practical purposes, and the branch-and-bound procedure is conceptually and computationally much simpler.

Example 5.3 Suppose that: $X = \{s_1, \ldots, s_5\}$, $U = \{c_1, c_2, c_3\}$, $W = \{s_4, s_5\}$, $x_0 = s_1$, the system under control is described by the following state transition equation:

$$x_{t+1} = \begin{array}{c|ccccc} & x_t = s_1 & s_2 & s_3 & s_4 & s_5 \\ \hline u_t = c_1 & s_1 & s_1 & s_4 & s_4 & s_5 \\ c_2 & s_2 & s_4 & s_3 & s_4 & s_5 \\ c_3 & s_4 & s_3 & s_2 & s_4 & s_5 \end{array}$$

the fuzzy constraints are

$$C(s_1) = 0.7/c_1 + 1/c_2 + 0.3/c_3$$
$$C(s_2) = 0.4/c_1 + 0.8/c_2 + 1/c_3$$
$$C(s_3) = 1/c_1 + 0.7/c_2 + 0.4/c_3$$

and the fuzzy goal is

$$G^{\overline{N}} = 0.8/s_4 + 1/s_4$$

The solution process of this problem is shown in Figure 25. It is easy to see that we obtain the following two optimal solutions (optimal sequences of controls at the consecutive control stages) represented by two paths in the decision tree shown in bold arrows:

$$u_0^* = c_2 \quad u_1^* = c_3 \quad u_2^* = c_1$$

and

$$u_0^* = c_2 \quad u_1^* = c_2$$

and for both of them $\mu_D(. \mid x_0) = 0.8$.   □

This completes our discussion of a relevant and interesting case of an implicitly specified termination time. We have limited the scope of our discussion to the deterministic system under control which is the basic case. The analysis and solution techniques have been in general more sophisticated and difficult than in the case of a fixed and specified termination time considered in Chapter 4.

Concerning the case of a stochastic system under control, one encounters difficulties even while trying to formally state the control problem. On the other hand, the case with a fuzzy system under control may be solved relatively easily by using the approximation by reference fuzzy sets discussed in Sections 4.3.1, 4.3.3 or 7.3.1.

6

Control Processes with a Fuzzy Termination time

In the two previous chapters we discussed formulations of multistage control problems under fuzziness in which a crisp, i.e. nonfuzzy, termination condition was imposed. Namely, we assumed in Chapter 4 a fixed and specified, and in Chapter 5 an implicitly specified termination time. In practice, however, a crisp (nonfuzzy) termination time may often not represent the real perception of the very essence of the planning (control) horizon that might be appropriate for the problem considered. Many examples may here be cited as, e.g., in virtually all (longer term) planning problems even if we state the termination time (planning horizon) as, say, 25 years, then it is tacitly assumed that this is just a rough estimate, and the process should terminate (e.g., with the attainment of some socioeconomic goals) in *more or less* 10 years or, say, in *not much more than* 25 years. On the other hand, in many cases a small increase of the termination time can greatly improve the outcome or performance of the process, while a small decrease may have a negligible effect or none at all.

This all does suggest that it may often be expedient to allow a less crisp, and "softer" definition of the termination time of the control process by allowing its formulation as a fuzzy set such as *more or less 25 years, much less than 10 stages*, etc.

The idea of such a *fuzzy termination time* was proposed in 1977 in Fung and Fu (1977), and Kacprzyk (1977). Though these two approaches were similar conceptually, and with respect to a general problem perception and formulation, they differed in details and solution procedures.

We will now present the idea of a fuzzy termination time and the resulting formulations of the control processes, for the deterministic, stochastic, and fuzzy system under control. We will only sketch Fung and Fu's (1977) approach as it is less relevant for the multistage control processes considered in this book, and consider Kacprzyk's (1977) approach in detail as it is much more consistent with the general line of reasoning adopted here.

First, let us denote the set of all possible control stages by $S = \{0, 1, \ldots, K - 1, K, K + 1, \ldots, N\}$; N is here some fixed, highest possible nonfuzzy termination time (control stage), and $N \leq \infty$. The fuzzy termination time of the control process is now given as a fuzzy set T defined in the set of control stages S, characterized by its membership function $\mu_T : S \longrightarrow [0, 1]$ such that $\mu_T(t) \in [0, 1]$ is viewed as a measure of how preferable $t \in S$ is as the termination time, from $\mu_T(t) = 1$ for the most preferable to $\mu_T(t) = 0$ for the least preferable (unacceptable, or impossible)

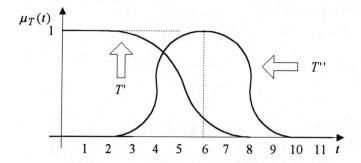

Figure 26   Some examples of a fuzzy termination time

through all intermediate values, and the higher the value the more preferable t as the termination time.

The control process should therefore terminate at some time (control stage) M in the support [cf. (2.13), page 25] of the fuzzy termination time T, i.e. $M \in \operatorname{supp} T = \{t \in S : \mu_T(t) > 0\}$. Moreover, suppose that $\operatorname{supp} T = \{K, K+1, \ldots, N\}$, i.e. that the termination time should occur at "later" control stages. This is a technical assumption which will simplify our discussion but will not limit generality; in fact, such an assumption is quite natural.

Some examples of fuzzy termination times are shown in Figure 26. Notice that T' corresponds to the case of time optimal control as it may be viewed as a requirement to finish the control process as soon as possible, while T'' is more applicable for planning processes in which it is required to attain some goal after, say, "about" n years.

As in the previous cases of fixed and specified and implicit termination times, the fuzzy constraints, $\mu_{C^t}(u_t)$, are imposed on the controls $u_t \in U$ at the consecutive control stages $t = 0, 1, \ldots, N-1$, and the fuzzy goal is imposed on the final state attained, $x_N \in X$, i.e. $\mu_{G^N}(x_N)$; we assume that the fuzzy goal is imposed only on the final state and not on the intermediate states, which will simplify further notation but will not limit the generality of our discussion as fuzzy goals at all intermediate control stages may readily be assumed. The systems under control may be, as in the previous sections, deterministic, stochastic, and fuzzy. The fuzzy decision is also defined analogously as before.

Assuming now for simplicity the basic case of a deterministic system under control, the fuzzy termination time can be introduced into the control problem formulation (i.e. into the fuzzy decision) in the following two basic ways:

- due to Fung and Fu (1977)

$$\mu_D(u_0, \ldots, u_{M-1} \mid x_0) =$$
$$= \mu_{C^0}(u_0) \wedge \ldots \wedge \mu_{C^{M-1}}(u_{M-1}) \wedge \mu_{G^M}(x_M) \wedge \mu_T(M) \quad (6.1)$$

i.e. the fuzzy termination time concerns the whole fuzzy decision, and
- due to Kacprzyk (1977, 1978b, c)

$$\mu_D(u_0, \ldots, u_{M-1} \mid x_0) =$$
$$= \mu_{C^0}(u_0) \wedge \ldots \wedge \mu_{C^{M-1}}(u_{M-1}) \wedge [\mu_T(M) \cdot \mu_{G^M}(x_M)] \quad (6.2)$$

i.e. the fuzzy termination time concerns the fuzzy goal only.

These two problem formulations are similar in respect to both the control problem formulation and its solution, and may lead in many cases to similar results. We will assume in the following the latter, that is (6.2).

For both the above approaches, i.e. (6.1) and (6.2), the control problem is now formulated as to find an *optimal termination time* M^* and an *optimal sequence of controls* $u_0^*, \ldots, u_{M^*-1}^*$ such that

$$\mu_D(u_0^*, \ldots, u_{M^*-1}^* \mid x_0) = \max_{M, u_0, \ldots, u_{M-1}} \mu_D(u_0, \ldots, u_{M-1} \mid x_0) \qquad (6.3)$$

Now, we will consecutively consider the cases of the deterministic, stochastic and fuzzy systems under control, and their respective control problems and solution techniques.

6.1   CONTROL OF A DETERMINISTIC SYSTEM

As always in our prior discussions, the (time-invariant) deterministic system under control is described by the state transition equation (2.130), page 59, i.e.

$$x_{t+1} = f(x_t, u_t), \qquad t = 0, 1, \ldots \qquad (6.4)$$

where $x_t, x_{t+1} \in X = \{s_1, \ldots, s_n\}$ are (nonfuzzy) states at control stages t and $t + 1$, respectively, and $u_t \in U = \{c_1, \ldots, c_m\}$ is a (nonfuzzy) control at time t; $t = 0, 1, \ldots, M$.

At each control stage t, the control u_t is subjected to a fuzzy constraint $\mu_{C^t}(u_t)$, and on the final state to be attained, $x_M \in X$, a fuzzy goal $\mu_{G^M}(x_M)$ is imposed. As we have already mentioned, the fuzzy decision in Kacprzyk's (1977, 1978c) sense, i.e. (6.2), is assumed so that the fuzzy termination time $\mu_T(t)$ concerns the fuzzy goal only.

To simplify further notation, we introduce a *modified fuzzy goal* $\mu_{\overline{G}^M}(x_M)$ given as

$$\mu_{\overline{G}^M}(x_M) = \mu_T(M) \cdot \mu_{G^M}(x_M), \qquad \text{for each } x_M \in X \qquad (6.5)$$

The fuzzy decision (6.2) may now be therefore rewritten as

$$\mu_D(u_0, \ldots, u_{M-1} \mid x_0) =$$
$$= \mu_{C^0}(u_0) \wedge \ldots \wedge \mu_{C^{M-1}}(u_{M-1}) \wedge \mu_{\overline{G}^M}(x_M) \qquad (6.6)$$

and the control problem considered is to find an optimal termination time M^* and an optimal sequence of controls $u_0^*, \ldots, u_{M^*-1}^*$ such that

$$\mu_D(u_0^*, \ldots, u_{M^*-1}^* \mid x_0) =$$
$$= \max_{M, u_0, \ldots, u_{M-1}} [\mu_{C^0}(u_0) \wedge \ldots \wedge \mu_{C^{M-1}}(u_{M-1}) \wedge \mu_{\overline{G}^M}(x_M)] \qquad (6.7)$$

where $x_M \in X$ is determined from x_0 and $u_0, \ldots, u_{M-1} \in U$ through the state transition equation (6.4).

It can readily be noticed that the very structure of this problem is virtually the same as that of problem (4.3), page 86, for the case of a fixed and specified termination time. This implies that the same two basic solution techniques:

- dynamic programming, and
- branch-and-bound

can be used, and they will now be consecutively presented.

6.1.1 Solution by dynamic programming

The first approach to the solution of problem (6.7) was proposed by Kacprzyk (1977, 1978c) and was based on dynamic programming. Stein's (1980) later improvement employed a dynamic programming scheme too. These two approaches will be presented below.

6.1.1.1 Kacprzyk's approach

First, let us recall that $\operatorname{supp} T = \{M\} = \{t \in S : \mu_T(t) > 0\} = \{K, K+1, \ldots, N-1, N\}$, i.e. K is the earliest possible and N is the latest possible termination time. This clearly implies that the sequence of controls $u_0, \ldots, u_{N-1}$ may be partitioned into two parts:

- $u_0, \ldots, u_{K-2}$, i.e. the controls that do not lead to the termination of the control process, and
- $u_{K-1}, \ldots, u_{M-1}$, i.e. the controls that may lead to the termination of the process.

As we can remember from Chapter 4, due to Bellman's optimality principle, in an optimal sequence of controls from the initial to the final control stage, i.e. $u_0^*, \ldots, u_{M^*-1}^*$, its part from the $t = K - 1$ to the final control stage $t = M^* - 1$, i.e. $u_{K-1}^*, \ldots, u_{M^*-1}^*$, must be itself optimal.

The problem considered (6.7) may be therefore rewritten as

$$
\begin{aligned}
\mu_D(u_0^*, &\ldots, u_{M^*-1}^* \mid x_0) = \\
&= \max_{u_0, \ldots, u_{K-2}} \{\mu_{C^0}(u_0) \wedge \ldots \wedge \mu_{C^{K-2}}(u_{K-2}) \wedge \\
&\wedge \max_{M, u_{K-1}, \ldots, u_{M-1}} [\mu_{C^{K-1}}(u_{K-1}) \wedge \ldots \wedge \mu_{C^{M-1}}(u_{M-1}) \wedge \mu_{\overline{G}^M}(x_M)]\} \quad (6.8)
\end{aligned}
$$

Now, if we denote, for $i = 1, \ldots, M - K + 1$,

$$
\begin{aligned}
\overline{\mu}_{G^{M-i}}(x_{M-i}, M) &= \\
&= \max_{u_{M-i}} [\mu_{C^{M-i}}(u_{M-i}) \wedge \overline{\mu}_{G^{M-i+1}}(x_{M-i+1}, M)] \quad (6.9)
\end{aligned}
$$

where $\overline{\mu}_{G^M}(x_M, M) = \mu_{\overline{G}^M}(x_M)$, and perform the dynamic programming backward iterations analogous to (4.5) and (4.6) on page 87, then we will obtain the following set of dynamic programming recurrence equations, analogous to (4.7):

$$
\left\{
\begin{aligned}
&\overline{\mu}_{G^{M-i}}(x_{M-i}, M) = \\
&\qquad = \max_{u_{M-i}} [\mu_{C^{M-i}}(u_{M-i}) \wedge \overline{\mu}_{G^{M-i+1}}(x_{M-i+1}, M)] \\
&x_{M-i+1} = f(x_{M-i}, u_{M-i}) \\
&i = 1, \ldots, M - K + 1; M = K, K+1, \ldots, N
\end{aligned}
\right. \quad (6.10)
$$

By solving this set of recurrence equations we determine $\bar{\mu}_{G^{K-1}}(x_{K-1}, M)$ and its corresponding optimal sequence of controls (or, actually, an optimal sequence of control policies) $u^*_{K-1}, \ldots, u^*_{M-1}$, for all $M \in \{K, K+1, \ldots, N\}$.

Then, the $\mu_{G^{K-1}}(x_{K-1})$ sought, and the corresponding optimal termination time M^* and optimal sequence of controls $u^*_{K-1}, \ldots, u^*_{M^*-1}$ are determined by solving

$$\mu_{G^{K-1}}(x_{K-1}) = \max_M \bar{\mu}_{G^{K-1}}(x_{K-1}, M) \tag{6.11}$$

The first part of the optimal sequence of controls, $u^*_0, \ldots, u^*_{K-2}$, is now found as in the case of the fixed and specified termination time considered in Section 4.1.1.1, i.e. by solving the set of recurrence equations of type (4.7), page 87, i.e.

$$\begin{cases} \mu_{K-1-i}(x_{K-1-i}) = \max_{u_{K-1-i}}[\mu_{C^{K-1-i}}(u_{K-1-i}) \wedge \mu_{G^{K-i}}(x_{K-i})] \\ x_{K-i} = f(x_{K-1-i}, u_{K-1-i}), \qquad i = 1, \ldots, K-1 \end{cases} \tag{6.12}$$

where $\mu_{G^{K-1}}(x_{K-1})$ is determined from (6.11).

Example 6.1 Suppose that: $X = \{s_1, \ldots, s_5\}$, $U = \{c_1, c_2, c_3\}$, $S = \{0, 1, \ldots, 6\}$, the system under control be given by the following state transition equation (table) [cf. (6.4)]:

$$x_{t+1} = \begin{array}{c|ccccc} x_t = s_1 & s_2 & s_3 & s_4 & s_5 \\ \hline u_t = c_1 & s_2 & s_4 & s_1 & s_5 & s_3 \\ c_2 & s_3 & s_5 & s_2 & s_1 & s_4 \\ c_3 & s_5 & s_3 & s_4 & s_2 & s_1 \end{array}$$

the fuzzy constraints are

$$C^0 = 0.5/c_1 + 1/c_2 + 0.7/c_3 \quad C^1 = 0.6/c_1 + 1/c_2 + 0.8/c_3$$
$$C^2 = 1/c_1 + 0.7/c_2 + 0.3/c_3 \quad C^3 = 0.4/c_1 + 0.6/c_2 + 1/c_3$$
$$C^4 = 1/c_1 + 0.7/c_2 + 0.5/c_3 \quad C^5 = 0.7/c_1 + 1/c_2 + 0.7/c_3$$

the fuzzy goal is

$$G^M = 0.2/s_1 + 0.4/s_2 + 0.6/s_3 + 0.8/s_4 + 1/s_5$$

and the fuzzy termination time is

$$T = 0.7/4 + 1/5 + 0.8/6$$

First, we solve the set of dynamic programming recurrence equations (6.10) for the consecutive possible termination times $M \in \operatorname{supp} T = \{4, 5, 7\}$, and obtain the following optimal policies for the particular M's:

- for $M = 6$

$$\begin{array}{lllll} a^*_5(s_1) = c_3 & a^*_5(s_2) = c_2 & a^*_5(s_3) = c_3 & a^*_5(c_1) = c_1 & a^*_5(s_5) = c_2 \\ a^*_4(s_1) = c_1 & a^*_4(s_2) = c_1 & a^*_4(s_3) \in \{c_1, c_2\} & a^*_4(s_4) = c_2 & a^*_4(s_5) = c_2 \\ a^*_3(s_1) = c_3 & a^*_3(s_2) = c_3 & a^*_3(s_3) = c_3 & a^*_3(s_4) = c_3 & a^*_3(s_5) = c_3 \end{array}$$

- for $M = 5$

$$\begin{array}{lllll} a^*_4(s_1) = c_2 & a^*_4(s_2) = c_1 & a^*_4(s_3) = c_3 & a^*_4(s_4) = c_1 & a^*_4(s_5) = c_2 \\ a^*_3(s_1) = c_3 & a^*_3(s_2) = c_2 & a^*_3(s_3) = c_3 & a^*_3(s_4) = c_3 & a^*_3(s_5) \in \{c_2, c_3\} \end{array}$$

- for $M = 4$

$$a_3^*(s_1) = c_3 \quad a_3^*(s_2) = c_2 \quad a_3^*(s_3) = c_3 \quad a_3^*(s_4) = c_2 \quad a_3^*(s_5) = c_2$$

Now, by solving (6.11) we obtain the following values of $\mu_{G^3}(x_3)$ for the particular $x_3 = s_1, \ldots, s_5$, and their corresponding optimal termination times M^*:

$$
\begin{aligned}
\mu_{G^3}(s_1) &= 0.7 & M^* &\in \{4, 5, 6\} \\
\mu_{G^3}(s_2) &= 0.7 & M^* &= 6 \\
\mu_{G^3}(s_3) &= 1 & M^* &= 6 \\
\mu_{G^3}(s_4) &= 0.8 & M^* &= 5 \\
\mu_{G^3}(s_5) &= 0.8 & M^* &= 6
\end{aligned}
$$

with the optimal policies determined as previously for the termination times given above.

Then, from the set of recurrence equations (6.12) we obtain the following optimal policies for $t = 2, 1, 0$:

$$
\begin{aligned}
a_2^*(s_1) &\in \{c_1, c_2\} & a_2^*(s_2) &= c_1 & a_2^*(s_3) &\in \{c_1, c_2\} & a_2^*(s_4) &= c_1 & a_2^*(s_5) &= c_1 \\
a_1^*(s_1) &= c_3 & a_1^*(s_2) &= c_2 & a_1^*(s_3) &\in \{c_2, c_3\} & a_1^*(s_4) &= c_3 & a_1^*(s_5) &= c_2 \\
a_0^*(s_1) &= c_2 & a_0^*(s_2) &= c_2 & a_0^*(s_3) &= c_2 & a_0^*(s_4) &= c_2 & a_0^*(s_5) &= c_2
\end{aligned}
$$

Thus, for the particular initial states $x_0 = s_1, \ldots s_5$, the optimal termination times M^* and the optimal sequences of controls $u_0^*, u_1^*, \ldots, u_{M^*}^*$ are:

$$
\begin{aligned}
\text{for } x_0 = s_1: \quad & M^* = 5 \quad c_2 c_2 c_1 c_3 c_1 \\
& M^* = 6 \quad c_2 c_3 c_1 c_3 c_1 c_2 \\
\text{for } x_0 = s_2: \quad & M^* = 6 \quad c_2 c_2 c_1 c_3 c_1 c_3 \\
\text{for } x_0 = s_3: \quad & M^* = 5 \quad c_2 c_2 c_1 c_3 c_2 \\
\text{for } x_0 = s_4: \quad & M^* = 5 \quad c_2 c_3 c_1 c_3 c_1 \\
\text{for } x_0 = s_5: \quad & M^* = 5 \quad c_2 c_3 c_1 c_3 c_1
\end{aligned}
$$

$\square$

It may readily be seen that many other types of the fuzzy decision given in Section 3.1.2 satisfy the conditions which make possible the use of the dynamic programming recurrence equations similar to (6.10)–(6.12). Namely, to mention only the most relevant types of the fuzzy decision, the sets of dynamic programming equations are:

- for the general t-norm-type fuzzy decision [cf. (3.11), page 72]

$$
\begin{cases}
\overline{\mu}_{G^{M-i}}(x_{M-i}, M) = \\
\quad = \max_{u_{M-i}} [\mu_{C^{M-i}}(u_{M-i}) \, t \, \overline{\mu}_{G^{M-i+1}}(x_{M-i+1}, M)] \\
x_{M-i+1} = f(x_{M-i}, u_{M-i}) \\
i = 1, \ldots, M - K + 1; M = K, K + 1, \ldots, N
\end{cases}
\tag{6.13}
$$

$$
\mu_{G^{K-1}}(x_{K-1}) = \max_M \overline{\mu}_{G^{K-1}}(x_{K-1}, M)
\tag{6.14}
$$

$$
\begin{cases}
\mu_{K-1-i}(x_{K-1-i}) = \\
\quad = \max_{u_{K-1-i}} [\mu_{C^{K-1-i}}(u_{K-1-i}) \, t \, \mu_{G^{K-i}}(x_{K-i})] \\
x_{K-i} = f(x_{K-1-i}, u_{K-1-i}), \qquad i = 1, \ldots, K - 1
\end{cases}
\tag{6.15}
$$

- for the product-type fuzzy decision [cf. (3.12), page 72]

$$\begin{cases} \overline{\mu}_{G^{M-i}}(x_{M-i}, M) = \\ \quad = \max_{u_{M-i}} [\mu_{C^{M-i}}(u_{M-i}) \cdot \overline{\mu}_{G^{M-i+1}}(x_{M-i+1}, M)] \\ x_{M-i+1} = f(x_{M-i}, u_{M-i}) \\ i = 1, \ldots, M - K + 1; M = K, K + 1, \ldots, N \end{cases} \quad (6.16)$$

$$\mu_{G^{K-1}}(x_{K-1}) = \max_M \overline{\mu}_{G^{K-1}}(x_{K-1}, M) \quad (6.17)$$

$$\begin{cases} \mu_{K-1-i}(x_{K-1-i}) = \\ \quad = \max_{u_{K-1-i}} [\mu_{C^{K-1-i}}(u_{K-1-i}) \cdot \mu_{G^{K-i}}(x_{K-i})] \\ x_{K-i} = f(x_{K-1-i}, u_{K-1-i}), \qquad i = 1, \ldots, K - 1 \end{cases} \quad (6.18)$$

- for the weighted-sum-type fuzzy decision [cf. (3.13), page 72]

$$\begin{cases} \overline{\mu}_{G^M}(x_M, M) = r_M \, \mu_{\overline{G}^M}(x_M) \\ \overline{\mu}_{G^{M-i}}(x_{M-i}, M) = \\ \quad = \max_{u_{M-i}} [r_{M-1} \, \mu_{C^{M-i}}(u_{M-i}) + \overline{\mu}_{G^{M-i+1}}(x_{M-i+1}, M)] \\ x_{M-i+1} = f(x_{M-i}, u_{M-i}), \qquad M = K, K + 1, \ldots, N \end{cases} \quad (6.19)$$

where $r_{K-1} + r_K + \cdots + r_M = r(M) < 1$,

$$\mu_{G^{K-1}}(x_{K-1}) = \max_M \overline{\mu}_{G^{K-1}}(x_{K-1}, M) \quad (6.20)$$

$$\begin{cases} \mu_{K-1-i}(x_{K-1-i}) = \\ \quad = \max_{u_{K-1-i}} [r_{K-i-1} \, \mu_{C^{K-1-i}}(u_{K-1-i}) + \mu_{G^{K-i}}(x_{K-i})] \\ x_{K-i} = f(x_{K-1-i}, u_{K-1-i}), \qquad i = 1, \ldots, K - 1 \end{cases} \quad (6.21)$$

where $r_0 + r_1 + \cdots + r_{K-2} = 1 - r(M^*)$;

- for the max-type fuzzy decision [cf. (3.19), page 73]

$$\begin{cases} \overline{\mu}_{G^{M-i}}(x_{M-i}, M) = \\ \quad = \max_{u_{M-i}} [\mu_{C^{M-i}}(u_{M-i}) \vee \overline{\mu}_{G^{M-i+1}}(x_{M-i+1}, M)] \\ x_{M-i+1} = f(x_{M-i}, u_{M-i}) \\ i = 1, \ldots, M - K + 1; M = K, K + 1, \ldots, N \end{cases} \quad (6.22)$$

$$\mu_{G^{K-1}}(x_{K-1}) = \max_M \overline{\mu}_{G^{K-1}}(x_{K-1}, M) \quad (6.23)$$

$$\begin{cases} \mu_{K-1-i}(x_{K-1-i}) = \\ \quad = \max_{u_{K-1-i}} [\mu_{C^{K-1-i}}(u_{K-1-i}) \vee \mu_{G^{K-i}}(x_{K-i})] \\ x_{K-i} = f(x_{K-1-i}, u_{K-1-i}), \qquad i = 1, \ldots, K - 1 \end{cases} \quad (6.24)$$

The general approach presented above, the essence of which is to partition the control sequence into two parts, one starting just before the control stage when the termination time is possible and ending at the final stage, and one starting from the initial control stage $t = 0$ until the first control stage of the former part, is certainly effective as it leads to two sets of dynamic programming recurrence equation, solving the problem considered. It is, however, possible to slightly modify it to increase its efficiency. This will be presented below.

6.1.1.2   Stein's approach

In a later article, Stein (1980) proposed an improved dynamic programming based scheme for solving the problem considered (6.7), page 191, that increased the numerical efficiency, in particular for large supp T, i.e. for the case of a large number of possible termination times.

The very idea of Stein's (1980) approach is based on the following reasoning. If the process under control is at the $(N-i)$-th control stage, $i = 1, \ldots, N$, then we face two possible options.

First, we can immediately stop and attain

$$\mu_{\overline{G}^{N-i}}(x_{N-i}) = \mu_T(N-i) \cdot \mu_{G^{N-i}}(x_{N-i}) \tag{6.25}$$

or, second, we can apply control u_{N-i} and obtain

$$\mu_{C^{N-i}}(u_{N-i}) \wedge \mu_{\overline{G}^{N-i}}(x_{N-i}) \tag{6.26}$$

Evidently, the better [in the sense of whether the value of (6.25) or (6.26) is higher] alternative should be chosen (at this particular control stage).

This line of reasoning leads to the following set of recurrence equations:

$$\left\{ \begin{array}{l} \mu_{G^{N-i}}(x_{N-i}) = \\ \qquad = \mu_{\overline{G}^{N-i}}(x_{N-i}) \vee \max_{u_{N-i}}[\mu_{C^{N-i}}(u_{N-i}) \wedge \mu_{G^{N-i+1}}(x_{N-i+1})] \\ x_{N-i+1} = f(x_{N-i}, u_{N-i}), \qquad i = 1, \ldots, N \end{array} \right. \tag{6.27}$$

The optimal termination time sought, M^*, is here determined by such a control stage $N - i$ at which, in the optimal sequence of controls, the terminating control occurs, i.e. is when during the solution of the set of equations (6.27) the following inequality happens to hold:

$$\mu_{\overline{G}^{N-i}}(x_{N-i}) > \max_{u_{N-i}}[\mu_{C^{N-i}}(u_{N-i}) \wedge \mu_{G^{N-i+1}}(x_{N-i+1})] \tag{6.28}$$

The solution of the set of equations (6.27) usually requires less computational effort than the solution of (6.10)–(6.12), page 192, or similar sets of equations for other types of the fuzzy decision [cf. (6.13)–(6.24)].

For illustration of Stein's (1980) approach, we will solve below the same example as in the case of Kacprzyk's approach (Example 6.1, page 193).

Example 6.2 For convenience of the reader, let us repeat the description of the problem. Suppose that: $X = \{s_1, \ldots, s_5\}$, $U = \{c_1, c_2, c_3\}$, $S = \{0, 1, \ldots, 6\}$, the system under control be given by the following state transition equation (table) [cf. (6.4), page 191]:

$$x_{t+1} =$$

$u_t = c_1$	$x_t = s_1$	s_2	s_3	s_4	s_5
$u_t = c_1$	s_2	s_4	s_1	s_5	s_3
c_2	s_3	s_5	s_2	s_1	s_4
c_3	s_5	s_3	s_4	s_2	s_1

the fuzzy constraints are

$$C^0 = 0.5/c_1 + 1/c_2 + 0.7/c_3 \quad C^1 = 0.6/c_1 + 1/c_2 + 0.8/c_3$$
$$C^2 = 1/c_1 + 0.7/c_2 + 0.3/c_3 \quad C^3 = 0.4/c_1 + 0.6/c_2 + 1/c_3$$
$$C^4 = 1/c_1 + 0.7/c_2 + 0.5/c_3 \quad C^5 = 0.7/c_1 + 1/c_2 + 0.7/c_3$$

the fuzzy goal is

$$G^M = 0.2/s_1 + 0.4/s_2 + 0.6/s_3 + 0.8/s_4 + 1/s_5$$

and the fuzzy termination time is

$$T = 0.7/4 + 1/5 + 0.8/6$$

First, using (6.27) we determine

$$G^5 = 0.7/s_1 + 0.8/s_2 + 0.64/s_3 + 0.8/s_4 + 1/s_5$$
$$G^4 = 0.8/s_1 + 0.8/s_2 + 0.7/s_3 + 1/s_4 + 0.7/s_5$$
$$G^3 = 0.7/s_1 + 0.7/s_2 + 1/s_3 + 0.8/s_4 + 0.8/s_5$$
$$G^2 = 0.7/s_1 + 0.8/s_2 + 0.7/s_3 + 0.8/s_4 + 1/s_5$$
$$G^1 = 0.8/s_1 + 1/s_2 + 0.8/s_3 + 0.8/s_4 + 0.8/s_5$$
$$G^0 = 0.8/s_1 + 0.8/s_2 + 1/s_3 + 0.8/s_4 + 0.8/s_5$$

Hence, the optimal control policies at the particular control stages are:

$$a_5^*(s_1) = c_3 \quad a_5^*(s_2) = c_2 \quad a_5^*(s_3) = c_3) \quad a_5^*(s_4) = c_1 \quad a_5^*(s_5) = c_2$$
$$a_4^*(s_1) = c_1 \quad a_4^*(s_2) = c_1 \quad a_4^*(s_3) \in \{c_1, c_2\} \quad a_4^*(s_4) = c_1 \quad a_4^*(s_5) = c_3$$
$$a_3^*(s_1) = c_3 \quad a_3^*(s_2) = c_3 \quad a_3^*(s_3) = c_3 \quad a_3^*(s_4) = c_3 \quad a_3^*(s_5) = c_3$$
$$a_2^*(s_1) \in \{c_1, c_2\} \quad a_2^*(s_2) = c_1 \quad a_2^*(s_3) \in \{c_1, c_2\} \quad a_2^*(s_4) = c_1 \quad a_2^*(s_5) = c_1$$
$$a_1^*(s_1) = c_3 \quad a_1^*(s_2) = c_2 \quad a_1^*(s_3) \in \{c_2, c_3\} \quad a_1^*(s_4) = c_3 \quad a_1^*(s_5) = c_2$$
$$a_0^*(s_1) = c_2 \quad a_0^*(s_2) = c_2 \quad a_0^*(s_3) = c_2 \quad a_0^*(s_4) = c_2 \quad a_0^*(s_5) = c_2$$

The optimal termination times M^*, i.e. the control stages when while solving (6.27) the inequality (6.28) occurs, and the corresponding optimal sequences of controls, $u_0^*, \ldots, u_{M^*-1}^*$, found from (6.27) are therefore as follows:

$$\text{for } x_0 = s_1 : \quad M^* = 6 \quad c_2 c_3 c_1 c_3 c_1 c_2$$
$$M^* = 5 \quad c_2 c_2 c_1 c_3 c_1$$
$$\text{for } x_0 = s_2 : \quad m^* = 6 \quad c_2 c_2 c_1 c_3 c_1 c_3$$
$$\text{for } x_0 = s_3 : \quad M^* = 5 \quad c_2 c_2 c_1 c_3 c_1$$
$$\text{for } x_0 = s_4 : \quad M^* = 5 \quad c_2 c_3 c_1 c_3 c_1$$
$$\text{for } x_0 = s_5 : \quad M^* = 5 \quad c_2 c_3 c_1 c_3 c_1$$

Thus, the results are the same as those obtained by using Kacprzyk's approach (cf. Example 6.1, page 193) though they have been obtained with less computational effort.   □

6.1.2   Solution by branch-and-bound

Now we will present a different approach to the solution of the problem considered (6.7), page 191, that is based on a branch-and-bound scheme. This approach was proposed by Kacprzyk (1983b), and is similar to the branch-and-bound algorithm for solving the cases of a fixed and specified (cf. Section 4.1.2) and implicit (cf. Section 5.3) termination times.

Namely, it is easy to see that the fuzzy decision [cf. (6.6), page 191]

$$\mu_D(u_0, \ldots, u_{M-1} \mid x_0) = \mu_{C^0}(u_0) \wedge \ldots \wedge \mu_{C^{M-1}}(u_{M-1}) \wedge \mu_{\overline{G}^M}(x_M)$$

satisfies the following conditions analogous to (4.17)–(4.19), page 92, that make the application of branch-and-bound possible: if

$$
\begin{cases}
v_0 = \mu_{C^0}(u_0) \\
\dots\dots\dots\dots\dots\dots\dots\dots\dots\dots\dots\dots\dots\dots \\
v_k = \mu_{C^0}(u_0) \wedge \dots \wedge \mu_{C^k}(u_k) \\
\dots\dots\dots\dots\dots\dots\dots\dots\dots\dots\dots\dots\dots \\
v_M = \mu_{C^0}(u_0) \wedge \dots \wedge \mu_{C^{M-1}}(u_{M-1}) \wedge \mu_{\overline{G}^M}(x_M)
\end{cases}
\tag{6.29}
$$

for each $M \in \{K, K+1, \dots, N\}$, then

$$
v_k \geq v_w \tag{6.30}
$$

and, in particular,

$$
v_k \geq v_M = \mu_D(u_0, \dots, u_{M-1} \mid x_0) \tag{6.31}
$$

for each $k < w < M$.

Therefore, the problem considered (6.7), page 191, i.e. to find an optimal termination time M^* and an optimal sequence of controls $u_0^*, \dots, u_{M^*-1}^*$ such that

$$
\mu_D(u_0^*, \dots, u_{M^*-1}^* \mid x_0) = \max_{M, u_0, \dots, u_{M-1}} \mu_D(u_0, \dots, u_{M-1} \mid x_0)
$$

with $\mu_D(u_0, \dots, u_{M-1} \mid x_0)$ given by (6.6), may be solved by using a branch-and-bound scheme analogous to that presented in Section 4.1.2. Its block diagram is shown in Figure 27. The parameters K and N denote here the earliest and latest possible termination times, respectively, and the meaning of other symbols, format of tables, etc. is the same as for the branch-and-bound scheme in Section 4.1.2 (Figure 18 on page 94).

Example 6.3 For illustration, let us solve the same example as in the case of using dynamic programming (Example 6.1, page 193) whose specification will be repeated here for convenience.

Suppose that: $X = \{s_1, \dots, s_5\}$, $U = \{c_1, c_2, c_3\}$, $S = \{0, 1, \dots, 6\}$, the system under control be given by the following state transition equation (table) [cf. (6.4), page 191]:

$x_{t+1} =$	$u_t = c_1$	$x_t = s_1$	s_2	s_3	s_4	s_5
	$u_t = c_1$	s_2	s_4	s_1	s_5	s_3
	c_2	s_3	s_5	s_2	s_1	s_4
	c_3	s_5	s_3	s_4	s_2	s_1

the fuzzy constraints are

$$
\begin{aligned}
C^0 &= 0.5/c_1 + 1/c_2 + 0.7/c_3 & C^1 &= 0.6/c_1 + 1/c_2 + 0.8/c_3 \\
C^2 &= 1/c_1 + 0.7/c_2 + 0.3/c_3 & C^3 &= 0.4/c_1 + 0.6/c_2 + 1/c_3 \\
C^4 &= 1/c_1 + 0.7/c_2 + 0.5/c_3 & C^5 &= 0.7/c_1 + 1/c_2 + 0.7/c_3
\end{aligned}
$$

the fuzzy goal is

$$
G^M = 0.2/s_1 + 0.4/s_2 + 0.6/s_3 + 0.8/s_4 + 1/s_5
$$

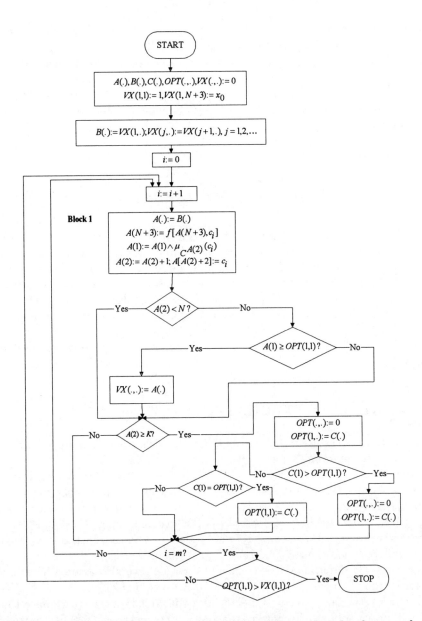

Figure 27 Block diagram of the branch-and-bound procedure for the control
of a deterministic system with a fuzzy termination time

and the fuzzy termination time is

$$T = 0.7/4 + 1/5 + 0.8/6$$

Suppose that the initial state is $x_0 = s_1$. Then, the decision tree is shown as in Figure 28, page 201. Notice that in this case, as opposed to dynamic programming, we need to specify the initial state to start from, and we determine the optimal controls rather than the optimal policies; this is, however, not relevant in practice.

As it can be easily found, and as is shown in the decision tree in Figure 28, the optimal solutions (for the initial state $x_0 = s_1$!) is evidently the same as the one obtained using dynamic programming (cf. Example 6.1), i.e.

$$N^* = 5 \quad c_2 c_2 c_1 c_3 c_1$$
$$M^* = 6 \quad c_2 c_3 c_1 c_3 c_1 c_2$$

The optimal solutions for other initial states $x_0 = s_2, \ldots, s_5$ can be found in a similar way.   □

Finally, let us remark that the branch-and-bound procedure presented above can also be employed for solving the problem considered for other types of fuzzy decisions that satisfy conditions like (6.29)–(6.31), page 198, notably for the product-type fuzzy decision [cf. (3.12), page 72]. The block diagram is in this case the same as in Figure 27 with an obvious replacement of "∧" (for the min-type fuzzy decision) by "·" (for the product-type fuzzy decision), or another operation in case of another type of fuzzy decision.

Now we will proceed to an interesting and challenging case of a stochastic system under control.

6.2   CONTROL OF A STOCHASTIC SYSTEM

In this section we will consider a challenging case of a stochastic system under control operating in a fuzzy environment, and with a fuzzy termination time.

The system under control is assumed to be (4.117), page 138, i.e. a Markov chain whose dynamics (state transitions) are described by a conditional probability $p(x_{t+1} \mid x_t, u_t)$ where $x_t, , x_{t+1} \in X = \{s_1, \ldots, s_n\}$ are states at control stages t and $t + 1$, respectively, and $u_t \in U = \{c_1, \ldots, c_m\}$ is the control at stage t, $t = 0, 1, \ldots$.

At each control stage t, the control u_t is subjected to a fuzzy constraint $\mu_{C^t}(u_t)$, and on the final state x_M a fuzzy goal $\mu_{G^M}(x_M)$ is imposed.

As in the case of a deterministic system under control, we also assume here that the fuzzy termination time, $\mu_T(t)$, defined as a fuzzy set in $S = \{1, \ldots, K-1, K, K+1, \ldots, N\}$, concerns the fuzzy goal only, i.e. the fuzzy decision is of type (6.2), page 190. The modified fuzzy goal, to be entered into the fuzzy decision (6.2) is therefore $\mu_{\overline{G}^M}(x_M) = \mu_T(M) \cdot \mu_{G^M}(x_M)$.

In order to account for randomness in the state transitions, we consider the fuzzy goal to be a fuzzy event in Zadeh's (1968b) sense (cf. Section 2.1.8.1), i.e. the expected value of the membership function of the fuzzy decision is

$$E\mu_{G^M}(x_M) = E\mu_{G^M}(x_M \mid x_{M-1}, u_{M-1}) =$$

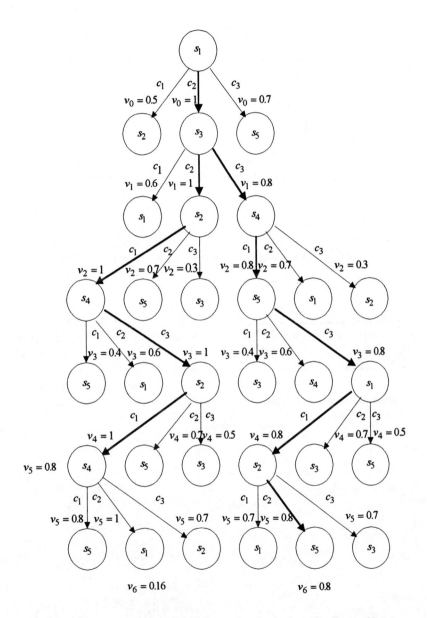

Figure 28 Decision tree with the optimal solutions for Example 6.3

$$= \sum_{x_M \in X} p(x_M \mid x_{M-1}, u_{M-1}) \mu_{G^M}(x_M) \qquad (6.32)$$

so that

$$E\mu_{\overline{G}^M}(x_M) = \mu_T(M) \cdot \mu_{G^M}(x_M) \qquad (6.33)$$

The problem is to find an optimal termination time M^* and an optimal sequence of controls $u_0^*, \ldots, u_{M^*-1}^*$ such that

$$\mu_D(u_0^*, \ldots, u_{M^*-1}^* \mid x_0) =$$
$$= \max_{M, u_0, \ldots, u_{M-1}} [\mu_{C^0}(u_0) \wedge \ldots \wedge \mu_{C^{M-1}}(u_{M-1}) \wedge E\mu_{\overline{G}^M}(x_M)] \qquad (6.34)$$

This problem was first formulated and solved by Kacprzyk (1978b, c), and then Stein (1980) proposed a modification to improve numerical efficiency. Both these approaches are based on dynamic programming, and will be presented below.

Finally, let us make some remark concerning the above problem specifications and formulation. Since the state transitions are random and the membership grades of the particular crisp termination times in the fuzzy termination time are specified in advance, then the results obtained may be regarded as suboptimal (cf. Stein, 1980).

6.2.1 Kacprzyk's approach

The method for solving problem (6.34) proposed by Kacprzyk (1978b, c) is analogous to that given in Section 6.1.1.1 for the case of a deterministic system under control, and also to the method presented in Section 4.2.1 for the stochastic system under control with a fixed and specified termination time.

First, since $\operatorname{supp} T = \{M\} = \{t \in S : \mu_T(t) > 0\} = \{K, K+1, \ldots, N\}$, then the sequence of controls $u_0, \ldots, u_{M-1}$ can evidently be partitioned into two parts:

- $u_0, \ldots, u_{K-2}$, and
- $u_{K-1}, \ldots, u_{M-1}$.

In an optimal sequence of controls $u_0^*, \ldots, u_{M^*-1}^*$, its part $u_{K-1}^*, \ldots, u_{M^*-1}^*$ must obviously be itself optimal, which is due to the principle of optimality, so that (6.34) may be rewritten as

$$\mu_D(u_0^*, \ldots, u_{M^*-1}^* \mid x_0) =$$
$$= \max_{u_0, \ldots, u_{K-2}} \{\mu_{C^0}(u_0) \wedge \ldots \wedge \mu_{C^{K-2}}(u_{K-2}) \wedge$$
$$\wedge \max_{M, u_{K-1}, \ldots, u_{M-1}} [\mu_{C^{K-1}}(u_{K-1}) \wedge \ldots$$
$$\ldots \wedge \mu_{C^{M-1}}(u_{M-1}) \wedge E\mu_{\overline{G}^M}(x_M)]\} \qquad (6.35)$$

It is easy to see that the very structure of (6.35) is virtually the same as that of, e.g., (6.8), page 192, and makes it possible to employ dynamic programming for the solution. Taking into account the partitioning of the sequence of controls into two parts as shown above, and proceeding as in the case of (6.10)–(6.12), we obtain the following sets of dynamic programming recurrence equations:

- for the control stages $K - 1, \ldots, M$

$$
\begin{cases}
\overline{\mu}_{GM-i}(x_{M-i}, M) = \\
\quad = \max_{u_{M-i}}[\mu_{CM-i}(u_{M-i}) \wedge E\overline{\mu}_{GM-i+1}(x_{M-i+1}, M)] \\
E\overline{\mu}_{GM-i+1}(x_{M-i+1}, M) = \\
\quad = \sum_{x_{M-i+1} \in X} p(x_{M-i+1} \mid x_{M-i}, u_{M-i}) \times \\
\qquad \times \overline{\mu}_{GM-i+1}(x_{M-i+1}, M) \\
i = 1, \ldots, M - K + 1; M = K, K+1, \ldots, N
\end{cases}
\tag{6.36}
$$

where $\overline{\mu}_{GM}(x_M, M) = \mu_{\overline{G}M}(x_M)$, and

$$
\mu_{GK-1}(x_{K-1}) = \max_{M \in \{K, \ldots, N\}} \overline{\mu}_{GK-1}(x_{K-1}, M)
\tag{6.37}
$$

which give M^* and $u_K^*, u_{K+1}^*, \ldots, u_{M^*-1}^*$;
- for the control stages $0, 1, \ldots, K - 2$

$$
\begin{cases}
\mu_{GK-i-1}(x_{K-i-1}, M) = \\
\quad = \max_{u_{K-i-1}}[\mu_{CK-i-1}(u_{K-i-1}) \wedge E\overline{\mu}_{GK-i}(x_{K-i})] \\
E\mu_{GK-i}(x_{K-i})] = \\
\quad = \sum_{x_{K-i} \in X} p(x_{K-i} \mid x_{K-i-1}, u_{K-i-1}) \cdot \mu_{GK-i}(x_{K-i}) \\
i = 1, \ldots, K - 1
\end{cases}
\tag{6.38}
$$

which gives $u_0^*, \ldots, u_{K-2}^*$.

Example 6.4 Let: $X = \{s_1, \ldots, s_5\}$, $U = \{c_1, c_1, c_2\}$, $S = \{, 1, \ldots, 6\}$, and the system under control be given by the following conditional probability:

$$
p(x_{t+1} \mid x_t, u_t) =
$$

		$x_{t+1} = s_1$	s_2	s_3	s_4	s_5
$u_t = c_1$	$x_t = s_1$	0.3	0.7	0	0	0
	s_2	0	0	0.2	0.6	0.2
	s_3	0.5	0.3	0.2	0	0
	s_4	0	0	0	0.1	0.9
	s_5	0	0.1	0.8	0.1	0

		$x_{t+1} = s_1$	s_2	s_3	s_4	s_5
$u_t = c_2$	$x_t = s_1$	0	0.4	0.6	0	0
	s_2	0	0	0.2	0.3	0.5
	s_3	0.1	0.7	0.2	0	0
	s_4	0.8	0.1	0.1	0	0
	s_5	0	0	0	0.9	0.1

		$x_{t+1} = s_1$	s_2	s_3	s_4	s_5
$u_t = c_3$	$x_t = s_1$	0	0	0.1	0.3	0.6
	s_2	0	0.2	0.5	0.2	0.1
	s_3	0	0	0.1	0.9	0
	s_4	0.3	0.7	0	0	0
	s_5	0.8	0.2	0	0	0

The fuzzy constraints are

$$C^0 = 0.5/c_1 + 1/c_2 + 0.7/c_3 \qquad C^1 = 0.6/c_1 + 1/c_2 + 0.7/c_3$$
$$C^2 = 1/c_1 + 0.7/c_2 + 0.3/c_3 \qquad C^3 = 0.4/c_1 + 0.6/c_2 + 1/c_3$$
$$C^4 = 1/c_1 + 0.7/c_2 + 0.5/c_3 \qquad C^5 = 0.7/c_1 + 1/c_2 + 0.7/c_3$$

the fuzzy goal is

$$G^M = 0.2/s_1 + 0.4/s_2 + 0.6/s_3 + 0.8/s_4 + 1/s_5$$

and the fuzzy termination time is

$$T = 0.7/4 + 1/5 + 0.8/6$$

First, by solving the set of recurrence equations (6.36) for the consecutive $M \in$ supp $T = \{4, 5, 6\}$ we obtain

- for $M = 6$

$$a_5^*(s_1) = c_3 \quad a_5^*(s_2) = c_2 \quad a_5^*(s_3) = c_3 \quad a_5^*(s_4) = c_2 \quad a_5^*(s_5) = c_2$$
$$a_4^*(s_1) = c_1 \quad a_4^*(s_2) = c_1 \quad a_4^*(s_3) = c_1 \quad a_4^*(s_4) = c_2 \quad a_4^*(s_5) = c_2$$
$$a_3^*(s_1) = c_3 \quad a_3^*(s_2) = c_3 \quad a_3^*(s_3) = c_3 \quad a_3^*(s_4) = c_3 \quad a_3^*(s_5) = c_3$$

- for $M = 5$

$$a_4^*(s_1) = c_2 \quad a_4^*(s_2) = c_1 \quad a_4^*(s_3) = c_3 \quad a_4^*(s_4) = c_1 \quad a_4^*(s_5) = c_2$$
$$a_3^*(s_1) = c_3 \quad a_3^*(s_2) = c_2 \quad a_3^*(s_3) = c_3 \quad a_3^*(s_4) = c_3 \quad a_3^*(s_5) = c_2$$

- for $M = 4$

$$a_3^*(s_1) = c_3 \quad a_3^*(s_2) = c_3 \quad a_3^*(s_3) = c_3 \quad a_3^*(s_4) = c_1 \quad a_3^*(s_5) = c_2$$

Next, via (6.37) we obtain the following values of $\mu_{G^3}(x_3)$ and their corresponding optimal termination times:

$$\mu_{G^3}(s_1) = 0.7640 \qquad M^* = 5$$
$$\mu_{G^3}(s_2) = 0.6836 \qquad M^* = 6$$
$$\mu_{G^3}(s_3) = 0.9320 \qquad M^* = 5$$
$$\mu_{G^3}(s_4) = 0.7160 \qquad M^* = 5$$
$$\mu_{G^3}(s_5) = 0.6884 \qquad M^* = 6$$

Finally, solving the set of recurrence equation (6.38), we obtain the following remaining optimal policies:

$$a_2^*(s_1) = c_1 \quad a_2^*(s_2) = c_1 \quad a_2^*(s_3) = c_1 \quad a_2^*(s_4) = c_2 \quad a_2^*(s_5) = c_1$$
$$a_1^*(s_1) = c_3 \quad a_1^*(s_2) = c_2 \quad a_1^*(s_3) = c_3 \quad a_1^*(s_4) = c_3 \quad a_1^*(s_5) = c_2$$
$$a_0^*(s_1) = c_2 \quad a_0^*(s_2) = c_2 \quad a_0^*(s_3) = c_2 \quad a_0^*(s_4) = c_2 \quad a_0^*(s_5) = c_2$$

$$\square$$

As in the case of a deterministic system under control, here also the min-type fuzzy decision assumed above can be replaced by other types of fuzzy decision, and the resulting dynamic programming recurrence equations are:

- for the general t-norm type fuzzy decision [cf. (3.11), page 72]

$$
\begin{cases}
\overline{\mu}_{GM-i}(x_{M-i}, M) = \\
\quad = \max_{u_{M-i}}[\mu_{CM-i}(u_{M-i})\, t\, E\overline{\mu}_{GM-i+1}(x_{M-i+1}, M)] \\
E\overline{\mu}_{GM-i+1}(x_{M-i+1}, M) = \\
\quad = \sum_{x_{M-i+1}\in X} p(x_{M-i+1} \mid x_{M-i}, u_{M-i})\times \\
\quad\quad \times \overline{\mu}_{GM-i+1}(x_{M-i+1}, M) \\
i = 1, \ldots, M - K + 1; M = K, K + 1, \ldots, N
\end{cases}
\tag{6.39}
$$

$$
\mu_{GK-1}(x_{K-1}) = \max_{M} \overline{\mu}_{GK-1}(x_{K-1}, M)
\tag{6.40}
$$

$$
\begin{cases}
\mu_{K-1-i}(x_{K-1-i}) = \\
\quad = \max_{u_{K-1-i}}[\mu_{CK-1-i}(u_{K-1-i})\, t\, E\mu_{GK-i}(x_{K-i})] \\
E\overline{\mu}_{GK-i}(x_{K-i}, M) = \\
\quad = \sum_{x_{K-i}\in X} p(x_{K-i} \mid x_{K-i-1}, u_{K-i-1})\times \\
\quad\quad \times \overline{\mu}_{GK-i}(x_{K-i}, M) \\
i = 1, \ldots, K - 1
\end{cases}
\tag{6.41}
$$

- for the product-type fuzzy decision [cf. (3.12), page 72]

$$
\begin{cases}
\overline{\mu}_{GM-i}(x_{M-i}, M) = \\
\quad = \max_{u_{M-i}}[\mu_{CM-i}(u_{M-i}) \cdot E\overline{\mu}_{GM-i+1}(x_{M-i+1}, M)] \\
E\overline{\mu}_{GM-i+1}(x_{M-i+1}, M) = \\
\quad = \sum_{x_{M-i+1}\in X} p(x_{M-i+1} \mid x_{M-i}, u_{M-i})\times \\
\quad\quad \times \overline{\mu}_{GM-i+1}(x_{M-i+1}, M) \\
i = 1, \ldots, M - K + 1; M = K, K + 1, \ldots, N
\end{cases}
\tag{6.42}
$$

$$
\mu_{GK-1}(x_{K-1}) = \max_{M} \overline{\mu}_{GK-1}(x_{K-1}, M)
\tag{6.43}
$$

$$
\begin{cases}
\mu_{K-1-i}(x_{K-1-i}) = \\
\quad = \max_{u_{K-1-i}}[\mu_{CK-1-i}(u_{K-1-i}) \cdot E\mu_{GK-i}(x_{K-i})] \\
E\overline{\mu}_{GK-i}(x_{K-i}, M) = \\
\quad = \sum_{x_{K-i}\in X} p(x_{K-i} \mid x_{K-i-1}, u_{K-i-1})\times \\
\quad\quad \times \overline{\mu}_{GK-i}(x_{K-i}, M) \\
i = 1, \ldots, K - 1
\end{cases}
\tag{6.44}
$$

- for the weighted-sum-type fuzzy decision [cf. (3.13), page 72]

$$
\begin{cases}
\overline{\mu}_{GM}(x_M, M) = r_M E\mu_{\overline{G}M}(x_M) \\
\overline{\mu}_{GM-i}(x_{M-i}, M) = \\
\quad = \max_{u_{M-i}}[r_{M-1}\, \mu_{CM-i}(u_{M-i}) + E\overline{\mu}_{GM-i+1}(x_{M-i+1}, M)] \\
E\overline{\mu}_{GM-i+1}(x_{M-i+1}, M) = \\
\quad = \sum_{x_{M-i+1}\in X} p(x_{M-i+1} \mid x_{M-i}, u_{M-i})\times \\
\quad\quad \times \overline{\mu}_{GM-i+1}(x_{M-i+1}, M) \\
i = 1, \ldots, N - K + 1; M = K, K + 1, \ldots, N
\end{cases}
\tag{6.45}
$$

where $r_{K-1} + r_K + \cdots + r_M = r(M) < 1$,

$$
\mu_{GK-1}(x_{K-1}) = \max_{M} \overline{\mu}_{GK-1}(x_{K-1}, M)
\tag{6.46}
$$

$$\begin{cases} \mu_{K-1-i}(x_{K-1-i}) = \\ \quad = \max_{u_{K-1-i}}[r_{K-i-1}\,\mu_{C^{K-1-i}}(u_{K-1-i}) + E\mu_{G^{K-i}}(x_{K-i})] \\ E\overline{\mu}_{G^{K-i}}(x_{K-i}, M) = \\ \quad = \sum_{x_{K-i}\in X} p(x_{K-i} \mid x_{K-i-1}, u_{K-i-1}) \cdot \overline{\mu}_{G^{K-i}}(x_{K-i}, M) \\ i = 1, \ldots, K-1 \end{cases}$$

$$(6.47)$$

where $r_0 + r_1 + \cdots + r_{K-2} = 1 - r(M^*)$;

- for the max-type fuzzy decision [cf. (3.19), page 73]

$$\begin{cases} \overline{\mu}_{G^{M-i}}(x_{M-i}, M) = \\ \quad = \max_{u_{M-i}}[\mu_{C^{M-i}}(u_{M-i}) \vee E\overline{\mu}_{G^{M-i+1}}(x_{M-i+1}, M)] \\ E\overline{\mu}_{G^{M-i+1}}(x_{M-i+1}, M) = \\ \quad = \sum_{x_{M-i+1}\in X} p(x_{M-i+1} \mid x_{M-i}, u_{M-i}) \times \\ \qquad \times \overline{\mu}_{G^{M-i+1}}(x_{M-i+1}, M) \\ i = 1, \ldots, M-K+1; M = K, K+1, \ldots, N \end{cases}$$

$$(6.48)$$

$$\mu_{G^{K-1}}(x_{K-1}) = \max_M \overline{\mu}_{G^{K-1}}(x_{K-1}, M) \qquad (6.49)$$

$$\begin{cases} \mu_{K-1-i}(x_{K-1-i}) = \\ \quad = \max_{u_{K-1-i}}[\mu_{C^{K-1-i}}(u_{K-1-i}) \vee E\mu_{G^{K-i}}(x_{K-i})] \\ E\overline{\mu}_{G^{K-i}}(x_{K-i}, M) = \\ \quad = \sum_{x_{K-i}\in X} p(x_{K-i} \mid x_{K-i-1}, u_{K-i-1}) \times \\ \qquad \times \overline{\mu}_{G^{K-i}}(x_{K-i}, M) \\ i = 1, \ldots, K-1 \end{cases}$$

$$(6.50)$$

6.2.2 Stein's approach

An improved dynamic programming based approach for solving the problem considered (6.35) was also proposed by Stein (1980), similarly as for the case of a deterministic system under control (Section 6.1.1.2). Stein's (1980) approach has increased numerical efficiency, in particular for large supp T, i.e. when the number of possible termination times is high.

The idea of this approach is practically the same as in the case of a deterministic system under control (Section 6.1.1.2), and it is based on the following reasoning.

If the process under control is at the $(N-i)$-th control stage, $i = 1, \ldots, N$, then we face two possibilities. First, we can immediately stop and attain

$$E\mu_{\overline{G}^{N-i}}(x_{N-i}) = \mu_T(N-i) \cdot E\mu_{G^{N-i}}(x_{N-i}) \qquad (6.51)$$

or, second, we can apply control u_{N-i} and obtain

$$\mu_{C^{N-i}}(u_{N-i}) \wedge E\mu_{\overline{G}^{N-i}}(x_{N-i}) \qquad (6.52)$$

Evidently, the better alternative [in the sense of which value is higher: (6.51) or (6.52)] should be chosen.

This line of reasoning leads to the following set of recurrence equations:

$$\begin{cases} \mu_{G^{N-i}}(x_{N-i}) = \mu_{\overline{G}^{N-i}}(x_{N-i}) \vee \\ \quad \vee \max_{u_{N-i}}[\mu_{C^{N-i}}(u_{N-i}) \wedge E\,\mu_{G^{N-i+1}}(x_{N-i+1})] \\ x_{N-i+1} = f(x_{N-i}, u_{N-i}), \qquad i = 1, \ldots, N \end{cases}$$

$$(6.53)$$

with the optimal termination time sought, $M^* \in \operatorname{supp} T$, determined by such a control stage $N - i$ at which, in the optimal sequence of controls, the terminating control occurs, i.e. when during the solution of the set of equations (6.53) the following inequality happens to hold:

$$\mu_{\overline{G}^{N-i}}(x_{N-i}) > \max_{u_{N-i}}[\mu_{C^{N-i}}(u_{N-i}) \wedge E\mu_{G^{N-i+1}}(x_{N-i+1})] \qquad (6.54)$$

Example 6.5 The example is the same as the one used for illustrating Kacprzyk's approach presented in Section 6.2.1, i.e. Example 6.4, page 203. For convenience of the reader, let us repeat its specifications.

Let $X = \{s_1, \ldots, s_5\}$, $U = \{c_1, c_1, c_2\}$, and $S = \{1, \ldots, 6\}$. The system under control is given by the following conditional probability:

$$p(x_{t+1} \mid x_t, u_t) =$$

		$x_{t+1} = s_1$	s_2	s_3	s_4	s_5
$u_t = c_1$	$x_t = s_1$	0.3	0.7	0	0	0
	s_2	0	0	0.2	0.6	0.2
	s_3	0.5	0.3	0.2	0	0
	s_4	0	0	0	0.1	0.9
	s_5	0	0.1	0.8	0.1	0

		$x_{t+1} = s_1$	s_2	s_3	s_4	s_5
$u_t = c_2$	$x_t = s_1$	0	0.4	0.6	0	0
	s_2	0	0	0.2	0.3	0.5
	s_3	0.1	0.7	0.2	0	0
	s_4	0.8	0.1	0.1	0	0
	s_5	0	0	0	0.9	0.1

		$x_{t+1} = s_1$	s_2	s_3	s_4	s_5
$u_t = c_3$	$x_t = s_1$	0	0	0.1	0.3	0.6
	s_2	0	0.2	0.5	0.2	0.1
	s_3	0	0	0.1	0.9	0
	s_4	0.3	0.7	0	0	0
	s_5	0.8	0.2	0	0	0

The fuzzy constraints are

$$C^0 = 0.5/c_1 + 1/c_2 + 0.7/c_3 \qquad C^1 = 0.6/c_1 + 1/c_2 + 0.7/c_3$$
$$C^2 = 1/c_1 + 0.7/c_2 + 0.3/c_3 \qquad C^3 = 0.4/c_1 + 0.6/c_2 + 1/c_3$$
$$C^4 = 1/c_1 + 0.7/c_2 + 0.5/c_3 \qquad C^5 = 0.7/c_1 + 1/c_2 + 0.7/c_3$$

the fuzzy goal is

$$G^M = 0.2/s_1 + 0.4/s_2 + 0.6/s_3 + 0.8/s_4 + 1/s_5$$

and the fuzzy termination time is

$$T = 0.7/4 + 1/5 + 0.8/6$$

First, by solving the set of recurrence equations (6.53) we obtain

$$G^5 = 0.7/s_1 + 0.688/s_2 + 0.624/s_3 + 0.8/s_4 + 1/s_5$$
$$G^4 = 0.6916/s_1 + 0.8048/s_2 + 0.6812/s_3 + 0.98/s_4 + 0.7/s_5$$
$$G^3 = 0.7821/s_1 + 0.7676/s_2 + 0.9501/s_3 + 0.7708/s_4 + 0.7142/s_5$$
$$G^2 = 0.7719/s_1 + 0.7954/s_2 + 0.8114/s_3 + 0.7199/s_4 + 0.9139/s_5$$
$$G^1 = 0.805/s_1 + 0.8352/s_2 + 0.7962/s_3 + 0.7883/s_4 + 0.7766/s_5$$
$$G^0 = 0.8118/s_1 + 0.7841/s_2 + 0.8244/s_3 + 0.8071/s_4 + 0.7872/s_5$$

and the optimal policies at the particular control stages are

$$
\begin{array}{lllll}
a_5^*(s_1) = c_3 & a_5^*(s_2) = c_2 & a_5^*(s_3) = c_3 & a_5^*(s_4) = c_2 & a_5^*(s_5) = c_2 \\
a_4^*(s_1) = c_1 & a_4^*(s_2) = c_1 & a_4^*(s_3) = c_1 & a_4^*(s_4) = c_1 & a_4^*(s_5) = c_2 \\
a_3^*(s_1) = c_3 & a_3^*(s_2) = c_3 & a_3^*(s_3) = c_3 & a_3^*(s_4) = c_3 & a_3^*(s_5) = c_3 \\
a_2^*(s_1) = c_1 & a_2^*(s_2) = c_1 & a_2^*(s_3) = c_1 & a_2^*(s_4) = c_1 & a_2^*(s_5) = c_1 \\
a_1^*(s_1) = c_2 & a_1^*(s_2) = c_2 & a_1^*(s_3) = c_2 & a_1^*(s_4) = c_3 & a_1^*(s_5) = c_3 \\
a_0^*(s_1) = c_2 & a_0^*(s_2) = c_2 & a_0^*(s_3) = c_2 & a_0^*(s_4) = c_2 & a_0^*(s_5) = c_2
\end{array}
$$

Now, using the condition (6.54), we find that if x_5 attained is s_4 or s_5, then we should terminate, i.e. the optimal termination time is $M^* = 5$. Otherwise, i.e. if x_5 is s_1, s_2 or s_3, then we should continue until the end (that is, until the highest possible termination time), so that the optimal termination time is $M^* = 6$. In both these cases we evidently apply the consecutive optimal policies as shown above.   □

The results obtained are different than those obtained using Kacprzyk's approach (Example 6.4, page 203). However, in both the approaches the optimal policies are of type (3.56) [or (3.57)], page 81, i.e. they relate an optimal control to a current state only.

6.2.3 Using fuzzy probability of a fuzzy event

In both the approaches to the control of a probabilistic system in a fuzzy environment presented above in Sections 6.2.1 and 6.2.2, the probability of a fuzzy event in the classical sense of Zadeh (1968b) was assumed, i.e. as a real number from the unit interval (cf. Section 2.1.8.1). Since, as it may readily be noticed, the structure of the problem considered, (6.34), is virtually the same as that of (4.122), page 140, for the fixed and specified termination time, then fuzzy probability of a fuzzy event in the sense of Yager (1979), i.e. as a fuzzy set defined in the unit interval, can also be introduced into problem (6.34) similarly as for problem (4.122).

For convenience of the reader, let us briefly repeat the line of reasoning presented in Section 4.2.1.2.

The stochastic system under control is governed by the conditional probability $p(x_{t+1} \mid x_t, u_t)$, where $x_t, x_{t+1} \in X = \{s_1, \ldots, s_n\}$ are the states at control stages t and $t+1$, respectively, and $u_t \in U = \{c_1, \ldots, c_m\}$ is the control at stage t, $t = 0, 1, \ldots$ [cf. (4.117), page 138].

At each control stage t, u_t is subjected to a fuzzy constraint $\mu_{C^t}(u_t)$ and on the final state x_M a fuzzy goal $\mu_{G^M}(x_M)$ is imposed.

The fuzzy termination time is T defined as a fuzzy set in $S = \{1, \ldots, K - 1, K, K + 1, \ldots, N\}$, and the process should terminate at some termination time $M \in \operatorname{supp} T = \{t \in: \mu_T(t) > 0\} = \{K, K + 1, \ldots, N\}$.

If we view the modified fuzzy goal $\mu_{\overline{G}^M}(x_M) = \mu_T(M) \cdot \mu_{G^M}(x_M)$ as a fuzzy event, then its fuzzy probability – assumed in Yager's (1979) sense (cf. Section 2.1.8.2) – is equal to [cf. (4.129), page 142]

$$E_f \mu_{\overline{G}^M}(x_M) = E_f \mu_{\overline{G}^M}(x_M \mid x_{M-1}, u_{M-1}) = \sum_{\alpha \in (0,1]} \alpha / p(\overline{G}_\alpha^M) \qquad (6.55)$$

where $\overline{G}_\alpha^M$ is the α-cut of $\overline{G}^M$ expressed by (4.127), page 142.

The problem is to find an optimal termination time M^* and an optimal sequence of controls $u_0^*, \ldots, u_{M^*-1}^*$ such that

$$\mu_D(u_0^*, \ldots, u_{M^*-1}^* \mid x_0) =$$
$$= \max_{M, u_0, \ldots, u_{M-1}} [\mu_{C^0}(u_0) \underline{\wedge} \ldots \underline{\wedge} \mu_{C^{M-1}}(u_{M-1}) \underline{\wedge} E_f \mu_{\overline{G}^M}(x_M)] \qquad (6.56)$$

and since for each particular realization of $u_0, \ldots, u_{M-1}$, the $\mu_{C^t}(u_t)$'s assume real values in $[0,1]$, $t = 0, 1, \ldots, M$, but $E_f \mu_{\overline{G}^M}(x_M)$] is a fuzzy set in $[0,1]$, they are evidently directly incompatible, and the normal "$\wedge$" cannot be used directly; hence "$\underline{\wedge}$" is written to denote that some modified "$\wedge$" is to be employed.

Basically, the simplest and most straightforward solution would be to use some "trickery" to be able to express all the terms in the right-hand side of (6.56) as real numbers from the unit interval. A subsuming function F defined by (4.132, page 143, may be used because it equates with a fuzzy set in the unit interval its subsuming ("equivalent") real number (from the unit interval).

Problem (6.56) becomes in such a case to find an optimal termination time M^* and an optimal sequence of controls $u_0^*, \ldots, u_{M^*-1}^*$ such that

$$\mu_D(u_0^*, \ldots, u_{M^*-1}^* \mid x_0) =$$
$$= \max_{M, u_0, \ldots, n_{M-1}} [\mu_{C^0}(u_0) \wedge \ldots \wedge \mu_{C^{M-1}}(u_{M-1}) \wedge F(E_f \mu_{\overline{G}^M}(x_M))] \qquad (6.57)$$

where, obviously, both $F(E_f \mu_{\overline{G}^M}(x_M)) \in [0,1]$ and $\mu_{C^t}(u_t) \in [0,1]$, $t = 0, 1, \ldots, M$, hence "$\wedge$" can be used.

The structure of this problem is evidently virtually the same as of problem (6.34), page 202, hence dynamic programming, and both Kacprzyk's (cf. Section 6.2.1) and Stein's (cf. Section 6.2.2) approach can be used. Since the line of reasoning is the same, for brevity we will only present the respective dynamic programming recurrence equations.

6.2.3.1 Kacprzyk's approach

The fuzzy dynamic programming recurrence equations for this case, which are analogous to (6.36)–(6.38), page 203, are now:

- for the control stages $K - 1, K \ldots, M$

$$\begin{cases} \overline{\mu}_{G^{M-i}}(x_{M-i}, M) = \max_{u_{M-i}} [\mu_{C^{M-i}}(u_{M-i}) \wedge \\ \qquad \wedge F(E_f \overline{\mu}_{G^{M-i+1}}(x_{M-i+1}, M))] \\ E_f \overline{\mu}_{G^{M-i+1}}(x_{M-i+1}, M) = \sum_{\alpha \in (0,1]} \alpha / p(G_\alpha^{M-i+1}) \\ i = 1, \ldots, M - K + 1; M = K, K + 1, \ldots, N \end{cases} \qquad (6.58)$$

where $\bar{\mu}_{GM}(x_M, M) = \mu_{\overline{G}^M}(x_M)$, and

$$\mu_{G^{K-1}}(x_{K-1}) = \max_{M \in \{K, \dots, N\}} \bar{\mu}_{G^{K-1}}(x_{K-1}, M) \qquad (6.59)$$

which gives M^* and $u_K^*, u_{K+1}^*, \dots, u_{M^*-1}^*$; and

• for the control stages $0, 1, \dots, K-2$

$$\begin{cases} \mu_{G^{K-i-1}}(x_{K-i-1}, M) = \max_{u_{K-i-1}}[\mu_{C^{K-i-1}}(u_{K-i-1}) \wedge \\ \qquad \wedge F(E_f \bar{\mu}_{G^{K-i}}(x_{K-i}))] \\ E_f \mu_{G^{K-i}}(x_{K-i}) = \sum_{\alpha \in (0,1]} \alpha / p(G_\alpha^{K-i}) \\ i = 1, \dots, K-1 \end{cases} \qquad (6.60)$$

By employing (6.58) and (6.59) we obtain therefore M^* and $u_{K-1}^*, \dots, u_{M^*-1}^*$, and (6.60) yields $u_0^*, \dots, u_{K-2}^*$; we obtain in fact respective optimal control policies, and they are again in the sense of (3.56), i.e. relate the optimal control at a particular control stage to the current state only.

6.2.3.2 Stein's approach

The recurrence equations in this case – analogous to (6.53), page 206 – are

$$\begin{cases} \mu_{G^{N-i}}(x_{N-i}) = \mu_{\overline{G}^{N-i}}(x_{N-i}) \vee \max_{u_{N-i}}[\mu_{C^{N-i}}(u_{N-i}) \wedge \\ \qquad \wedge F(E_f \mu_{G^{N-i+1}}(x_{N-i+1}))] \\ E_f \mu_{G^{N-i+1}}(x_{N-i+1}) = \sum_{\alpha \in (0,1]} \alpha / p(G_\alpha^{N-i+1}) \\ i = 1, \dots, N \end{cases} \qquad (6.61)$$

where $\mu_{G^N}(x_N) = \mu_{G^M}(x_M)$, and M^* is given by such a control stage $t = N - i \in \{K, K+1, \dots, N-1\}$ when there occurs

$$\mu_T(N-i) \cdot \mu_{G^{N-i}}(x_{N-i}) >$$
$$> \max_{u_{N-i}}[\mu_{C^{N-i}}(u_{N-i}) \wedge F(E_f \mu_{G^{N-i+1}}(x_{N-i+1}))] \qquad (6.62)$$

The results obtained using the above two formulations are in general different. Moreover, they are also different than those obtained using the nonfuzzy probability of a fuzzy event in the sense of Zadeh (1968b) as presented in Sections 6.2.1 and 6.2.2.

6.3   CONTROL OF A FUZZY SYSTEM

The case with a fuzzy system under control can be obviously dealt with by combining the idea of a fuzzy termination time (see the beginning of Chapter 6), and the two approaches to the control of a fuzzy system with a fixed and specified termination time presented in Section 4.3.1 (using dynamic programming) and Section 4.3.2 (using branch-and-bound). Moreover, for improving numerical efficiency, one can use the idea of an interpolative-reasoning-based approach presented in Section 4.3.3.

6.3.1   Solution by dynamic programming

The use of dynamic programming is here an extension of Baldwin and Pilsworth's (1982) approach presented in Section 4.3.1.

The fuzzy system under control is as previously described by a fuzzy state transition equation [cf. (2.143), page 61], i.e.

$$X_{t+1} = F(X_t, U_t), \qquad t = 0, 1, \ldots$$

which is equivalent to a conditioned fuzzy set $\mu_{X_{t+1}}(x_{t+1} \mid x_t, u_t)$, where X_t, X_{t+1} are fuzzy states at control stages t and $t+1$, respectively, defined as fuzzy sets in $X = \{s_1, \ldots, s_n\}$, and U_t is the fuzzy control at control stage t defined as a fuzzy set in $U = \{c_1, \ldots, c_m\}$.

At each control stage t, the fuzzy control U_t is subjected to a fuzzy constraint $\mu_{C^t}(u_t)$, and the final fuzzy state to be attained is subjected to a fuzzy goal $\mu_{G^M}(x_M)$.

The fuzzy termination time is T defined as a fuzzy set in $S = \{1, \ldots, K-1, K, K+1, \ldots, N\}$. The control process is therefore to terminate at some $M \in \operatorname{supp} T = \{K, K+1, \ldots, N\}$.

The line of reasoning that leads to the problem formulation is virtually the same as that for the fixed and specified termination time presented in Section 6.1.1, and will be omitted for brevity.

We arrive at the following problem formulation: find and optimal termination time M^* and an optimal sequence of fuzzy controls $U_0^*, \ldots, U_{M^*-1}^*$ such that

$$\mu_D(u_0^*, \ldots, U_{M^*-1}^* \mid X_0) =$$
$$= \max_{M, U_0, \ldots, U_{M-1}} [\max_{u_0}(\mu_{U_0}(u_0) \wedge \mu_{C^0}(u_0)) \wedge \ldots$$
$$\ldots \max_{u_{M-1}}(\mu_{U_{M-1}}(u_{M-1}) \wedge \mu_{C^{M-1}}(u_{M-1})) \wedge$$
$$\wedge \max_{x_M}(\mu_{X_M}(x_M) \wedge \mu_T(M) \cdot \mu_{G^M}(x_M))] \tag{6.63}$$

The sets of dynamic programming recurrence equations, analogous to (6.10)–(6.12), page 192, are:

$$\begin{cases} \overline{\mu}_{G^M}(X_M, M) = \max_{x_M}[\mu_{X_M}(x_M) \wedge \mu_T(M) \cdot \mu_{G^M}(x_M)] \\ \overline{\mu}_{G^{M-i}}(X_{M-i}, M) = \max_{x_{M-i}}[\max_{u_{M-i}}(\mu_{U_{M-i}}(u_{M-i}) \wedge \\ \qquad \wedge \overline{\mu}_{G^{M-i+1}}(X_{M-i+1}, M))] \\ \mu_{X_{M-i+1}}(x_{M-i+1}) = \max_{x_{M-i}}[\max_{u_{M-i+1}}(\mu_{U_{M-i}}(u_{M-i}) \wedge \\ \qquad \wedge \mu_{X_{M-i+1}}(x_{M-i+1} \mid x_{M-i}, u_{M-i})) \wedge \mu_{X_{M-i}}(x_{M-i}))] \\ i = 1, \ldots, M-K+1; M = K, \ldots, N \end{cases} \tag{6.64}$$

$$\mu_{G^{K-1}}(X_{K-1}) = \max_M \overline{\mu}_{G^{K-1}}(X_{K-1}, M) \tag{6.65}$$

$$\begin{cases} \mu_{G^{K-1-i}}(X_{K-i-1}) = \max_{U_{K-i-1}}[\max_{U_{K-i-1}}(\mu_{U_{K-i-1}}(u_{K-i-1}) \wedge \\ \qquad \wedge \mu_{C^{K-i-1}}(u_{K-i-1})) \wedge \mu_{G^{K-i}}(X_{K-i})] \\ \mu_{X_{K-i}}(x_{K-i}) = \max_{x_{K-i-1}}[\max_{u_{K-i-1}}(\mu_{U_{K-i-1}}(u_{K-i-1}) \wedge \\ \qquad \wedge \mu_{X_{K-i-1}}(x_{K-i} \mid x_{K-i-1}, u_{K-i-1})) \wedge \mu_{X_{K-i-1}}(x_{K-i-1}))] \\ i = 1, \ldots, K-1 \end{cases} \tag{6.66}$$

To solve the above sets of fuzzy dynamic programming recurrence equations we can employ the approach based on the use of reference fuzzy states and controls presented in Section 4.3.1 or an interpolative reasoning based approach presented in Section 4.3.3. Because of lack of space we will not give details on how to apply the above two approaches for the problem considered in this section which is analogous to the above-mentioned solution procedures for a fixed and specified termination time.

6.3.2   Solution by branch-and-bound

This solution method is an extension of Kacprzyk's (1979) branch-and-bound approach presented in Section 4.3.2 for the case of a fuzzy system under control and a fixed and specified termination time.

The fuzzy system under control is again given as the following fuzzy state transition equation [cf. (2.143), page 61]

$$X_{t+1} = F(X_t, u_t), \qquad t = 0, 1, \ldots$$

which is also equivalent to a conditioned fuzzy set $\mu_{X_{t+1}}(x_{t+1} \mid x_t, u_t)$, where X_t, X_{t+1} are fuzzy states (defined in $X = \{s_1, \ldots, s_n\}$) at control stages t and $t+1$, respectively, and $u_t \in U = \{c_1, \ldots, c_m\}$ is a nonfuzzy control at control stage t. We assume here nonfuzzy controls, but the idea of the method may also be applied in the case of fuzzy controls by assuming first some reference fuzzy controls, and then constructing the decision tree (cf. Figure 22, page 165) in Section 4.3.2, with edges associated with those reference fuzzy controls similarly as discussed in Section 4.3.1.

At each control stage t, the control u_t is subjected to a fuzzy constraint $\mu_{C^t}(u_t)$ and on the final fuzzy state X_M a fuzzy goal $\mu_{G^M}(x_M)$ is imposed.

The fuzzy termination time is T defined in $S = \{1, \ldots, K-1, K, K+1, \ldots, N\}$; the process should terminate at some $M \in \operatorname{supp} T = \{K, K+1, \ldots, N\}$.

Since the final state is now fuzzy, then – as already shown in Section 4.3.2 – it cannot directly enter the fuzzy constraint, and we can define a measure of closeness (distance) between the fuzzy goal G^M and the final fuzzy state X_M, denoted by $\mu_{\overline{G}^M}(X_M)$, which will play the role of a fuzzy goal [cf. (4.160), page 153], i.e.

$$\mu_{\overline{G}^M}(X_M) = 1 - d(X_M, G^M) \tag{6.67}$$

where $d(X_M, G^M)$ is a normalized distance between the two fuzzy sets X_M and G^M exemplified by a linear (4.161), page 154, or quadratic (4.162), page 154, distance, or even some other form as

$$\mu_{\overline{G}^M}(X_M) = e(X_M, G^M) \tag{6.68}$$

where $e(X_M, G^M)$ is the degree to which X_M and G^M are equal given by, say, one of (2.7)–(2.10), page 24.

The problem is now to find an optimal termination time M^* and an optimal sequence of controls $u_0^*, \ldots, u_{M^*-1}^*$ such that

$$\mu_D(u_0^*, \ldots, u_{M^*-1}^* \mid X_0) = \max_{M, u_0, \ldots, u_{M-1}} [\mu_{C^0}(u_0) \wedge \ldots$$
$$\ldots \wedge \mu_{C^{M-1}}(u_{M-1}) \wedge \mu_T(M) \cdot \mu_{\overline{G}^M}(X_M)] \tag{6.69}$$

It may readily be seen that the structure of the fuzzy decision, and of the control problem (6.69), immediately implies the satisfaction of conditions similar to (4.18)–(4.19) that make it possible to devise a branch-and-bound algorithm.

Problem (6.69) may therefore be solved by the branch-and-bound procedure of the type presented in Section 4.1.2. Its block diagram is the same as shown in Figure 18 on page 94, with the following obvious replacements: X_M instead of x_M, and $\mu_{\overline{G}^M}(X_M)$ instead of $\mu_{\overline{G}^M}(x_M)$. All the tables (arrays) are the same but the format of their rows is different to account for the fuzziness of the state (the same as shown in Section 4.1.2).

See that, as in the case of the fixed and specified and implicitly given termination times, the branch-and-bound procedure can also be employed in the case of other types of fuzzy decision, notably the product-type fuzzy decision (cf. Section 4.1.2).

As we have already mentioned, the case with fuzzy controls may also be solved by branch-and-bound but the reference fuzzy controls should be used (cf. Section 4.3.1).

This concludes our discussion of the case of a fuzzy termination time. We will proceed now to the last case of an infinite termination time.

7

Control Processes with an Infinite Termination Time

In all the control problems considered in the previous chapters the termination time (planning horizon) was assumed finite, i.e. $t = 0, 1, \ldots, N$, $N < \infty$, no matter whether it was fixed and specified, implicit or fuzzy.

The solution process of the resulting problem formulation required some iterations over the consecutive control stages within the above (finite) time span from 0 to N. This all may be fully justified when the number of control stages is not too high, i.e. when the control process is not too long, and – above all – when the process itself exhibits some reasonably intensive variability over time.

There are, however, many control problems that are assumed to proceed over a very long time span, and vary little, their object being just to maintain the present level of activity. Many examples of such control processes may be found in technological, economic, social, etc. systems as there may be industries that just cannot plan any expansion since the demand for their products is saturated, there may be regions (e.g., rural) that can at best experience stagnation and avoid decay because of, say, the out-migration of the younger population to neighboring urban areas and the resulting aging of the existing population, etc.

In the cases mentioned above, the use of models presented in the previous chapters, with iterations over the consecutive control stages, may evidently be inefficient and a conceptually different approach may be needed. In such an approach, basically, one should assume an *infinite termination time*, and an *optimal stationary strategy* is to be sought. Needless to say that the iterations over the consecutive infinitely numerous control stages should be replaced by some sort of finite iteration scheme, i.e., over a finite number of control stages. This line of reasoning has a long tradition in multistage decision making (and control) under uncertainty (randomness), probably best exemplified by Markov decision processes. These started in about the mid-1950's and the fundamental work by Howard appeared in 1960. This work was followed by a vast literature, and in the case of books the reader is referred to, say, Bertsekas and Shreve (1978) or Puterman (1994).

Basically, the essence of Howard's (1960, 1971) works was the introduction of a *policy iteration technique* that made it possible to solve infinite termination time problems by a finite sequence of iterations.

In the case of multistage decision making (control) under fuzziness, the infinite termination time problem was formulated and solved by Kacprzyk and Staniewski

(1980b, 1982, 1983a), and then extended in, e.g., Kacprzyk, Safteruk and Staniewski (1981a, b); see also Kacprzyk's (1983b) book for a comprehensive analysis. These works concerned all the systems under control (deterministic, stochastic and fuzzy), and showed that the solution process might proceed by employing some sort of a policy iteration type scheme.

The particular problem formulations and solution techniques depend here to a large extent on the type of system under control. We will therefore present below first what might be considered to be a general problem formulation, not related explicitly to any system under control, and then tailor this formulation to each particular system under control.

At each control stage t, $t = 0, 1 \ldots$, the control $u_t \in U = \{c_1, \ldots, c_m\}$ is subjected to a *state-dependent fuzzy constraint* $C(x_t)$ in U characterized by its membership function $\mu_C(u_t \mid x_t)$. The fuzzy goal G in $X = \{s_1, \ldots, s_n\}$, characterized by its membership function $\mu_G(x_t)$, is clearly the same for all the control stages. The state-dependency of the fuzzy constraints and the assumption of the same fuzzy goal stem evidently from the very nature of the infinite termination time problem.

The fuzzy decision is

$$
\begin{aligned}
\mu_D(u_0, u_1, \ldots \mid x_0) = \\
= \quad [\mu_C(u_0 \mid x_o) \wedge \mu_G(x_1)] \wedge [\mu_C(u_1 \mid x_1) \wedge \mu_G(x_2)] \wedge \ldots = \\
= \quad \lim_{N \to \infty} \bigwedge_{t=0}^{N} [\mu_C(u_t \mid x_t) \wedge \mu_G(x_{t+1})]
\end{aligned} \tag{7.1}
$$

It is often expedient, as in nonfuzzy infinite termination time control problems, to introduce a *discount factor* $b > 1$, and then

$$
\begin{aligned}
\mu_D(u_0, u_1, \ldots \mid x_0, b) = \\
= \quad [\mu_C(u_0 \mid x_0) \wedge \mu_G(x_1)] \wedge b[\mu_C(u_1 \mid x_1) \wedge \mu_G(x_2)] \wedge \\
\wedge b^2 [\mu_C(u_2 \mid x_2) \wedge \mu_G(x_3)] \wedge \ldots = \\
= \quad \lim_{N \to \infty} \bigwedge_{t=0}^{N} b^t [\mu_C(u_t \mid x_t) \wedge \mu_G(x_{t+1})]
\end{aligned} \tag{7.2}
$$

The role of the discount factor may be diverse. First, it may be used to express the greater importance of earlier control stages, i.e. of what is to occur in a near than in a distant future. Second, it may in some way subdue the impact of later constraints and goals that are evidently less sure as they concern a distant future. Third, its introduction may be beneficial for the efficiency, or even sometimes the possibility, of solution of the control problem considered.

The introduction of the discount factor $b > 1$ in the way as assumed in (7.2) is very natural as the multiplication by $b^t > 1$, for $b > 1$ and $t > 1$ increases the value of $b^t[\mu_G(x_1) \wedge \mu_C(u_1 \mid x_1) \wedge \mu_G(x_2)]$, hence – due to the "$\wedge$" operation employed – diminishes its impact on the value of $\mu_D(u_0, u_1, \ldots \mid x_0, b)$ in (7.2), which yields the effect intended. For more details, and different forms and aspects of discounting (importances assigned to the particular control stages), see Section 3.1.2.

The problem is now to find an optimal sequence of controls $u_0^*, u_1^*, \ldots$ such that

$$\mu_D(u_0^*, u_1^*, \ldots \mid x_0) = \max_{u_0, u_1, \ldots} \lim_{N \to \infty} \bigwedge_{t=0}^{N} [\mu_C(u_t \mid x_t) \wedge \mu_G(x_{t+1})] \qquad (7.3)$$

and in the case of discounting by $b > 1$

$$\mu_D(u_0^*, u_1^*, \ldots \mid x_0, b) = \max_{u_0, u_1, \ldots} \lim_{N \to \infty} \bigwedge_{t=0}^{N} b^t [\mu_C(u_t \mid x_t) \wedge \mu_G(x_{t+1})] \qquad (7.4)$$

We are in fact interested in the determination of an optimal stationary policy, possibly in the sense of (3.63), page 82, i.e. $a^* : X \to U$, relating an optimal control to the current state only, such that

$$u_t^* = a_t^*(x_t), \qquad t = 0, 1, \ldots$$

If it is not possible to find such a simple optimal stationary policy, then we will seek an optimal stationary policy in the sense of (3.67), page 82, i.e. $a^* : [0,1] \times X \to U$ such that [cf. (3.68)]

$$u_t^* = a^*[w(x_0, u_0, x_1, u_1, \ldots, x_{t-1}, u_{t-1}), x_t], \qquad t = 0, 1, \ldots$$

i.e. relating an optimal control to the current state, x_t, and a "summary" of the past trajectory $x_0, u_0, x_1, u_1, x_{t-1}, u_{t-1}$, $w(x_0, u_0, x_1, u_1, \ldots, x_{t-1}, u_{t-1})$ [cf. (3.69)].

In terms of the stationary policies and stationary strategy problems (7.3) and (7.4) become, respectively, to find an optimal stationary strategy $a_\infty^* = (a^*, a^*, \ldots)$ such that

$$\mu_D(a_\infty^* \mid x_0) =$$

$$= \max_{a_\infty} \mu_D(a_\infty \mid x_0) = \max_{a_\infty} \lim_{N \to \infty} \bigwedge_{t=0}^{N} [\mu_C(a(x_t) \mid x_t) \wedge \mu_G(x_{t+1})] \qquad (7.5)$$

and, in the case of discounting by $b > 1$, to find an optimal stationary strategy $a_\infty^* = (a^*, a^*, \ldots)$ such that

$$\mu_D(a_\infty^* \mid x_0, b) =$$

$$= \max_{a_\infty} \mu_D(a_\infty \mid x_0) = \max_{a_\infty} \lim_{N \to \infty} \bigwedge_{t=0}^{N} b^t [\mu_C(a(x_t) \mid x_t) \wedge \mu_G(x_{t+1})] \qquad (7.6)$$

with the ordering "$\succeq$" between the strategies given by

$$a_\infty' \succeq a_\infty'' \iff \mu_D(a_{\infty}' \mid x_0) \geq \mu_D(a_{\infty}'' \mid x_0) \qquad (7.7)$$

for each $x_0 \in X$, and, evidently, $a_\infty^* \succeq a_\infty$, for each a_∞.

Now we will proceed to the analysis of cases of the particular systems under control.

7.1   CONTROL OF A DETERMINISTIC SYSTEM

The deterministic system under control is given, as in all previous cases, by a state transition equation (2.130), page 59, i.e.

$$x_{t+1} = f(x_t, u_t), \qquad t = 0, 1, \ldots \tag{7.8}$$

where $x_t, x_{t+1} \in X = \{s_1, \ldots, s_n\}$ are the (nonfuzzy) states at control stage t and $t+1$, respectively, and $u_t \in U = \{c_1, \ldots, c_m\}$ is the (nonfuzzy) control at stage t. The initial state is $x_0 \in X$.

At each control stage t, the control u_t is subjected to a state-dependent fuzzy constraint $\mu_C(u_t \mid x_t)$, and on the state x_t a fuzzy goal $\mu_G(x_t)$ is imposed.

Using the fuzzy decision with the discount factor, i.e. (7.2), we obtain

$$\mu_D(a_\infty \mid x_0) = \lim_{N \to \infty} \bigwedge_{t=0}^{N} b^t [\mu_C(a(x_t) \mid x_t) \wedge \mu_G(f(x_t, a(x_t)))] \tag{7.9}$$

and the problem is to find an optimal stationary strategy $a_\infty^* = (a^*, a^*, \ldots)$ such that

$$\mu_D(a_\infty^* \mid x_0) = \max_{a_\infty} \mu_D(a_\infty \mid x_0) =$$

$$= \lim_{N \to \infty} \bigwedge_{t=0}^{N} b^t [\mu_C(a(x_t) \mid x_t) \wedge \mu_G(f(x_t, a(x_t)))] \tag{7.10}$$

As already mentioned, the discount factor $b > 1$ increases the value of further terms (for higher t's) so that, due to "$\wedge$," they have less and less impact on the value of (7.9). Moreover, in this case the discount factor will also simplify the analysis and solution of the problem.

Problem (7.10) was first formulated and solved by Kacprzyk and Staniewski (1980b, 1982, 1983a) [cf. in particular Kacprzyk and Staniewski (1983a)], and we will follow that approach below.

Initially, we assume some arbitrary, not necessarily stationary strategy $A = (a_0, a_1, \ldots)$.

The first step is the determination of a *functional equation* that relates the values of the fuzzy decision from the subsequent control stages on as, e.g., its value from stage t on (i.e. until the end of the process, $N = \infty$) to its value from stage $t + 1$ on (i.e. until $N = \infty$).

Due to the time-invariance of the system under control, the fuzzy decision (7.9) may be written as

$$\mu_D(A \mid x_0) = \mu_C(a_0(x_0) \mid x_0) \wedge$$
$$\wedge \mu_G[f(x_0, a_0(x_0))] \wedge b^t \mu_D[TA \mid f(x_0, a_0(x_0))] \tag{7.11}$$

where $TA = T(a_0, a_1, \ldots) = (a_1, a_2, \ldots)$, i.e. T is a one-stage time shift operator. For a stationary strategy a_∞, we evidently have $Ta_\infty = a_\infty$. This expression (7.11) serves clearly the purpose of a functional equation.

And, more generally, we have

$$\mu_D(T^t A \mid x_t) = \mu_C(a_t(x_t) \mid x_t) \wedge$$
$$\wedge \mu_G[f(x_t, a_t(x_t))] \wedge b \mu_D[T^{t+1} A \mid f(x_t, a_t(x_t))] \tag{7.12}$$

where $T^t A = T^t(a_0, a_1, \ldots) = (a_t, a_{t+1}, \ldots); t = 0, 1, \ldots$.

The functional equation (7.11) should clearly be meant as the following set of equations, for $x_0 = s_1, \ldots, s_n$:

$$
\begin{cases}
\mu_D(A \mid s_1) = \mu_C(a_0(s_1) \mid s_1) \wedge \\
\qquad \wedge \mu_G[f(s_1, a_0(s_1))] \wedge b\, \mu_D[TA \mid f(s_1, a_0(s_1))] \\
\cdots\cdots\cdots\cdots\cdots\cdots\cdots\cdots\cdots\cdots\cdots\cdots\cdots\cdots\cdots \\
\mu_D(A \mid s_n) = \mu_C(a_0(s_n) \mid s_n) \wedge \\
\qquad \wedge \mu_G[f(s_n, a_0(s_n))] \wedge b\, \mu_D[TA \mid f(s_n, a_0(s_n))]
\end{cases}
\tag{7.13}
$$

It is now easy to see that since the sets of states and controls, X and U, are finite, then

$$
\mu_D[TA \mid f(x_0, a_0(x_0))] = H(a_0)\, b\, \mu_D(A \mid x_0)
\tag{7.14}
$$

where $H(a_0)$ is an $m \times n$ 0–1 transition matrix that somehow transforms the values (from a finite set) of $\mu_D(A \mid .)$ into the values (from a finite set too) of $\mu_D(TA \mid .)$.

Now we introduce some operator $Z : [0,1]^n \longrightarrow [0,1]^n$ such that

$$
Z(a_t)w = \mu_C(a_t(x_t) \mid x_t) \wedge \mu_G[f(x_t, a_t(a_t(x_t)))] \wedge H(a_t)bw
\tag{7.15}
$$

So, if we denote $(a, A) = (a, a_1, a_2, \ldots)$, then

$$
\mu_D((a, A) \mid x_0) = Z(a)\mu_D(A \mid x_0)
\tag{7.16}
$$

We have now the following important relationships and properties.

Lemma 7.1 *The operator $Z(a)$ defined by (7.15) is monotone, i.e.*

$$
w_1 \geq w_2 \implies Z(a)w_1 \geq Z(a)w_2
\tag{7.17}
$$

which is evidently a direct consequence of "$\wedge$" used in (7.15).

Among the consequences of Lemma 7.1, the following three important properties will be employed in the following.

Proposition 7.1 *If, for each policy $v : X \longrightarrow U$, there holds*

$$
\mu_D(A^* \mid x_0) \geq \mu_D((v, A^*) \mid x_0)
\tag{7.18}
$$

then the strategy A^* is optimal.

Proof. From the assumptions, i.e. (7.18), and (7.16), we have $Z(a)\mu_D(A^* \mid x_0) \leq \mu_D(A^* \mid x_0)$, for each policy $a : X \longrightarrow U$. Therefore, taking an arbitrary strategy $A = (a_0, a_1, \ldots)$ and assuming $v = a_k$, we obtain $Z(a_k)\mu_D(A^* \mid x_0) \leq \mu_D(A^* \mid x_0)$.

Since, due to Lemma 7.1, $Z(.)$ is monotone, then the superposition $Z(a_0) \cdot \ldots \cdot Z(a_{k-1})$ is obviously monotone too, and hence

$$
[Z(a_0) \cdot \ldots \cdot Z(a_k)]\mu_D(A^* \mid x_0) \leq [Z(a_0) \cdot \ldots \cdot Z(a_{k-1})]\mu_D(A^* \mid x_0)
$$

and, if we apply the above inequality for the consecutive values of k, then we finally obtain

$$
\mu_D(A^* \mid x_0) \geq [Z(a_0) \cdot \ldots \cdot Z(a_k)]\mu_D(A^* \mid x_0) = \mu_D((a_0, \ldots, a_k, A^*) \mid x_0)
$$

for each $k > 0$.

Thus, if $k \longrightarrow \infty$, then we obtain $\mu_D(A^* \mid x_0) \geq \mu_D(A \mid x_0)$, for each A, i.e. the strategy A^* is optimal. □

Proposition 7.2 *If, for any policy a, there holds*

$$\mu_D((a, A) \mid x_0) > \mu_D(A \mid x_0) \qquad (7.19)$$

then

$$\mu_D(a_\infty \mid x_0) > \mu_D(A \mid x_0) \qquad (7.20)$$

Proof. From the assumption (7.19), and from (7.16), we have that $Z(a)\mu_D(A \mid x_0) \geq \mu_D(A \mid x_0)$. Then, in an analogous way as in the proof of Proposition 7.1, we obtain that, for each $k > 0$,

$$\underbrace{[Z(a) \cdot \ldots \cdot Z(a)]}_{k} \mu_D(A \mid x_0) \geq \underbrace{[Z(a) \cdot \ldots \cdot Z(a)]}_{k-1} \mu_D(A \mid x_0)$$

Now, it is easy to see that by mathematical induction we arrive at the following inequality satisfied for each $k > 0$:

$$\underbrace{[Z(a) \cdot \ldots \cdot Z(a)]}_{k} \mu_D(A \mid x_0) \geq \mu_D((a, A) \mid x_0) > \mu_D(A \mid x_0)$$

which, for $k \longrightarrow \infty$, implies $\mu_D(a_\infty \mid x_0) > \mu_D(A \mid x_0)$. $\qquad \square$

The essense of Proposition 7.2 is that if we can only improve a strategy, then a better stationary strategy may be found.

And, finally, we have the main theorem that will provide a basis for an algorithm to determine an optimal stationary strategy sought.

Theorem 7.1 *Denote first by $B(i, a)$, for each $i \in \{1, \ldots, n\}$ and some stationary policy $a : X \longrightarrow U$, the set of all such policies $v : X \longrightarrow U$ that*

$$\mu_D(a_\infty \mid s_i) < $$
$$< \mu_C(v(s_i) \mid s_i) \wedge \mu_G[f(s_i, v(s_i))] \wedge b\, \mu_D[a_\infty \mid f(s_i, a(s_i))] \qquad (7.21)$$

If now:

1. *$B(i, a) = \emptyset$, for each $i \in \{1, \ldots, n\}$, then a_∞ is optimal;*
2. *$B(i, a) \neq \emptyset$, then for any policy $z : X \longrightarrow U$ such that:*

    a. *$z(s_i) \in B(i, a)$, for some $i \in \{1, \ldots, m\}$, and*

    b. *$z(s_i) = a(s_i)$, for all $z(s_i) \notin B(i, a)$,*

    there holds

$$\mu_D(z_\infty \mid x_0) > \mu_D(a_\infty \mid x_0) \qquad (7.22)$$

Proof. From (7.16) and (7.15) we obviously have that

$$\mu_D((z, a_\infty) \mid s_i) = \mu_C(z(s_i) \mid s_i) \wedge \mu_G[f(s_i, z(s_i))] \wedge b\, \mu_D[a_\infty \mid f(s_i, a(s_i))]$$

Now, by the assumption (7.21), the inequality $\mu_D((z, a_\infty) \mid s_i) > \mu_D(a_\infty \mid s_i)$ holds if and only if $z(s_i) \in B(i, a)$, and $\mu_D((z, a_\infty) \mid s_i) = \mu_D(a_\infty \mid s_i)$ if and only if $z(s_i) = a(s_i)$.

Thus, if $B(i, a) = \emptyset$, for each s_i, then $\mu_D(a_\infty \mid x_0) \geq \mu_D((z, a_\infty) \mid x_0)$, for each z. This implies, due to Proposition 7.1, that a_∞ is optimal.

However, if for any policy z satisfying the conditions 2a and 2b of this theorem the inequality $\mu_D((z, a_\infty) \mid x_0) > \mu_D(a_\infty \mid x_0)$ holds, then due to Proposition 7.1 the inequality $\mu_D(z_\infty \mid x_0) > \mu_D(a_\infty \mid x_0)$ holds. $\square$

The essence of Theorem 7.1 is that an arbitrary strategy is either optimal, which corresponds to the case $B(i, a) = \emptyset$, or there exists its improvement, which corresponds to the case $B(i, a) \neq \emptyset$. Since the number of stationary strategies is finite, which is evidently implied by the finiteness of the set of states and controls, X and U, then the following result is an immediate consequence.

Corollary 7.1 *There exists an optimal stationary strategy.*

An optimal stationary strategy, whose existence is ascertained by Corollary 7.1 need not be unique, however. Moreover, see that the optimal stationary strategy is here in the sense of (3.63), page 82, that is, it relates a current optimal control to the current state only.

The properties stated above, in particular Theorem 7.1, may be easily shown to lead to the following policy iteration type algorithm for the determination of an optimal stationary strategy solving the problem considered (7.10), page 218:

Step 1. Assume an arbitrary stationary strategy $a_\infty = (a, a, \ldots)$.

Step 2. Solve in the $\mu_D(a_\infty \mid s_i)$'s the following set of n equations:

$$\begin{cases} \mu_D(a_\infty \mid s_1) = \mu_C(a(s_1) \mid s_1) \wedge \\ \qquad \wedge \mu_G[f(s_1, a(s_1))] \wedge b \, \mu_D[a_\infty \mid f(s_1, a(s_1))] \\ \cdots\cdots\cdots\cdots\cdots\cdots\cdots\cdots\cdots\cdots\cdots\cdots\cdots\cdots\cdots \qquad (7.23) \\ \mu_D(a_\infty \mid s_n) = \mu_C(a(s_n) \mid s_n) \wedge \\ \qquad \wedge \mu_G[f(s_n, a(s_n))] \wedge b \, \mu_D[a_\infty \mid f(s_n, a(s_n))] \end{cases}$$

If the values of $\mu_D(a_\infty \mid s_j)$'s obtained are not unique, as it often happens, then the highest one should be taken. This is clearly implied by the use of "$\wedge$."

Step 3. Improve the strategy, i.e. using the values of $\mu_D(a_\infty \mid s_i)$'s determined in Step 2, find for each $s_i \in X = \{s_1, \ldots, s_n\}$ a maximizing policy $z^* : X \longrightarrow U$ such that

$$\mu_C(z^*(s_i) \mid s_i) \wedge \mu_G[f(s_i, z^*(s_i))] \wedge$$
$$b \, \mu_D[a_\infty \mid f(s_i, a(s_i))] = \max_z \{ \mu_C(z(s_i) \mid s_i) \wedge$$
$$\wedge \mu_G[f(s_i, z(s_i))] \wedge b \, \mu_D[a_\infty \mid f(s_i, a(s_i))] \} \qquad (7.24)$$

Step 4. If the maximizing strategy found, $z_\infty^* = (z^*, z^*, \ldots)$, is the same as the previous one, i.e. if $z_\infty^* = a_\infty$, then it is an optimal stationary strategy to be determined. Otherwise, assume $a_\infty = z_\infty^*$ and return to Step 2.

Example 7.1 Let $X = \{s_1, s_2, s_3\}$, $U = \{c_1, c_2, c_3\}$, $b = 1.01$, the fuzzy constraints being:

$$C(s_1) = 1/c_1 + 0.7/c_2 + 0.3/c_2$$
$$C(s_2) = 0.7/c_1 + 1/c_2 + 0.5/c_3$$
$$C(s_3) = 0.4/c_1 + 0.8/c_2 + 1/c_3$$

the fuzzy goal is

$$G = 0.3/s_1 + 0.7/s_2 + 1/s_3$$

and the state transition table of the system under control is

$$x_{t+1} = \qquad \begin{array}{c|ccc} & x_t = s_1 & s_2 & s_3 \\ \hline u_t = c_1 & s_2 & s_1 & s_2 \\ c_2 & s_1 & s_2 & s_3 \\ c_3 & s_3 & s_2 & s_1 \end{array}$$

The consecutive steps of the algorithm are now as follows:

Step 1. We chose, for instance, $a_\infty = (a, a, \ldots)$ such that $a(s_1) = c_1$, $a(s_2) = c_2$, and $a(s_3) = c_2$.

Step 2. We solve (7.23), i.e.

$$\begin{cases} \mu_D(a_\infty \mid s_1) = 1 \wedge 0.7 \wedge 1.01\,\mu_D(a_\infty \mid s_2) = 0.7 \wedge 1.01\,\mu_D(a_\infty \mid s_2) \\ \mu_D(a_\infty \mid s_2) = 1 \wedge 0.7 \wedge 1.01\,\mu_D(a_\infty \mid s_2) = 0.7 \wedge 1.01\,\mu_D(a_\infty \mid s_2) \\ \mu_D(a_\infty \mid s_3) = 0.4 \wedge 0.7 \wedge 1.01\,\mu_D(a_\infty \mid s_2) = 0.4 \wedge 1.01\,\mu_D(a_\infty \mid s_2) \end{cases}$$

and obtain the solution:

$$\mu_D(a_\infty \mid s_1) = 0.7 \qquad \mu_D(a_\infty \mid s_2) = 0.7 \qquad \mu_D(a_\infty \mid s_1) = 0.3$$

Step 3. By solving (7.24), we obtain: $z^*(s_1) = c_1$, $z^*(s_2) = c_2$, and $z^*(s_3) = c_1$. Since this is different than the initially assumed a_∞, then we take $a_\infty = z_\infty^*$ and return to Step 2.

Step 2. We solve (7.23), i.e.

$$\begin{cases} \mu_D(a_\infty \mid s_1) = 1 \wedge 0.7 \wedge 1.01\,\mu_D(a_\infty \mid s_2) = 0.7 \wedge 1.01\,\mu_D(a_\infty \mid s_2) \\ \mu_D(a_\infty \mid s_2) = 1 \wedge 0.7 \wedge 1.01\,\mu_D(a_\infty \mid s_2) = 0.7 \wedge 1.01\,\mu_D(a_\infty \mid s_2) \\ \mu_D(a_\infty \mid s_3) = 0.4 \wedge 0.7 \wedge 1.01\,\mu_D(a_\infty \mid s_2) = 0.4 \wedge 1.01\,\mu_D(a_\infty \mid s_2) \end{cases}$$

and obtain the solution:

$$\mu_D(a_\infty \mid s_1) = 0.7 \qquad \mu_D(a_\infty \mid s_2) = 0.7 \qquad \mu_D(a_\infty \mid s_1) = 0.3$$

Step 3. Solving (7.24), we obtain: $z^*(s_1) = c_1$, $z^*(s_2) = c_2$ and $z^*(s_3) = c_2$. Since this is different than the previous a_∞, then we assume $a_\infty = z_\infty^*$ and return to Step 2.

Step 2. We solve (7.23), i.e.

$$\begin{cases} \mu_D(a_\infty \mid s_1) = 1 \wedge 0.7 \wedge 1.01 \, \mu_D(a_\infty \mid s_2) = 0.7 \wedge 1.01 \, \mu_D(a_\infty \mid s_2) \\ \mu_D(a_\infty \mid s_2) = 1 \wedge 0.7 \wedge 1.01 \, \mu_D(a_\infty \mid s_2) = 0.7 \wedge 1.01 \, \mu_D(a_\infty \mid s_2) \\ \mu_D(a_\infty \mid s_3) = 0.8 \wedge 0.7 \wedge 1.01 \, \mu_D(a_\infty \mid s_3) = 0.8 \wedge 1.01 \, \mu_D(a_\infty \mid s_3) \end{cases}$$

and obtain the solution:

$$\mu_D(a_\infty \mid s_1) = 0.7 \qquad \mu_D(a_\infty \mid s_2) = 0.7 \qquad \mu_D(a_\infty \mid s_1) = 0.8$$

Step 3. Solving (7.24), we obtain: $z^*(s_1) = c_1$, $z^*(s_2) = c_2$ and $z^*(s_3) = c_2$. Since this is the same policy as the previous one, then z^*_∞ is an optimal stationary strategy to be found, i.e. $a^*_\infty = (a^*, a^*, \ldots)$ where

$$a^*(s_1) = c_1 \qquad a^*(s_2) = c_2 \qquad a^*(s_3) = c_2$$

$\square$

This concludes our discussion of the case of a deterministic system under control operating in the presence of fuzzy constraints and fuzzy goal. We have assumed the min-type fuzzy decision [cf. (3.24), page 74]. As to other types of a fuzzy decision [cf. (3.25)–(3.29), page 74], some information can be found at the end of the next section concerned with the stochastic system under control since the deterministic system is clearly a special case of a stochastic system.

Let us proceed now to the presumably most interesting, difficult and challenging case of a stochastic system under control operating in a fuzzy environment.

7.2   CONTROL OF A STOCHASTIC SYSTEM

We assume now that the stochastic system under control is a Markov chain whose dynamics (state transitions) are governed by the conditional probability (4.117), page 138, $p(x_{t+1} \mid x_t, u_t) \in [0, 1]$, where $x_t, x_{t+1} \in X = \{s_1, \ldots, s_n\}$ are the states at control stage t and $t + 1$, respectively, and $u_t \in U = \{c_1, \ldots c_m\}$ is the control at stage t, $t = 0, 1, \ldots$; $p(x_{t+1} \mid x_t, u_t)$ specifies the probability of attaining state x_{t+1} from state x_t and under control u_t.

At each control stage t, the control u_t is subjected to a state-dependent fuzzy constraint $\mu_C(u_t \mid x_t)$, and on the resulting state x_{t+1} a fuzzy goal $\mu_G(x_{t+1})$ is imposed.

The fuzzy decision is assumed to be without the discount factor [cf. (7.1), page 216], i.e.

$$\begin{aligned} \mu_D(u_0, u_1, \ldots \mid x_0) &= \\ &= [\mu_C(u_0 \mid x_0) \wedge \mu_G(x_1)] \wedge [\mu_C(u_1 \mid x_1) \wedge \mu_G(x_2)] \wedge \ldots = \\ &= \lim_{N \to \infty} \bigwedge_{t=0}^{N} [\mu_C(u_t \mid x_t) \wedge \mu_G(x_{t+1})] \end{aligned} \qquad (7.25)$$

The problem is now to find an optimal sequence of controls $u_0^*, u_1^*, \ldots$ such that

$$
\begin{aligned}
E\mu_D(u_0^*, u_1^*, \ldots \mid x_0) &= \\
&= \max_{u_0, u_1, \ldots} E\mu_D(u_0, u_1, \ldots \mid x_0) = \\
&= \max_{u_0, u_1, \ldots} E\{\lim_{N \longrightarrow \infty} \bigwedge_{t=0}^{N}[\mu_C(u_t \mid x_t) \wedge \mu_G(x_{t+1})]\}
\end{aligned}
\tag{7.26}
$$

where $E\mu_D(. \mid .)$ is the expected value of the membership function of fuzzy decision.

Notice that we apply here Kacprzyk and Staniewski's (1980a) formulation of the control of a stochastic system under fuzziness in which the maximization of the expected value of the membership function of fuzzy decision is involved – see Section 4.2.2. The second, Bellman and Zadeh's (1970) approach, presented in Section 4.2.1, the essence of which is the maximization of the probability of attaining the fuzzy goal under fuzzy constraints loses its relevance somewhat in the case of an infinite termination time.

We are evidently interested in finding an optimal stationary strategy $a_\infty^* = (a^*, a^*, \ldots)$ such that

$$
E\mu_D(a_\infty^* \mid x_0) = \max_{a_\infty} E\mu_D(a_\infty \mid x_0)
\tag{7.27}
$$

with a natural ordering between the stationary strategies, "$\succeq$," given as

$$
a_\infty' \succeq a_\infty'' \iff E\mu_D(a_\infty' \mid x_0) \geq E\mu_D(a'' \mid x_0), \qquad \text{for each } x_0 \in X
\tag{7.28}
$$

and, evidently,

$$
a_\infty^* \succeq a_\infty, \qquad \text{for each } a_\infty
\tag{7.29}
$$

Looking now at the control problem considered (7.27), one can readily recognize that the solution process virtually consists of the following two issues:

- the determination of $E\mu_D(a_\infty \mid x_0)$, and
- the determination of an optimal stationary strategy a_∞^*,

which will now be considered in detail.

7.2.1 Determination of $E\mu_D(a_\infty \mid x_0)$

It is quite natural that the determination of the expected value of the membership function of the min-type fuzzy decision for an infinite number of control stages (being equivalent to an infinite number of terms in the fuzzy decision), and even under a fixed stationary strategy, is not a trivial problem. A solution was proposed by Kacprzyk, Safteruk and Staniewski (1981a, b), and is as follows.

First, denote by $R = [r_{ij}]$, $i, j = 1, \ldots, n$, an $n \times n$ matrix whose elements are equal to

$$
r_{ij} = \mu_C(a(s_i) \mid s_i) \wedge \mu_G(s_j)
\tag{7.30}
$$

where $s_i, s_j \in X$, and $s_j = f(s_i, a(s_i))$, with a being a given stationary policy.

Since X and U are finite, then the r_{ij}'s may evidently only take on a finite number, say M, of distinct values, and they can be – without loss of generality –

ordered as $r^1 < \ldots < r^M$. Therefore, for any realization $x_0, x_1, \ldots$, we obviously have $\mu_D(a_\infty \mid x_0) \in \{r^1, \ldots, r^M\}$ which is clearly implied by "$\wedge$" in (7.30) and (7.26).

Now we will calculate the probability that

$$\mu_D(a_\infty \mid x_0) \leq r^k, \qquad k = 1, \ldots, M \tag{7.31}$$

First, we add to the state space X an additional trapping state s_{n+1} to which all the state transitions with $\mu_D(a_\infty \mid x_0) \leq r^k$ are directed. The new augmented state space, $\overline{X}$, is therefore $\overline{X} = \{s_1, \ldots, s_n, s_{n+1}\}$.

Then, it is easy to derive the augmented transition probability matrix for this new case, i.e. the one governing the state transitions for the states in $\overline{X}$, which is denoted by $Q^{(k)} = [q_{ij}^{(k)}], i, j = 1, \ldots, n+1$.

Let us now calculate

$$\underline{Q}^{(k)} = \lim_{l \to \infty} (Q^{(k)})^l = [\underline{q}_{ij}^{(k)}] \tag{7.32}$$

that is, the elements of this matrix are the probabilities that starting from state $x_0 = s_i \in X$ at least one state transition with $r_{ij} \leq r^k$ has been performed.

We introduce a matrix $B^{(k)} = [b_{ij}^{(k)}], i, j = 1, \ldots, n+1$, defined as

$$B^{(k)} = \underline{Q}^{(k)} - \underline{Q}^{(k-1)}, \qquad k = 1, \ldots, M \tag{7.33}$$

with $\underline{Q}^{(0)} = 0$, by definition. Evidently, $b_{i,n+1}^{(k)}$ is the probability that $\mu_D(a_\infty \mid s_i) = r^k$.

Therefore

$$E\mu_D(a_\infty \mid s_i) = \sum_{k=1}^{M} r^k \cdot b_{i,n+1}^{(k)} \tag{7.34}$$

Now we proceed to the determination of $\underline{Q}^{(k)}$. We assume first that the Markov process considered has no cyclic chains which implies the existence of $\underline{Q}^{(k)}$. On the other hand, if the cyclic chains exist, then they should be removed, for instance by adding auxiliary transitions to some states belonging to the cyclic chains. This does not influence the value of $E\mu_D(a_\infty \mid x_0)$.

Now, suppose that $b_1, \ldots, b_w \neq 1$ are the eigenvalues of $\underline{Q}^{(k)}$. Then, it may be shown that

$$(Q^{(k)})^l = \underline{Q}^{(k)} + b_1^l Q_1 + \cdots + b_w^l Q_w, \qquad \text{for each } l = 1, 2, \ldots \tag{7.35}$$

where $Q_1, \ldots, Q_w$ are matrices whose elements belong to the open interval $(-1, 1)$ and sum up in their rows to zero.

Hence, e.g., for $l = 21$, we obtain

$$\mid (Q^{(k)})^{21} - \underline{Q}^{(k)} \mid \leq b_1^{21} + \cdots + b_w^{21} \tag{7.36}$$

and, on the other hand,

$$tr\underline{Q}^{(21)} = b_1^{21} + \cdots + b_w^{21} + W \tag{7.37}$$

where W is the number of eigenvalues of $\underline{Q}^{21}$ that are equal to 1, and "tr" is the trace of a matrix.

It is therefore sufficient to calculate subsequently $Q^{(2)}, Q^{(4)}, \ldots, Q^{(2s)}$ until the following condition is satisfied:

$$tr\underline{Q}^{(2s)} - W \leq e \tag{7.38}$$

where e is a desired accuracy.

Then, knowing the $\underline{Q}^{(k)}$'s, we may calculate the $B^{(k)}$'s by using (7.33), and finally we may obtain $E\mu_D(a_\infty \mid x_0)$ via (7.34).

The method presented is relatively simple and efficient.

The determination of $E\mu_D(a_\infty \mid x_0)$ is the first, preliminary but indispensable condition to be able to solve the problem considered (7.27). However, the main and most challenging issue is here the determination of an optimal control strategy to be presented below.

7.2.2 Determination of an optimal stationary strategy

For convenience of the reader let us first repeat the formulation of the problem considered (7.26) and (7.27): find an optimal stationary strategy $a_\infty^* = (a^*, a^*, \ldots)$ such that

$$E\mu_D(a_\infty^* \mid x_0) = \max_{a_\infty} E\mu_D(a_\infty \mid x_0) =$$

$$= \max_{a_\infty} E\{ \lim_{N \longrightarrow \infty} \bigwedge_{t=0}^{N} [\mu_C(a(x_t) \mid x_t) \wedge \mu_G(x_{t+1})] \} \tag{7.39}$$

where the state transitions are governed by the conditional probability $p(x_{t+1} \mid x_t, u_t)$; $x_t, x_{t+1} \in X = \{s_1, \ldots, s_m\}$, and $u_t \in U = \{c_1, \ldots, c_m\}$.

The problem is, however, difficult to solve. The algorithm was proposed by Kacprzyk, Safteruk and Staniewski (1981a, b) [see also Kacprzyk (1983a, b) for an extensive discussion]. The idea of the analysis and the solution algorithm will now be presented.

First, we introduce the following notation:

$$h_t = (x_0, u_0, x_1, u_1, \ldots, x_{t-1}, u_{t-1}, x_t), \qquad t = 0, 1, \ldots \tag{7.40}$$

where $h_0 = (x_0)$, by definition, and

$$h = (x_0, u_0, x_1, u_1, x_2, \ldots) \tag{7.41}$$

The h_t represents therefore the trajectory ("history" of the control process) from the very beginning (i.e. $t = 0$) up to the control stage t, while the h represents the "total" trajectory, i.e. the history of the whole control process over an infinite horizon. Moreover, denote by $H_t = \{h_t\}$, for each t, the set of realizations of the trajectory h_t.

In general, we define a *policy* as [cf. (3.58), page 82]

$$a_t : H_t \longrightarrow U, \qquad t = 0, 1, \ldots \tag{7.42}$$

such that

$$u_t = a_t(h_t) = a_t(x_0, u_0, x_1, \ldots, x_{t-1}, u_{t-1}, x_t), \qquad t = 0, 1, \ldots \tag{7.43}$$

where $u_0 = a_0(h_0) = a_0(x_0)$.

Hence, the strategy is $A = (a_0, a_1, \ldots, a_t)$. We will therefore initially deal with the most general type of policy, i.e. which relates the control at the current control stage not only to the current state but also to all the previous states attained and controls applied.

Now, to simplify our further discussion, let us introduce the following notation:

$$r_{t-1} = r(x_{t-1}, a_{t-1}(h_{t-1}), x_t) = \mu_C(a_{t-1}(h_{t-1}) \mid x_{t-1}) \wedge \mu_G(x_t) \tag{7.44}$$

$$R(h_t) = r_0 \wedge \ldots \wedge r_{t-1} \tag{7.45}$$

$$R(h) = r_0 \wedge r_1 \wedge \ldots = \lim_{t \to \infty} R(h_t) \tag{7.46}$$

Let us now denote by $v^T(h_t, A)$ the expected value of $R(h_t)$, for some $T < \infty$, given some h_t and some $A = (a_0, a_1, \ldots, a_t)$, for $t < T$. This is evidently equal to

$$v^T(h_t, A) =$$
$$= \sum_{(x_{t+1}, x_{t+2} \ldots, x_T)} [(R(h_t) \wedge r_t \wedge \ldots \wedge r_{T-1}) \cdot p(x_{t+1} \mid x_t, a(h_t)) \times \cdots$$
$$\cdots \times p(x_T \mid x_{T-1}, a_{T-1}(h_{T-1}))] =$$
$$= \sum_{(x_{t+1}, x_{t+2}, \ldots, x_T)} [(R(h_t) \wedge \bigwedge_{k=t}^{T-1} r_k) \cdot \prod_{q=t}^{T-1} p(x_{q+1} \mid x_q, a_q(h_q))] \tag{7.47}$$

Moreover, let us introduce the following notations:

$$v(h_T, A) = \lim_{T \to \infty} v^T(h_T, A) \tag{7.48}$$

$$f^T(h_t) = \max_A v^T(h_t, A) \tag{7.49}$$

$$f(h_t) = \lim_{T \to \infty} f^T(h_t) \tag{7.50}$$

and, evidently, in (7.49) the maximization over A proceeds only over the set of policies $\{a_t, a_{t+1}, \ldots, a_{T-1}\}$ because the policies $a_0, a_1, \ldots, a_{t-1}$ are fixed as h_t has been assumed to be given.

Since, what is a natural consequence of the "$\wedge$" operation used, the sequences $(v^T(h_t, A))_{t=0,1,\ldots}$ and $(f^T(h_t))_{t=0,1,\ldots}$ are nonincreasing, then the limits in (7.48) and (7.50) exist.

Now, the A^* is called an *optimal strategy* if

$$v(h_t, A^*) = f(h_t), \qquad \text{for each } t = 0, 1, \ldots \tag{7.51}$$

and, needless to say, this is somewhat rigid and strong a definition of an optimal strategy.

As it is common in multistage control and decision making processes, we need for further analysis some *functional equation*. In the case considered it should relate the expected value of $R(h_t)$ [cf. (7.45)], for a given h_t, to the expected value of $R(h_{t+1})$, for a given h_{t+1}. Thus [cf. (7.47)], it should relate $v(h_t, A)$ to $v(h_{t+1}, A)$, under some given strategy A.

The following lemma gives the functional equation sought.

Lemma 7.2 *For each* $t = 0, 1, \ldots,$ h_t *and* A, *there holds*

$$v(h_t, A) = \sum_{x_{t+1}} v(x_{t+1}, A) \cdot p(x_{t+1} \mid x_t, a_t(h_t)) \qquad (7.52)$$

Proof. First, see that due to (7.47) we have, for each $T = 1, 2, \ldots$

$$v^T(h_t, A) = \sum_{x_{t+1}} v^{T+1}(h_{t+1}, A) \cdot p(x_{t+1} \mid x_t, a_t(h_t))$$

and since the summation is over a finite set X, the conditional probability $p(x_{t+1} \mid x_t, a_t(h_t))$ takes on its values in the unit interval $[0, 1]$, and the sequence $(v^T(h_t, A))_{t=0,1,\ldots}$ is convergent (which is a consequence of using "$\wedge$"), then the limit $\lim_{T \to \infty} v^T(h_t, A)$ exists, and by taking it we clearly obtain (7.52). $\qquad \square$

Now we have the following important property that is actually the *optimality condition* for the problem considered.

Theorem 7.2 *For each* t *and* h_t, *there holds*

$$f(h_t) = \max_{a_t} \sum_{x_{t+1}} f(h_{t+1}) \cdot p(x_{t+1} \mid x_t, a_t(h_t)) \qquad (7.53)$$

Proof. Due to the definition of an optimal strategy (7.51), we have $v(h_{t+1}, A) \le f_{t+1}(h_{t+1})$, for each A. From Lemma 7.2 we have

$$f(h_t) = \max_A v(h_t, A) =$$

$$= \max_A \sum_{x_{t+1}} v(h_{t+1}, A) \cdot p(x_{t+1} \mid x_t, a_t(h_t)) \le$$

$$\le \max_{a_t} \sum_{x_{t+1}} f(h_{t+1}) \cdot p(x_{t+1} \mid x_t, a_t(h_t)) \qquad (7.54)$$

On the other hand, since the sets X and U are finite, by assumption, then r_t given by (7.44) may evidently take on only a finite number of distinct values, i.e. $R(h_t)$ and $f(h_t)$ are bounded, for each h_t. The fact that $f(h_t)$ is an upper bound immediately implies that for each fixed t and $e > 0$ there exists a strategy $A' = (a'_0, a'_1, \ldots)$ such that $v(h_{t+1}, A') + a > f(h_{t+1})$, for each h_{t+1}.

Therefore, for each $A = (a_0, a_1, \ldots)$ and h_t we have

$$\sum_{x_{t+1}} f(h_{t+1}) \cdot p(x_{t+1} \mid x_t, a_t(h_t)) <$$

$$< \sum_{x_{t+1}} (v(h_{t+1}, A') + e) \cdot p(x_{t+1} \mid x_t, a_t(h_t)) =$$

$$= v(h_t, A'') + e \le f(h_t) + e$$

where $A'' = (a_0, a_1, \ldots, a_t, a'_{t+1}, a'_{t+2}, \ldots)$.

Since A and h_t are arbitrary, then the following inequality holds:

$$\max_A \sum_{x_{t+1}} f(h_{t+1}) \cdot p(x_{t+1} \mid x_t, a_t(h_t)) \leq \max_A v(h_{t+1}, A) = f(h_t) \qquad (7.55)$$

Therefore, "$\leq$" in (7.54) and "$\geq$" in (7.55) imply evidently (7.53).   □

We introduce now an important concept. Namely, $\hat{A} = (\hat{a}_0, \hat{a}_1, \ldots)$ is called a *conserving* strategy if

$$f(h_t) = \sum_{x_{t+1}} f(h_{t+1}) \cdot p(x_{t+1} \mid x_t, \hat{a}_t(h_t)) \qquad (7.56)$$

The following fact is then immediately implied by Theorem 7.2.

Corollary 7.2 *There always exists a conserving strategy.*

Then, we have the following important theorem.

Theorem 7.3 *Each conserving strategy is optimal.*

Proof. Since the strategy $\hat{A}$ is assumed to be conserving, then for each fixed t and h_t we have

$$v(h_t, \hat{A}) = \lim_{T \longrightarrow \infty} v^T(h_t, \hat{A}) =$$

$$= \lim_{T \longrightarrow \infty} \sum_{(x_{t+1}, x_{t+2}, \ldots, x_T)} [v^T(h_T, \hat{A})_{|h_t - \text{fixed}} \cdot \prod_{i=t}^{T-1} p(x_{i+1} \mid x_i, \hat{a}_i(h_i))] \geq$$

$$\geq \lim_{T \longrightarrow \infty} \sum_{(x_{t+1}, x_{t+2}, \ldots, x_T)} [f(h_T)_{|h_t - \text{fixed}} \cdot \prod_{i=t}^{T-1} p(x_{i+1} \mid x_i, \hat{a}_i(h_i))] = f(h_t)$$

We have therefore $v(h_t, \hat{A}) \geq f(h_t)$, and – on the other hand – $v(h_t, \hat{A}) \leq f(h_t)$, by definition, which implies that $v(h_t, \hat{A}) = f(h_t)$, i.e. that the strategy A is optimal. Since A is assumed to be arbitrary, then each conserving strategy is optimal.   □

An immediate consequence of the above theorem is the following corollary.

Corollary 7.3 *There exists an optimal strategy.*

We need to define now some important concept that will be relevant for our next discussion.

A strategy $\check{A} = (\check{a}_0, \check{a}_1, \ldots)$ is called *unimprovable* if

$$v(h_t, \check{A}) = \max_A \sum_{x_{t+1}} v(h_{t+1}, \check{A}) \cdot p(x_{t+1} \mid x_t, \check{a}_t(h_t)) \qquad (7.57)$$

and, obviously, this should be meant as unimprovable in a *single step* (at a *single control stage*), i.e. by a *single control*.

An unimprovable strategy is therefore the one that satisfies the optimality condition (7.53). Moreover, it is easy to see that each optimal strategy is unimprovable though, unfortunately, the opposite need not hold, in general.

If a strategy $A = (a_0, a_1, \ldots)$ is not unimprovable, then is clear that some strategy $\overline{A} = (\overline{a}_0, \overline{a}_1, \ldots)$ may be devised such that

$$v(h_t, A) \leq \sum_{x_{t+1}} v(h_{t+1}, A) \cdot p(x_{t+1} \mid x_t, \overline{a}_t(h_t)) \qquad (7.58)$$

and this strategy $\overline{A}$ is said to *improve* the strategy A. This is evidently again a single step strategy improvement, i.e. by a single control.

We have now an important result.

Theorem 7.4 *If a strategy $\overline{A}$ improves a strategy A, then*

$$v(h_t, A) \leq v(h_t, \overline{A}) \qquad (7.59)$$

for each t and h_t, and with the strict inequality "<" for at least one t and h_t.

Proof. Suppose that $\overline{A}$ improves A, and fix T. It will now be proved by backward induction that

$$v^T(h_t, \overline{A}) \geq v^T(h_t, A)$$

for all T and h_t, and $t < T$.

First, if $t = T$, then $v^T(h_T, \overline{A}) \geq v^T(h_T, A)$. The induction hypothesis is

$$v^T(h_{t+1}, \overline{A}) \geq v^T(h_{t+1}, A), \qquad \text{for each } h_{t+1}$$

Since it has been assumed that $\overline{A}$ improves A, and by the induction hypothesis, we obtain that

$$v^T(h_t, A) \leq \sum_{x_{t+1}} v^T(h_{t+1}, A) \cdot p(x_{t+1} \mid x_t, \overline{a}_t(h_t)) \leq$$

$$\leq \sum_{x_{t+1}} v^T(h_{t+1}, \overline{A}) \cdot p(x_{t+1} \mid x_t, \overline{a}_t(h_t)) = v^T(h_t, \overline{A})$$

which constitutes the induction step.

Now, for $T \longrightarrow \infty$ we obtain that $v(h_t, \overline{A}) \geq v(h_t, A)$, for each t and h_t, with the strict inequality for at least one t and h_t. $\square$

It should be apparent that the concept of a *strategy improvement* in the sense of (7.58) is of utmost importance. Namely, if an iteration procedure could be devised that determines consecutive strategies $A_0, A_1, \ldots$ such that A_{i+1} improves A_i, for $i = 0, 1, \ldots$, then such a procedure would terminate when an unimprovable strategy would be found, and cycles would be impossible due to Theorem 7.4. Such an unimprovable strategy would certainly be a very good solution though, strictly speaking, an unimprovable strategy need not, unfortunately, be optimal.

So far our discussion has concerned arbitrary strategies. Since, as has already been mentioned, we are in fact interested in stationary strategies, the concept of *stationarity* will be discussed now.

First, denote
$$w(h_t) = (x_t, R(h_t)) \tag{7.60}$$

where $R(h_t)$, given by (7.45), will be called a *summary* of the trajectory h_t, i.e. up to the control stage t.

Both x_t and $R(h_t)$ may evidently take on a finite number of distinct values as both the state and control sets, X and U, are finite; $w(h_t)$ may therefore take on a finite number of values (pairs of values in $X \times [0,1]$) too. Suppose that $w(h_t) \in W = \{y_1, \ldots, y_q\}$, where W is the set of (distinct) values of $w(h_t)$.

A strategy $A = (a_0, a_1, \ldots)$ will be called *stationary*, written $a_\infty = (a, a, \ldots)$, if

$$a_t(h_t) = a_{t'}(h_{t'}) \tag{7.61}$$

for each h_t and $h_{t'}$ such that $w(h_t) = w(h_{t'})$.

The policies given by (7.61) are clearly mappings $a : X \times [0,1] \longrightarrow U$, i.e. they relate the current control to the current state and a summary of the past trajectory, so that the stationarity is in the sense of (3.67), page 82.

From now on, until stated to the contrary, we will consider in Section 7.2 only the strategies and policies that are stationary in the above sense.

We have now the following important property of stationary policies:

Lemma 7.3 *For any stationary strategy a_∞, if $w(h_t) = w(h_{t'})$, then*

$$v(h_t, a_\infty) = v(h_{t'}, a_\infty) \tag{7.62}$$

$$f(h_t) = f(h_{t'}) \tag{7.63}$$

Proof. By the definition of $v(h_T, A)$ given by (7.48), there holds, for any a_∞, that if $w(h_t) = w(h_{t'})$, then $v^T(h_t, a_\infty) = v^T(h_{t'}, a_\infty)$. Thus, as $T \longrightarrow \infty$, we obtain (7.62).

To prove (7.63), suppose first that we have some arbitrary, not necessarily stationary, strategy $A = (a_0, a_1, \ldots)$. Define now a strategy $A' = (a'_0, a'_1, \ldots)$ such that $a'_{t'+k}(h_{t'+k}) = a_{t+k}(h_{t+k}), k = 1, 2, \ldots$, while the a_t's may be different for other h_t's.

By definition, we obtain therefore that $v(h_t, A) = v(h_{t'}, A) = v(h_{t'}, A') \leq f(h_{t'})$.

Since A has been assumed arbitrary, then $f(h_t) \leq f(h_{t'})$, and by a symmetric argument we obtain that $f(h_t) \geq f(h_{t'})$, i.e. $f(h_t) = f(h_{t'})$. $\qquad\square$

Now we have the following important consequences.

Corollary 7.4 *There exists a stationary conserving strategy.*

Proof. Suppose that $A = (a_0, a_1, \ldots)$ is a stationary strategy that obviously exists due to Corollary 7.2. If we now define a strategy $A\prime = (a'_0, a'_1, \ldots)$ such that if $w(h_t) = w(h_1)$, then $a'_t(h_t) = a'_1(h_1)$, for each t, then A' is evidently a stationary and conserving strategy. $\qquad\square$

Moreover, as an immediate consequence of Theorem 7.3 and this corollary, we also obtain the following result.

Corollary 7.5 *There exists an optimal stationary strategy.*

Now, we can summarize the main results obtained so far in the following way:

- an optimal stationary strategy exists,
- the (one-step) strategy improvement gives better and better strategies, and
- an optimal stationary strategy should be sought among equivalent strategies obtained as a result of strategy improvement (i.e. among unimprovable strategies).

These results imply the following policy iteration type (sub)optimization algorithm for solving the problem considered:

Step 1. Construct a new state space, $W = \{y_1, \ldots, y_q\}$ such that $y_i = (s_j, d_k)$, $s_j \in X = \{s_1, \ldots, s_n\}$, $d_k \in \{r : r = r(x_{t-1}, u_{t-1}, x_t) \in [0, 1]\}$, $x_{t-1}, x_t \in X$, and $u_{t-1} \in U = \{c_1, \ldots, c_m\}$.

Step 2. Construct a new conditional probability $p(w_t \mid w_{t-1}, u_{t-1})$, where $w_t, w_{t-1} \in W$ and $u_{t-1} \in U$; these new conditional probabilities govern the state transition among the newly introduced "states" w_t.

Step 3. Choose an arbitrary stationary strategy $a_\infty = (a, a, \ldots)$.

Step 4. Calculate for a_∞ the value of $E\mu_D(a_\infty \mid y_i)$ using the method presented in Section 7.2.1.

Step 5. For each $i = 1, \ldots, q$ find a stationary strategy $z'_\infty = (z'_0, z'_1, \ldots)$ that maximizes with respect to Z the following expression

$$\sum_{k=1}^{q} E\mu_D(z_\infty \mid y_k) \cdot p(y_k \mid y_i, z(y_i)) \tag{7.64}$$

Step 6. If $z'_\infty = a_\infty$, then STOP and assume z'_∞ as the (sub)optimal solution, i.e. $a^*_\infty = z'_\infty$. Otherwise, assume $a_\infty = z'_\infty$ and return to Step 4.

This algorithm yields as a result some unimprovable strategies, in the sense of (7.57), with no cycles. An optimal stationary strategy is among those unimprovable strategies found. If the number of such unimprovable strategies is not too high, we can employ full enumeration to find an optimal one, otherwise some heuristic technique is evidently preferable.

It may also be seen that if in Step 3 we obtain more than one policy maximizing (7.64), then a good practical rule would be to take the one that maximizes $r(w_t, z(w_t), w_{t+1})$.

The algorithm proposed will now be illustrated on the following simple example.

Example 7.2 Suppose that $X = \{s_1, \ldots, s_5\}$, $U = \{c_1, c_2, c_3\}$, and dynamics of the system under control is governed by the following conditional probability:

$$p(x_{t+1} \mid x_t, u_t) =$$

$u_t = c_1$	$x_t = s_1$	$x_{t+1} = s_1$	s_2	s_3	s_4	s_5
	$x_t = s_1$	0.5	0.5	0	0	0
	s_2	0.1	0.3	0.3	0.3	0
	s_3	0	0	1	0	0
	s_4	0	0	0	0	1
	s_5	0	0	0	0	1

$u_t = c_2$	$x_t = s_1$	$x_{t+1} = s_1$	s_2	s_3	s_4	s_5
	$x_t = s_1$	0.1	0.9	0	0	0
	s_2	0	0.3	0.4	0.3	0
	s_3	0	0.3	0.4	0.3	0
	s_4	0	0	1	0	0
	s_5	0	0	0	1	0

$u_t = c_3$	$x_t = s_1$	$x_{t+1} = s_1$	s_2	s_3	s_4	s_5
	$x_t = s_1$	0.5	0.5	0	0	0
	s_2	0	0	0.3	0.4	0.3
	s_3	0	0	1	0	0
	s_4	0	0	0	0	1
	s_5	0	0	0	0	1

Now, instead of listing the particular fuzzy constraints and fuzzy goals in explicit form, let us present them implicitly as follows:

$$r(x_t, u_t, x_{t+1}) = \mu_C(u_t \mid x_t) \wedge \mu_G(x_{t+1}) =$$

$u_t = c_1$	$x_t = s_1$	$x_{t+1} = s_1$	s_2	s_3	s_4	s_5
	$x_t = s_1$	0.1	0.9	0	0	0
	s_2	0.1	1	0.5	0.5	0
	s_3	0	0	0.7	0	0
	s_4	0	0	0	0	1
	s_5	0	0	0	0	1

$u_t = c_2$	$x_t = s_1$	$x_{t+1} = s_1$	s_2	s_3	s_4	s_5
	$x_t = s_1$	0.1	0.9	0	0	0
	s_2	0	0	0.9	0.5	0
	s_3	0	0	0.9	0	0
	s_4	0	0	0	0.4	0.4
	s_5	0	0	0	1	0

$u_t = c_3$	$x_t = s_1$	$x_{t+1} = s_1$	s_2	s_3	s_4	s_5
	$x_t = s_1$	0.8	0.8	0	0	0
	s_2	0	0	0.5	0.5	1
	s_3	0	0	0.3	0	0
	s_4	0	0	0	0.4	0.4
	s_5	0	0	0	0.4	0.4

The set of possible values of $r(x_t, u_t, x_{t+1})$ is therefore $\{0, 0.1, 0.3, 0.4, 0.5, 0.7, 0.8, 0.9, 1\}$, and the sets of possible values of $R(h_t)$ and $R(h)$, given as (7.45) and (7.46), respectively, are evidently the same.

We construct now the new state space $W = \{y\} = \{(x, d)\}$ such that $x \in \{s_1, \ldots, s_5\}$ and $d \in \{0.1, 0.3, 0.4, 0.5, 0.7, 0.8, 0.9, 1\}$. The stationary strategy is therefore $a_\infty = (a(y), a(y), \ldots)$.

The consecutive steps of the algorithm are now as follows (for simplicity we do not consider Steps 1 and 2 which are meant to provide the specification of the problem cosnidered):

Step 3. Initially, we choose, for example, the following stationary strategy:

$a(y) =$

	$d = 0.1$	0.3	0.4	0.5	0.7	0.8	0.9	1
$x = s_1$	c_1	c_1	c_1	c_1	c_1	c_1	c_1	c_1
s_2	c_1	c_1	c_1	c_1	c_1	c_1	c_1	c_1
s_3	c_1	c_1	c_1	c_1	c_1	c_1	c_1	c_1
s_4	c_1	c_1	c_1	c_1	c_1	c_1	c_1	c_1
s_5	c_1	c_1	c_1	c_1	c_1	c_1	c_1	c_1

Step 4. We obtain

$$E\mu_D(a_\infty \mid y) =$$

$=$

	$d = 0.1$	0.3	0.4	0.5	0.7	0.8	0.9	1
$x = s_1$	0.1	0.19	0.23	0.27	0.31	0.31	0.31	0.31
s_2	0.1	0.27	0.36	0.44	0.53	0.53	0.53	0.53
s_3	0.1	0.3	0.4	0.5	0.7	0.7	0.7	0.7
s_4	0.1	0.3	0.4	0.5	0.7	0.8	0.9	1
s_5	0.1	0.3	0.4	0.5	0.7	0.8	0.9	1

Step 5. The new stationary strategy z'_∞ that maximizes (7.64) is

$z'(y) =$

	$d = 0.1$	0.3	0.4	0.5	0.7	0.8	0.9	1
$x = s_1$	c_2	c_2	c_2	c_2	c_2	c_2	c_2	c_2
s_2	c_3	c_3	c_3	c_3	c_3	c_3	c_3	c_1
s_3	c_2	c_2	c_2	c_2	c_2	c_2	c_2	c_2
s_4	c_1	c_1	c_1	c_1	c_1	c_1	c_1	c_1
s_5	c_1	c_1	c_1	c_1	c_1	c_1	c_1	c_1

and since it is not the same as the previous strategy, the we return to Step 4 with $a_\infty = z'_\infty$.

Step 4. We obtain

$$E\mu_D(a_\infty \mid y) =$$

$=$

	$d = 0.1$	0.3	0.4	0.5	0.7	0.8	0.9	1
$x = s_1$	0.1	0.28	0.37	0.46	0.51	0.54	0.57	0.57
s_2	0.1	0.3	0.4	0.5	0.56	0.59	0.62	0.65
s_3	0.1	0.3	0.4	0.5	0.7	0.8	0.9	0.9
s_4	0.1	0.3	0.4	0.5	0.7	0.8	0.9	1
s_5	0.1	0.3	0.4	0.5	0.7	0.8	0.9	1

Step 5. The new z'_∞ that maximizes (7.64) is now

$$z'(y) = $$

	$d = 0.1$	0.3	0.4	0.5	0.7	0.8	0.9	1
$x = s_1$	c_2	c_2	c_2	c_2	c_2	c_2	c_2	c_2
s_2	c_3	c_3	c_3	c_3	c_3	c_3	c_3	c_1
s_3	c_2	c_2	c_2	c_2	c_2	c_2	c_2	c_2
s_4	c_1	c_1	c_1	c_1	c_1	c_1	c_1	c_1
s_5	c_1	c_1	c_1	c_1	c_1	c_1	c_1	c_1

and since this z'_∞ is the same as the previous one, then it is an optimal stationary strategy to be found, i.e. $a^*_\infty = z'_\infty$.

In this simple example the full enumeration is possible and it can be seen that this strategy is indeed the optimal solution.

$\square$

To conclude the discussion of a stochastic system under control, we will add some remarks on other types of the fuzzy decision [cf. (3.25)–(3.29), page 74]. For the weighted-sum-type fuzzy decision the problem amounts to the basic formulation in the stochastic setting (i.e. under uncertainty). Numerous algorithms have been proposed, and we refer the reader to, e.g., Howard (1960, 1971), Bertsekas and Shreve (1978) or Puterman (1994). The case of the product-type fuzzy decision (performance function) received much less attention in the literature, but here also some algorithms exist. The case of a maximum type performance function is evidently the most difficult one. In the stochastic setting it was considered by Kreps (1977a, b). His basic line of reasoning is very similar to that adopted in this book. Generally, what is no surprise, the case of a stochastic system under control over an infinite horizon in a fuzzy environment is quite similar to its respective formulations under uncertainty (randomness).

7.3 CONTROL OF A FUZZY SYSTEM

For the fuzzy system under control, its fuzzy state at control stage t is assumed to be a fuzzy set X_t defined in the state space $X = \{s_1, \ldots, s_n\}$ characterized by its membership function $\mu_{X_t}(x_t)$, and the fuzzy control at control stage t is also assumed to be a fuzzy set U_t defined in the control space $U = \{c_1, \ldots, c_m\}$ characterized by its membership function $\mu_{U_t}(u_t)$. We assume therefore a more general type of fuzzy system under control, i.e. the one with fuzzy states and fuzzy controls; the fuzzy system with nonfuzzy controls is evidently a special case here.

The temporal evolution of the fuzzy system under control considered is governed by the following fuzzy state transition equation [cf. (2.140), page 61]

$$X_{t+1} = F(X_t, U_t), \qquad t = 0, 1, \ldots \tag{7.65}$$

or in terms of membership functions [cf. (2.145), page 62]

$$\mu_{X_{t+1}}(x_{t+1}) = \max_{u_t} \max_{x_t} [\mu_{U_t}(u_t) \wedge$$
$$\wedge \mu_{X_t}(x_t) \wedge \mu_{X_{t+1}}(x_{t+1} \mid x_t, u_t)], \qquad \text{for each } x_{t+1} \in X \tag{7.66}$$

where $\mu_{X_{t+1}}(x_{t+1} \mid x_t, u_t)$ is an equivalent representation of the fuzzy state transition function F as mentioned in Section 2.4.3.

At each control stage t, the fuzzy control U_t is subjected to a fuzzy constraint $C''(X_t)$ defined as a fuzzy set in U and characterized by its membership function $\mu_{C''}(u_t \mid X_t)$, and on the fuzzy state X_t there is imposed a fuzzy goal G'' defined as a fuzzy set in X and characterized by its membership function $\mu_{G''}(x_t)$ that is evidently assumed to be the same for all the control stages.

We can immediately see the same inconsistency as in the case of controlling a fuzzy system with the previous types of the termination time [cf. Sections 4.3 and 6.3]. Namely, the fuzzy constraint $C''(X_t)$, i.e. $\mu_{C''}(u_t \mid X_t)$ specifies for each fuzzy state X_t a grade of membership of a particular $u_t \in U$. However, the control in our case [cf. (7.65) or (7.66)] is fuzzy, U_t, and the fuzzy constraint should be somehow reformulated to specify a grade of membership (in $[0, 1]$) for a particular fuzzy control U_t.

One can apply here the approach employed in Section 4.3.2 for the control of a fuzzy system with a fixed and specified termination time by employing branch-and-bound. Namely, one can reformulate $C''(X_t)$ using:

- a normalized distance between the two fuzzy sets $C''(X_t)$ and U_t, i.e.

$$\mu_{C'}(U_t \mid X_t) = 1 - d(C''(X_t), U_t), \qquad t = 0, 1, \ldots \qquad (7.67)$$

  where $d(.,.)$ is a normalized distance between two fuzzy sets exemplified by the linear (Hamming) (2.26), page 28, or quadratic (Euclidean) (2.27), page 28;

- a degree of equality of two fuzzy sets $C''(X_t)$ and U_t, i.e.

$$\mu_{C'}(U_t \mid X_t) = e(C''(X_t), X_t), \qquad t = 0, 1, \ldots \qquad (7.68)$$

  where $e(.,.)$ is a degree of equality of two fuzzy sets exemplified by (2.7)–(2.10), page 24.

- some other indices or measures of (dis)similarity as, e.g., Kaufmann and Gupta's (1985) dissimilarity index (4.202), page 170.

Using the above two reformulations of the initial fuzzy constraint, we obtain "auxiliary" fuzzy constraints that specify for each fuzzy control U_t its degree of membership in the (initial) fuzzy constraint.

The same line of reasoning applies to the fuzzy goal G'' characterized by its membership function $\mu_{G''}(x_t)$. Namely, it specifies the grade of membership of a particular nonfuzzy state x_t in the fuzzy goal G''. It has therefore to be reformulated into an "auxiliary" fuzzy goal G' which would specify a grade of membership of a particular fuzzy state X_t in the fuzzy goal. The same approach as for the fuzzy constraints can be used, i.e. by employing:

- a normalized distance between two fuzzy states

$$\mu_{G'}(X_t) = 1 - d(G'', X_t), \qquad t = 0, 1, \ldots \qquad (7.69)$$

- a degree of equality of two fuzzy sets

$$\mu_{G'}(X_t) = e(G'', X_t), \qquad t = 0, 1, \ldots \qquad (7.70)$$

The fuzzy decision [cf. (7.1), page 216], involving the above auxiliary fuzzy constraints and fuzzy goals, is therefore

$$
\begin{aligned}
\mu_D(U_0, U_1, \ldots \mid X_0) &= \\
&= [\mu_{C'}(U_0 \mid X_0) \wedge \mu_{G'}(X_1)] \wedge [\mu_{C'}(U_1 \mid X_1) \wedge \mu_{G'}(X_2)] \wedge \ldots = \\
&= \lim_{N \to \infty} \bigwedge_{t=0}^{N} [\mu_{C'}(U_t \mid X_t) \wedge \mu_{G^{t+1}}(X_{t+1})]
\end{aligned}
\tag{7.71}
$$

where the successive fuzzy states $X_1, X_2, \ldots$ are given by the fuzzy state transition equation (7.65).

The problem is then to find an optimal sequence of fuzzy controls $U_0^*, U_1^*, \ldots$ such that

$$
\mu_D(U_0^*, U_1^*, \ldots \mid X_0) = \max_{U_0, U_1, \ldots} \mu_D(U_0, U_1, \ldots \mid X_0)
\tag{7.72}
$$

with $\mu_D(U_0, U_1, \ldots \mid X_0)$ is given by (7.71). Needless to say that in this problem we are interested in an optimal stationary strategy.

At first glance an inherent difficulty in solving the problem (7.72) may be apparent. Namely, on the one hand, the state-dependent fuzzy constraints $\mu_{C'}(U_t \mid X_t)$, $t = 0, 1, \ldots$, should be specified for each fuzzy state. On the other hand, the maximization in the problem (7.72) should proceed over all possible sequences of fuzzy controls.

Unfortunately, the numbers of all possible fuzzy states and fuzzy controls are evidently infinite in theory, and at best very high in practice. So, a very high number of fuzzy constraints should be formulated, and eventually, to find a solution, the maximization over a very large set of fuzzy controls should be performed. Evidently, this can make the solution process ineffective, or at least highly inefficient.

A practical solution was proposed here by Kacprzyk and Staniewski (1982). Basically, it consists in an approximation of the source optimization problem by an auxiliary problem that is of a lower dimensionality and can be solved efficiently. Its solution should then be "readjusted." The idea is similar to that employed in Section 4.3.1 (and also in Section 4.3.3), i.e. consists of the use of some standard (called *reference* in the source paper by Kacprzyk and Staniewski, 1982) values of fuzzy states and controls. Now, we will present the essence of that approximation, and then show its application to the case considered.

7.3.1 Approximation by reference fuzzy sets

Basically, the idea of Kacprzyk and Staniewski's (1982) approach is to overcome the main obstacle in solving the problem (7.72), i.e. high numbers of possible fuzzy states and fuzzy controls, by somehow reducing these numbers. We have already presented a similar approach in the case of a fixed and specified termination time, both while using for the solution the branch-and-bound procedure (Kacprzyk, 1979 – cf. Section 4.3.2) and fuzzy dynamic programming (Baldwin and Pilsworth, 1982 – cf. Section 4.3.1).

Suppose, first, that we have some nonfuzzy set $Y = \{y_1, \ldots, y_k\}$, $k < \infty$, that serves the purpose of a universe of discourse in which fuzzy sets in Y are defined. Suppose that $\mathcal{Y} = \{W\} = \{W_1, W_2, \ldots\}$ is a family of all the fuzzy sets defined in Y.

Let us define now in Y a relatively small number, say w, of so-called *reference fuzzy sets* $\overline{W}_1, \ldots, \overline{W}_w$, and denote $\overline{\mathcal{Y}} = \{\overline{W}_1, \ldots, \overline{W}_w\}$, i.e. the family of all reference fuzzy sets. Obviously, $\overline{\mathcal{Y}} \subset \mathcal{Y}$, where $\mathcal{Y}$ is the family of all fuzzy sets defined in Y.

All the fuzzy sets $W \in \mathcal{Y}$ are now *approximated* by some reference fuzzy sets $\overline{W} \in \overline{\mathcal{Y}}$. The approximation may proceed in various ways. The first approach, which has quite a long tradition is closely related to the so-called *linguistic approximation* (cf. Kacprzyk, 1979, 1983b; Wenstøp, 1979; Zadeh, 1973). Its essence is basically the use of some normalized distance. The approximation considered amounts therefore to finding for an *approximated fuzzy set* $W \in \mathcal{Y}$ an *approximating reference fuzzy set* $\overline{W}^* \in \overline{\mathcal{Y}}$ such that

$$d(\overline{W}^*, W) = \min_{\overline{W} \in \overline{\mathcal{Y}}} d(\overline{W}, W) \tag{7.73}$$

where $d(.,.)$ is a normalized distance between two fuzzy sets [cf. (7.67)].

A similar role may be played by a degree of equality [cf. (7.68)] or another measure of similarity; evidently, we should employ then, e.g., "max" instead "min" in (7.73), i.e.

$$e(\overline{W}^*, W) = \max_{\overline{W} \in \overline{\mathcal{Y}}} e(\overline{W}, W) \tag{7.74}$$

where $e(.,.)$ is a degree of equality of two fuzzy sets.

The choice of the number and form of reference fuzzy sets is somewhat arbitrary, and some hints have already been given in Sections 4.3.1 and 4.3.2. Usually, they are represented by some triangular and/or trapezoid fuzzy numbers.

In the sequel the approximation by reference fuzzy sets will be generally denoted by $A(W)$.

Now let us briefly present how the approximation by reference fuzzy sets can be applied for the analysis and solution of the control problem considered.

7.3.2   Approximate formulation and solution of the problem

First, we will begin with the case of fuzzy states and fuzzy controls. If the state space is $X = \{s_1, \ldots, s_n\}$ and the control space is $U = \{c_1, \ldots, c_m\}$, then the fuzzy states are $X_t \in \mathcal{X}$ and the fuzzy controls are $U_t \in \mathcal{U}$, $t = 0, 1, \ldots$ We define now in X a relatively small number, say r, of *reference fuzzy states*, $\overline{S}_1, \ldots, \overline{S}_r$, i.e. $\overline{\mathcal{X}} = \{\overline{S}_1, \ldots, \overline{S}_r\}$. And analogously, in U we define a relatively small number, say s, of *reference fuzzy controls*, $\overline{C}_1, \ldots, \overline{C}_s$, i.e. $\overline{\mathcal{U}} = \{\overline{C}_1, \ldots, \overline{C}_s\}$. Then, all fuzzy states and fuzzy controls occurring in the course of control will be approximated by these reference fuzzy states and reference fuzzy controls.

The second step is the derivation of the fuzzy state transition equation, equivalent to (7.65), that would govern fuzzy state transitions in case of reference fuzzy states and reference fuzzy controls. Namely, for each reference fuzzy state $\overline{X}_t \in \overline{\mathcal{X}}$ and for each reference fuzzy control $\overline{U}_t \in \overline{\mathcal{U}}$, the fuzzy state transition equation (7.65) yields

$$X_{t+1} = F(\overline{X}_t, \overline{U}_t), \qquad t = 0, 1, \ldots \tag{7.75}$$

and, clearly, X_{t+1} need not be a reference fuzzy state, and therefore, we need to approximate X_{t+1} by a reference fuzzy state $\overline{X}_{t+1}$.

The fuzzy state transition equation (7.75) becomes therefore

$$\overline{X}_{t+1} = A(X_{t+1}) = A(F(\overline{X}_t, \overline{U}_t)), \qquad t = 0, 1, \ldots \tag{7.76}$$

which may be called a *reference fuzzy state transition equation*; it can be conveniently represented, for a sufficiently small number of reference fuzzy states and reference fuzzy control, in matrix form, as a *reference fuzzy state transition matrix* – cf. Example 4.22, page 159.

The third step is the reformulation of the fuzzy constraints and fuzzy goal to account for the use of reference fuzzy controls and reference fuzzy states only. The fuzzy constraint becomes therefore $\overline{C}(X_t)$, characterized by $\mu_{\overline{C}}(u_t \mid \overline{X}_t)$, i.e. we need to specify only a relatively small number of fuzzy constraints, for all possible $\overline{X}_t$'s, that is, $\overline{C}(\overline{S}_1), \ldots, \overline{C}(\overline{S}_r)$.

Finally, in the case of employing either a normalized distance (7.67) or a degree of equality (7.68), we obtain, respectively:

$$\mu_C(\overline{U}_t \mid \overline{X}_t) = 1 - d(\overline{C}(\overline{X}_t), \overline{U}_t), \qquad t = 0, 1, \ldots \qquad (7.77)$$

and

$$\mu_C(\overline{U}_t \mid \overline{X}_t) = e(\overline{C}(\overline{X}_t), \overline{U}_t), \qquad t = 0, 1, \ldots \qquad (7.78)$$

The same can be performed for the fuzzy goal, similarly as in (7.69) and (7.70), and we obtain, respectively:

$$\mu_G(\overline{X}_t) = 1 - d(G', \overline{X}_t), \qquad t = 0, 1, \ldots \qquad (7.79)$$

and

$$\mu_G(\overline{X}_t) = e(G', \overline{X}_t), \qquad t = 0, 1, \ldots \qquad (7.80)$$

The fuzzy decision is therefore

$$\begin{aligned}
\mu_d(\overline{U}_0, \overline{U}_1, \ldots \mid \overline{X}_0) &= \\
&= [\mu_C(\overline{U}_0 \mid \overline{X}_0) \wedge \mu_G(\overline{X}_1)] \wedge [\mu_C(\overline{U}_1 \mid \overline{X}_1) \wedge \mu_G(\overline{X}_2)] \wedge \ldots = \\
&= \lim_{N \to \infty} \bigwedge_{t=0}^{N} [\mu_C(\overline{U}_t \mid \overline{X}_t) \wedge \mu_G(\overline{X}_{t+1})]
\end{aligned} \qquad (7.81)$$

where $\overline{X}_1, \overline{X}_2, \ldots$ are obtained via the reference fuzzy state transition equation (7.76).

The problem is then to find an optimal sequence of reference fuzzy controls $\overline{U}_0^*, \overline{U}_1^*, \ldots$ such that

$$\mu_D(\overline{U}_0^*, \overline{U}_1^*, \ldots \mid \overline{X}_0) = \max_{\overline{U}_0, \overline{U}_1, \ldots} \mu_D(\overline{U}_0, \overline{U}_1, \ldots \mid \overline{X}_0) \qquad (7.82)$$

where $\mu_D(\overline{U}_0, \overline{U}_1, \ldots \mid \overline{X}_0)$ is given by (7.81). As always in the problem class considered, i.e. with an infinite termination time, we are in fact interested in finding an optimal stationary strategy.

Note that through an extensive use of approximation by reference fuzzy sets, the problem considered (7.82) becomes formally equivalent to that of controlling a deterministic system (with an infinite termination time) since the state ranges now over the finite state of reference fuzzy states $\{\overline{S}_1, \ldots, \overline{S}_r\}$, and the control ranges over the finite state of reference fuzzy controls $\{\overline{C}_1, \ldots, \overline{C}_s\}$. For such a case the solution algorithm has already been given in Section 7.1.

Moreover, if we additionally introduce a discount factor $b > 0$, the problem (7.82) becomes that of finding an optimal sequence of reference fuzzy controls $\overline{U}_0^*, \overline{U}_1^*, \ldots$ such that

$$\mu_D(\overline{U}_0^*, \overline{U}_1^*, \ldots \mid \overline{X}_0) =$$

$$= \max_{\overline{U}_0, \overline{U}_1, \ldots} \lim_{N \to \infty} \bigwedge_{t=0}^{N} b^t [\mu_C(\overline{U}_t \mid \overline{X}_t) \wedge \mu_G(\overline{X}_{t+1})] \qquad (7.83)$$

Obviously, we are in fact interested in finding an optimal stationary strategy $a_\infty^* = (a^*, a^*, \ldots)$, where $a^* : \overline{\mathcal{Y}} \longrightarrow \overline{\mathcal{U}}$ such that

$$\overline{U}_t^* = a^*(\overline{X}_t), \qquad t = 0, 1, \ldots \qquad (7.84)$$

that is, relating the current (at control stage t) optimal reference fuzzy control to the current reference fuzzy state.

As has been shown in Section 7.1, such an optimal stationary strategy exists and can be found in a finite number of iterations by using a policy-iteration-type algorithm.

Because of lack of space, we will not repeat here in terms of reference fuzzy states and reference fuzzy controls the whole line of reasoning that leads to that algorithm [cf. (7.11)–(7.22), page 218], but only present below the consecutive steps of the algorithm tailored to our particular purposes. These steps are as follows:

Step 1. Choose an arbitrary stationary strategy $a_\infty = (a, a, \ldots)$.

Step 2. Solve in the $\mu_D(a_\infty \mid \overline{S}_i)$'s the following set of r (functional) equations:

$$\begin{cases} \mu_D(a_\infty \mid \overline{S}_1) = \mu_C(a(\overline{S}_1) \mid \overline{S}_1) \wedge \mu_G[A(F(\overline{S}_1, a(\overline{S}_1)))] \wedge \\ \qquad \wedge b\mu_D[a_\infty \mid A(F(\overline{S}_1, a(\overline{S}_1)))] \\ \cdots\cdots\cdots\cdots\cdots\cdots\cdots\cdots\cdots\cdots\cdots\cdots\cdots\cdots\cdots\cdots\cdots\cdots \qquad (7.85) \\ \mu_D(a_\infty \mid \overline{S}_r) = \mu_C(a(\overline{S}_r) \mid \overline{S}_r) \wedge \mu_G[A(F(\overline{S}_r, a(\overline{S}_1)))] \wedge \\ \qquad \wedge b\mu_D[a_\infty \mid A(F(\overline{S}_r, a(\overline{S}_r)))] \end{cases}$$

Step 3. Improve the strategy. Using the $\mu_D(a_\infty \mid \overline{S}_i)$'s determined in Step 2, find for each $i \in \{1, \ldots, r\}$ a (stationary) policy $z' : \overline{X} \longrightarrow \overline{U}$, $\overline{U}_t = z'(\overline{X}_t)$, which maximizes with respect to z the following term

$$\mu_C(z(\overline{S}_i) \mid \overline{S}_i) \wedge$$
$$\wedge \mu_G[A(F(\overline{S}_i, a(\overline{S}_i)))] \wedge b\mu_D[a_\infty \mid A(F(\overline{S}_i, a(\overline{S}_i)))] \quad (7.86)$$

Step 4. If the strategy found in Step 3, $z'_\infty = (z', z', \ldots)$, is the same as the previous one, $a_\infty = (a, a, \ldots)$, then it is an optimal stationary strategy sought. Otherwise, assume $a_\infty = z'_\infty$ and return to Step 2.

The algorithm yields in the end an optimal stationary strategy $a_\infty^* = (a^*, a^*, \ldots)$ such that

$$\overline{U}_t^* = a^*(\overline{X}_t), \qquad t = 0, 1, \ldots \qquad (7.87)$$

This strategy relates optimal reference fuzzy controls to reference fuzzy states, because in our approach only the reference fuzzy states and controls "exist." The approximation has been, however, performed only to make the problem considered practically solvable, and in the source problem the fuzzy states and fuzzy controls need not evidently be only the reference ones. Therefore we have somehow to "adjust" the optimal stationary strategy found (7.87) so that it can be used for all possible fuzzy states and fuzzy controls, both reference and not reference.

A simple approach was proposed for this purpose by Kacprzyk and Staniewski (1982). Its idea is as follows. Assume that the particular reference fuzzy states $\overline{S}_1, \ldots, \overline{S}_r \in \overline{\mathcal{X}}$ are related through the optimal stationary policy (7.87) to the reference fuzzy controls $\overline{C}_u, \overline{C}_q, \ldots, \overline{C}_v \in \overline{\mathcal{U}}$ as follows:

$$\overline{C}_u = a^*(\overline{S}_1) \quad \overline{C}_q = a^*(\overline{S}_2) \quad \ldots \quad \overline{C}_v = a^*(\overline{S}_r) \tag{7.88}$$

This may be meant as:

If the reference fuzzy state is $\overline{S}_1$, then the optimal reference fuzzy control is $\overline{C}_u$

If the reference fuzzy state is $\overline{S}_2$, then the optimal reference fuzzy control is $\overline{C}_q$

. . .

If the reference fuzzy state is $\overline{S}_r$, then the optimal reference fuzzy control is $\overline{C}_v$

The optimal stationary policy may be therefore represented by the following fuzzy conditional statement [cf. (2.62), page 38]

$$\begin{cases} \text{IF } \overline{X}_t = \overline{S}_1 \text{ THEN } \overline{U}_t^* = \overline{C}_u \\ \text{ELSE } \ldots \text{ ELSE} \\ \text{IF } \overline{X}_t = \overline{S}_r \text{ THEN } \overline{U}_t^* = \overline{C}_v \end{cases} \tag{7.89}$$

Such a fuzzy conditional statement may be equated [cf. (2.62)] with a fuzzy relation R^* defined in $X \times X$ such that

$$R^* = \overline{S}_1 \times \overline{C}_u + \cdots + \overline{S}_r \times \overline{C}_v \tag{7.90}$$

where "$\times$" is the Cartesian product of two fuzzy sets defined by (2.56), page 37, and "$+$" is the union of two fuzzy sets given as (2.32), page 30.

Therefore, (7.90) is equivalent, in terms of membership functions, to

$$\mu_{R^*}(x_t, u_t) = [\mu_{\overline{S}_1}(x_t) \wedge \mu_{\overline{C}_u}(u_t)] \vee \ldots$$
$$\ldots \vee [\mu_{\overline{S}_r}(x_t) \wedge \mu_{\overline{C}_v}(u_t))], \qquad \text{for each } x_t \in X, u_t \in U \tag{7.91}$$

This fuzzy relation represents an *optimal fuzzy stationary policy*.

All the above has been done so far in terms of reference fuzzy controls and reference fuzzy states. Now, if we have a fuzzy state X_t, not necessarily a reference one, then the optimal fuzzy control U_t^*, not necessarily a reference one, that is related to this fuzzy state under the optimal stationary policy (7.88) [or, equivalently, (7.89)], is given by the max–min composition "∘" (2.53), page 36, of X_t and R^*, i.e.

$$U_t^* = X_t \circ R^* \tag{7.92}$$

or in terms of membership functions

$$\mu_{U_t^*}(u_t) = \max_{x_t \in X}[\mu_{X_t}(x_t) \wedge \mu_{R^*}(x_t, u_t)], \qquad \text{for each } u_t \in U \tag{7.93}$$

and, clearly, other types of composition can also be employed as, e.g., (2.54) or (2.55), page 36.

The result obtained, i.e. an optimal fuzzy control U_t^*, is therefore a fuzzy set (defined in the space of controls U). If we are interested in finding a nonfuzzy value of optimal control, $u_t^* \in U$, then we can choose various defuzzification methods outlined in Section 2.1.9 as, e.g.,

- the one with the highest grade of membership in U_t^* [cf. (2.84), page 46], i.e. such a u_t^* that

$$\mu_{U_t^*}(u_t^*) = \max_{u_t \in U} \mu_{U_t^*}(u_t) \tag{7.94}$$

- the one obtained by the center-of-gravity method [cf. (2.83), page 46], i.e.

$$u_t^* = \frac{\sum_{j=1}^m c_j \cdot \mu_{U_t^*}(c_j)}{\sum_{j=1}^m \mu_{U_t^*}(c_j)} \tag{7.95}$$

Now let us solve for illustration a simple example.

Example 7.3 Suppose that $X = \{s_1, \ldots, s_{10}\}$, $U = \{c_1, \ldots, c_5\}$, the reference fuzzy states are

$$\overline{S}_1 = 1/s_1 + 1/s_2 + 0.7/s_3 + 0.4/s_4 + 0.1/s_5$$
$$\overline{S}_2 = 0.4/s_1 + 0.71/s_2 + 1/s_3 + 0.8/s_4 + 0.3/s_5$$
$$\overline{S}_3 = 0.2/s_3 + 0.6/s_4 + 1/s_5 + 0.7/s_6 + 0.3/s_7$$
$$\overline{S}_4 = 0.3/s_5 + 0.7/s_6 + 1/s_7 + 0.8/s_8 + 0.4/s_9$$
$$\overline{S}_5 = 0.1/s_6 + 0.4/s_7 + 0.7/s_8 + 0.9/s_9 + 1/s_{10}$$

and the reference fuzzy controls are

$$\overline{C}_1 = 1/c_1 + 0.8/c_2 + 0.5/c_3 + 0.3/c_4 + 0.1/c_5$$
$$\overline{C}_2 = 0.3/c_1 + 0.7/c_2 + 1/c_3 + 0.7/c_4 + 0.2/c_5$$
$$\overline{C}_3 = 0.1/c_1 + 0.3/c_2 + 0.5/c_3 + 0.8/c_4 + 1/c_5$$

where, as always, the singletons with the grade of membership equal to 0 are omitted.

For clarity, we will present the fuzzy system under control already in the form of the following state transition table governing transitions between reference fuzzy states under reference fuzzy controls [cf. (7.76), page 238]

$$\overline{X}_{t+1} = $$

	$\overline{X}_t = \overline{S}_1$	$\overline{S}_2$	$\overline{S}_3$	$\overline{S}_4$	$\overline{S}_5$
$\overline{U}_t = \overline{C}_1$	$\overline{S}_1$	$\overline{S}_1$	$\overline{S}_2$	$\overline{S}_3$	$\overline{S}_4$
$\overline{C}_2$	$\overline{S}_2$	$\overline{S}_2$	$\overline{S}_3$	$\overline{S}_4$	$\overline{S}_5$
$\overline{C}_3$	$\overline{S}_2$	$\overline{S}_3$	$\overline{S}_3$	$\overline{S}_5$	$\overline{S}_5$

Next, suppose that the fuzzy constraints and fuzzy goal, given again in their final forms, i.e. involving the reference fuzzy controls and reference fuzzy states, are, respectively:

$$C(\overline{S}_1) = 0.2/\overline{C}_1 + 0.6/\overline{C}_2 + 1/\overline{C}_3$$
$$C(\overline{S}_2) = 0.3/\overline{C}_1 + 0.7/\overline{C}_2 + 1/\overline{C}_3$$
$$C(\overline{S}_3) = 0.4/\overline{C}_1 + 1/\overline{C}_2 + 0.6/\overline{C}_3$$
$$C(\overline{S}_4) = 0.7/\overline{C}_1 + 1/\overline{C}_2 + 0.6/\overline{C}_3$$
$$C(\overline{S}_5) = 1/\overline{C}_1 + 0.6/\overline{C}_2 + 0.2/\overline{C}_3$$

and

$$G = 0.3/\overline{S}_1 + 0.7/\overline{S}_2 + 1/\overline{S}_3 + 0.5/\overline{S}_4 + 0.1/\overline{S}_5$$

Finally, let the discount factor be $b = 1.01$.

The subsequent steps of the algorithm are now as follows:

Step 1. We choose, for instance, the following stationary policy a_∞:

$$a(\overline{S}_1) = \overline{C}_2 \quad a(\overline{S}_2) = \overline{C}_2 \quad a(\overline{S}_3) = \overline{C}_2 \quad a(\overline{S}_4) = \overline{C}_2 \quad a(\overline{S}_5) = \overline{C}_2$$

Step 2. We solve the set of equations (7.85), i.e.

$$
\begin{cases}
\mu_D(a_\infty \mid \overline{S}_1) = 0.6 \wedge 0.7 \wedge 1.01 \cdot \mu_D(a_\infty \mid \overline{S}_2) = \\
\qquad = 0.6 \wedge 1.01 \cdot \mu_D(a_\infty \mid \overline{S}_2) \\
\mu_D(a_\infty \mid \overline{S}_2) = 0.7 \wedge 0.7 \wedge 1.01 \cdot \mu_D(a_\infty \mid \overline{S}_2) = \\
\qquad = 0.7 \wedge 1.01 \cdot \mu_D(a_\infty \mid \overline{S}_2) \\
\mu_D(a_\infty \mid \overline{S}_3) = 1 \wedge 1 \wedge 1.01 \cdot \mu_D(a_\infty \mid \overline{S}_3) = \\
\qquad = 1 \wedge 1.01 \cdot \mu_D(a_\infty \mid \overline{S}_3) \\
\mu_D(a_\infty \mid \overline{S}_4) = 1 \wedge 0.5 \wedge 1.01 \cdot \mu_D(a_\infty \mid \overline{S}_4) = \\
\qquad = 0.5 \wedge 1.01 \cdot \mu_D(a_\infty \mid \overline{S}_4) \\
\mu_D(a_\infty \mid \overline{S}_5) = 0.6 \wedge 0.1 \wedge 1.01 \cdot \mu_D(a_\infty \mid \overline{S}_4) = \\
\qquad = 0.1 \wedge 1.01 \cdot \mu_D(a_\infty \mid \overline{S}_4)
\end{cases}
$$

and obtain the solution

$$\mu_D(a_\infty \mid \overline{S}_1) = 0.6$$
$$\mu_D(a_\infty \mid \overline{S}_2) = 0.7$$
$$\mu_D(a_\infty \mid \overline{S}_3) = 1$$
$$\mu_D(a_\infty \mid \overline{S}_4) = 0.5$$
$$\mu_D(a_\infty \mid \overline{S}_5) = 0.1$$

Step 3. We improve the strategy by using (7.86) and obtain the following stationary strategy z'_∞:

$$z'(\overline{S}_1) \in \{\overline{C}_2, \overline{C}_3\}$$
$$z'(\overline{S}_2) = \overline{C}_3$$
$$z'(\overline{S}_3) = \overline{C}_2$$
$$z'(\overline{S}_4) = \overline{C}_1$$
$$z'(\overline{S}_5) \in \{\overline{C}_1, \overline{C}_2, \overline{C}_3\}$$

which is different than the previous strategy, hence we return to Step 2. Since the strategy z'_∞ is not unique because for $\overline{S}_1$ and $\overline{S}_5$ various reference fuzzy controls can be used, then we assume, e.g., that $a_\infty = z'_\infty$ is

$$a(\overline{S}_1) = \overline{C}_3 \quad a(\overline{S}_2) = \overline{C}_3 \quad a(\overline{S}_3) = \overline{C}_2 \quad a(\overline{S}_4) = \overline{C}_1 \quad a(\overline{S}_5) = \overline{C}_1$$

Step 2. We solve the set of equations (7.85), i.e.

$$
\begin{cases}
\mu_D(a_\infty \mid \overline{S}_1) = 1 \wedge 0.7 \wedge 1.01 \cdot \mu_D(a_\infty \mid \overline{S}_2) = \\
\qquad = 0.7 \wedge 1.01 \cdot \mu_D(a_\infty \mid \overline{S}_2) \\
\mu_D(a_\infty \mid \overline{S}_2) = 1 \wedge 1 \wedge 1.01 \cdot \mu_D(a_\infty \mid \overline{S}_3) = \\
\qquad = 1 \wedge 1.01 \cdot \mu_D(a_\infty \mid \overline{S}_3) \\
\mu_D(a_\infty \mid \overline{S}_3) = 1 \wedge 1 \wedge 1.01 \cdot \mu_D(a_\infty \mid \overline{S}_3) = \\
\qquad = 1 \wedge 1.01 \cdot \mu_D(a_\infty \mid \overline{S}_3) \\
\mu_D(a_\infty \mid \overline{S}_4) = 0.7 \wedge 1 \wedge 1.01 \cdot \mu_D(a_\infty \mid \overline{S}_3) = \\
\qquad = 0.7 \wedge 1.01 \cdot \mu_D(a_\infty \mid \overline{S}_3) \\
\mu_D(a_\infty \mid \overline{S}_5) = 1 \wedge 0.5 \wedge 1.01 \cdot \mu_D(a_\infty \mid \overline{S}_3) = \\
\qquad = 0.5 \wedge 1.01 \cdot \mu_D(a_\infty \mid \overline{S}_3)
\end{cases}
$$

and obtain the solution

$$
\begin{aligned}
\mu_D(a_\infty \mid \overline{S}_1) &= 0.7 \\
\mu_D(a_\infty \mid \overline{S}_2) &= 1 \\
\mu_D(a_\infty \mid \overline{S}_3) &= 1 \\
\mu_D(a_\infty \mid \overline{S}_4) &= 0.7 \\
\mu_D(a_\infty \mid \overline{S}_5) &= 0.5
\end{aligned}
$$

Step 3. We improve the strategy via (7.86) and obtain

$$z'(\overline{S}_1) = \overline{C}_3 \quad z'(\overline{S}_2) = \overline{C}_3 \quad z'(\overline{S}_3) = \overline{C}_2 \quad z'(\overline{S}_4) = \overline{C}_1 \quad z'(\overline{S}_5) = \overline{C}_1$$

which is the same as the previous strategy, and hence is the optimal solution sought, i.e.

$$
\begin{aligned}
a^*(\overline{S}_1) &= \overline{C}_3 \\
a^*(\overline{S}_2) &= \overline{C}_3 \\
a^*(\overline{S}_3) &= \overline{C}_2 \\
a^*(\overline{S}_4) &= \overline{C}_1 \\
a^*(\overline{S}_5) &= \overline{C}_1
\end{aligned}
$$

Now, this optimal stationary strategy is represented due to (7.89) as the following fuzzy conditional statement:

$$
\begin{aligned}
&\text{IF } \overline{S}_1 \text{ THEN } \overline{C}_3 \\
&\quad \text{ELSE} \\
&\text{IF } \overline{S}_2 \text{ THEN } \overline{C}_3 \\
&\quad \text{ELSE} \\
&\text{IF } \overline{S}_3 \text{ THEN } \overline{C}_2 \\
&\quad \text{ELSE} \\
&\text{IF } \overline{S}_4 \text{ THEN } \overline{C}_1 \\
&\quad \text{ELSE} \\
&\text{IF } \overline{S}_5 \text{ THEN } \overline{C}_1
\end{aligned}
$$

which is in turn equivalent, according to (7.91) to the following fuzzy relation:

$$\overline{X}_{t+1} =$$

$$=$$

	$x_t = s_1$	s_2	s_3	s_4	s_5	s_6	s_7	s_8	s_9	s_{10}
$u_t = c_1$	0.1	0.1	0.2	0.3	0.3	0.7	1	0.8	0.9	1
c_2	0.3	0.3	0.3	0.6	0.7	0.7	0.8	0.8	0.8	0.8
c_3	0.5	0.5	0.5	0.6	1	0.7	0.5	0.5	0.5	0.5
c_4	0.8	0.8	0.7	0.6	0.7	0.7	0.3	0.3	0.3	0.3
c_5	1	1	0.7	0.4	0.2	0.2	0.2	0.1	0.1	0.1

If now we have, for instance, a fuzzy state (not a reference one) $X_t = 0.3/s_1 + 0.7/s_2 + 1/s_3 + 0.6/s_4 + 0.2/s_5$, then (7.92) yields the following optimal fuzzy control:

$$U_t^* = 0.3/c_1 + 0.6/c_2 + 0.6/c_3 + 0.7/c_4 + 0.7/c_5$$

and, if we use (7.94) to find the nonfuzzy optimal control, then we obtain the optimal nonfuzzy control $u_t^* \in \{c_4, c_5\}$. □

Notice that the method presented above is in fact applicable in the case of a dense set of reference fuzzy states and controls. If these sets are sparse, we need to use the interpolative reasoning based method presented in Section 4.3.3.

This concludes our analysis of a very relevant and interesting case of a fuzzy system controlled in a fuzzy environment over an infinite horizon. The solution of this, initially practically unsolvable problem has been obtained by using some approximations to transform the problem into one for which an efficient solution procedure is available.

This is also the last topic related to the control processes with an infinite termination time. Needless to say, these processes are certainly the most difficult to analyze and solve, in particular the case of a stochastic system under control. However, the case of a fuzzy system under control is presumably the most relevant for practice.

Finally, this chapter also concludes our analysis of the particular classes of prescriptive models of (multistage) fuzzy control in a fuzzy environment. Now, to illustrate some of the models presented, some relevant examples will be discussed in more detail.

8

Examples of Applications

The previous sections have provided a wide spectrum of models for formulating and solving diverse multistage control problems under fuzziness (imprecision). The models have involved fuzziness in various aspects as, e.g., the goals and constraints, dynamics of the system under control, the termination time, etc. Our discussion has been widely illustrated by examples, though their purpose has been mainly to show the particular steps of respective algorithms. In this chapter we will show in some "real world" applications of some "soft" problems.

Emphasis will be on how the problem is formulated, and how its fuzzy (imprecise) aspects are represented within our framework. This includes mainly the formulation of a model of the system under control, fuzzy constraints and fuzzy goals, and their proper aggregation (fuzzy decision).

A crucial step is here the identification of those elements, mainly of the system under control. As we have already mentioned in Section 2.4.3.1, we will not consider these issues in detail because in this book emphasis is more on algorithmic aspects. There may be another reason for such an attitude in regard to applications which will be discussed in this chapter. Namely, they concern very diverse areas (systems): socioeconomic, environmental, technological, medical, management, etc.

Identification procedures are clearly general, but in most cases they are tailored to a particular application area because, for instance, data that are available in one field are not available (or are not reliable) in another. Therefore, since our models are usually some fuzzifications of conventional models available in a specific field, the interested user will be able to use his or her available knowledge on how to identify (nonfuzzy) models that field, and "adjust" this expertise (and maybe also tools and techniques available) to take into account fuzzy elements to be introduced.

Then, a respective control problem will be formulated, and its solution will often be only outlined as it will be basically the application of consecutive steps of an appropriate solution algorithm presented in detail in one of the previous chapters of this book.

The exposition of basic elements of the theory of fuzzy sets provided in the previous chapters will be sufficient to follow our next discussions. Sometimes it will be necessary to introduce more application-specific concepts but these will be explained. Clearly, some basic knowledge on a particular area (regional development, research and development, power systems, etc.) will be needed to follow our discussion in detail, and then to eventually apply and further develop the models presented.

Needless to say, these "real-world" problems will be, by necessity, presented in much less detail than in their real analysis and solution (implementation). This is due, first, to our desire to maintain readability to a wide audience. Second, in many applications to be presented we have to rely on their authors' presentations which are by necessity short and incomplete. We will be unable therefore to cover in this chapter all issues that may be interesting to potential users, and which are certainly crucial, as, for example, sensitivity analyses. Emphasis will be on the essence of the problem and its modeling, i.e. how to express it via a fuzzy model of multistage control type.

The applications to be discussed may be divided into two groups. The first group comprises those applications which will be presented in much more detail, showing steps and ways of fuzzy problem formulation so that the reader obtain as illustrative and wide as possible an exposition.

In this group we will present in the consecutive sections the models which concern:

- socioeconomic regional development (Section 8.1),
- flood control (Section 8.2),
- research and development (R&D) planning (Section 8.3),
- scheduling of unit commitment in a power system (Section 8.4),
- anesthesia administration during surgery (Section 8.5),
- resource allocation (Section 8.6),
- inventory control (Section 8.7), and
- a review of some other applications in scheduling of power generator maintenance, design of a fuzzy controller, designing a distillation column in chemical engineering, determination of shortest paths for the transportation of hazardous waste, scheduling of autonomous guided vehicles in flexible manufacturing systems, and optimizing spare parts inventory in a power station (Section 8.8).

Most of the applications discussed in this book are non-technological, and concern broadly perceived decision (or policy) making. One reason is that we are deeply convinced that the present "fuzzy boom" which occurs in technological applications of fuzzy control only, and which often limits the interest of scholars and researchers to such applications, and even make many people believe that only these applications make sense, will evolve in the near future into non-technological areas, involving in particular decision and policy making issues as their importance to society is presumably higher than that of technology alone. It should be noted that such non-technological applications have been advocated since the very inception of fuzzy sets theory, as people have always thought that all decision making type problems are particularly susceptible to fuzzy analysis due to a crucial role played by human elements with their "soft" assessments, imprecise value systems, confused perception of rationality, etc. This fact is unfortunately not known to many newcomers to fuzzy sets who think that the applications may only be basically to fuzzy control of technological systems.

Therefore, this chapter should help the reader better understand the development of multistage fuzzy control models, and provide a background, hints, clues, etc. for the development of such models for his or her own problems.

8.1 SOCIOECONOMIC REGIONAL DEVELOPMENT

Regional development is one of more important issues facing many local and central authorities all over the world, independently of the political, economic and social systems. These issues will presumably be more and more important as the frontiers of countries erode, and a regional consciousness becomes more and more present and important for preserving some identity that is needed for the social groups to avoid, e.g., alienation and other dangerous social disasters that modern societies are very often plagued by.

Unfortunately, all regional problems are difficult as they involve various aspects (political, economic, social, environmental, technological, etc.), different parties (inhabitants, authorities of different levels, formal and informal groups, etc.), many entities that are difficult to precisely single out, define and quantify, etc. These problems are therefore typically "soft" in the sense of systems analysis (cf. Checkland, 1976, 1979).

The importance of regional development has evidently implied many attempts of its formal analysis and solution involving mathematical modeling, construction of decision support systems, etc. Unfortunately, conventional "hard" tools and techniques based on, say, traditional modeling and optimization have not always led to satisfactory results. One of the reasons is certainly that these means have not made it possible to properly express an inherent "softness" of the problems considered. As an attempt to overcome this difficulty, the use of fuzzy sets has been proposed in a series of papers by Kacprzyk and Straszak (1980–86) the nature of which will be outlined below.

The problem considered in these papers may be briefly stated as follows. We consider a region that is plagued by some severe difficulties. To be more specific, as considered in Kacprzyk and Straszak (1980–84), the region is rural plagued by out-migration of the younger population to neighboring urban centers, the resulting aging of the rural population, and – as a final consequence – a socioeconomic decay. This out-migration is mainly caused by a poor *life quality* perceived. To stop this decay, life quality should therefore be (considerably) improved to enhance its social perception, and in effect to stop the out-migration of younger population. It is further assumed that, quite obviously, the region does not possess adequate financial means, and some external outlays (investments, in practice) should be provided by higher level authorities.

The problem is then to find the amount of these investments, satisfying some constraints imposed on them, and their temporal distribution (over a specified planning horizon) to best attain goals of the development and their resulting social perception. It is clear that the nature of the problem stated as above indicates that a multistage control setting (under fuzziness) discussed in this book should be a proper framework for analysis.

8.1.1   A fuzzy model of socioeconomic regional development

The very essence of a socioeconomic regional development problem considered here is very clearly shown in Figure 29. The region is there represented by a socioeconomic dynamic system under control whose state at development (planning) stage $t - 1$, X_{t-1}, is characterized by some set of relevant socioeconomic life quality indicators. Then, the control (investment) at stage $t - 1$, u_{t-1}, changes the state (the values of

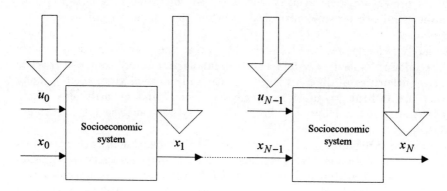

Figure 29   The essence of socioeconomic regional development

these indicators) to X_t. This is repeated for the particular stages $t = 1, \ldots, N$ where N is a planning horizon assumed.

The assessment of "goodness" of the development stage t is performed taking into account both the "goodness" of the control applied u_{t-1}, and the "goodness" of the state attained X_t; the former has to do with how well some constraints are satisfied, and the latter with how well some goals are attained. This assessment is repeated for $t = 1, \ldots, N$. Such an assessment, from another perspective, concerns both the "costs," i.e. the u_{t-1}'s, and "benefits," i.e. the X_t's.

The control problem is now to find such a sequence of investments (controls) for which the assessment of development given above is the most favorable.

The main problem is evidently how to properly formulate and evaluate those costs and benefits, and this will constitute in fact the major part of our next discussion.

First, let us present the socioeconomic system as a dynamic system under control shown in Figure 30. Its state (output) X_t is assumed to be equated with a *life quality index* that consists of the following seven *life quality indicators*:

- x_t^1 – economic quality (e.g., wages, salaries, income, ...),
- x_t^2 – environmental quality,
- x_t^3 – housing quality,
- x_t^4 – health service quality,
- x_t^5 – infrastructure quality,
- x_t^6 – work opportunity,
- x_t^7 – leisure time opportunity,

i.e. the life quality index may be represented as the following vector $X_t = [x_t^1, \ldots, x_t^7]$ whose entries are values of the seven life quality indicators (at time t).

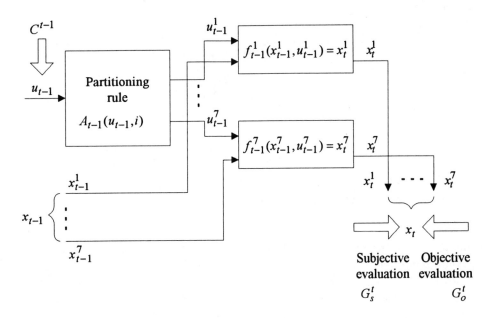

Figure 30   Socioeconomic system under control

The above seven life quality indicators have been chosen in Kacprzyk and Straszak (1980–86) as they cover virtually all practically relevant social, economic, environmental, infrastructural, etc. aspects of regional development. Needless to say these aspects are by no means separable, and are highly interwoven. This does complicate the analysis, and advocates the use of the fuzzy model to be presented.

One should also note that the choice of life quality indicators should be tailored to a particular situation, and is in itself a serious problem as not only the relevance and adequacy of the particular aspects should be taken into account, but also their implementability in the sense of, say, their quantifiablility.

The control at stage $t-1$, u_{t-1}, is some investment devoted to the development of the region. Clearly, as already mentioned, it consists of some external (from outside the region) funds, though the model does not differentiate between the external and internal (coming from the region's own sources) funds.

The control u_{t-1} is evidently subjected to some limitation. However, in virtually all practical cases such a limitation is not clear-cut and abrupt as, say, $\$1,000,000$ and nothing more. Even if such a crisp (definite) limit is set, it is tacitly assumed that it can be exceeded, if really needed, to some extent by say, a new application to higher level authorities motivated by an unexpected cost increase, a wider span of activities, etc.

We impose therefore on u_{t-1} a fuzzy constraint $\mu_{C^{t-1}}(u_{t-1})$ in a piecewise linear form as shown in Figure 31 to be read as follows. There is some investment u_{t-1}^p planned in advance. The investment (control) u_{t-1} may be fully utilized up to that amount u_{t-1}^p, and hence $\mu_{C^{t-1}}(u_{t-1}) = 1$ for $0 < u_{t-1} < u_{t-1}^p$. However, this planned investment will usually be insufficient and some additional contingency investment

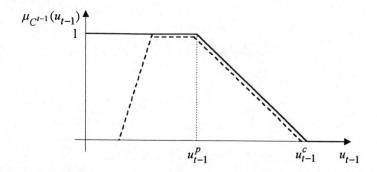

Figure 31 Fuzzy constraints on investment u_{t-1}

will be needed. It may be obtained from the authorities but, inevitably, the higher the amount requested, the more difficult to obtain. Therefore, the $\mu_{C^{t-1}}(u_{t-1})$ diminishes as u_{t-1} exceeds u_{t-1}^p. There is evidently some maximal contingency investment, i.e. more cannot be obtained, and this is represented by u_{t-1}^c (i.e. this maximal contingency investment is in fact equal to $u_{t-1}^c - u_{t-1}^p$). Thus, $\mu_{C^{t-1}}(u_{t-1}) = 0$ for $u_{t-1} \leq u_{t-1}^c$.

In many practical cases the fuzzy constraints are as shown in the dotted line in Figure 31 in that too low a use of available investments should also be avoided since, as experience confirms, this may well imply a reduction of funds allocated by higher level authorities in the future because, being unused, they may be considered unnecessary.

An obvious merit of such a representation of a constraint on the available investment is, first, its simplicity and an easy reflection of an inherent softness. Second, it may be elicited from experts in a relatively easy way as two points are only needed, u_{t-1}^p and u_{t-1}^c. Third, such a fuzzy constraint reflects clearly a satisfaction type attitude that seems to be very well suited for this class of practical problems.

The control (investment) at stage $t - 1$, u_{t-1}, is now partitioned into seven constituents, $u_{t-1}^1, \ldots, u_{t-1}^7$, devoted to improve the respective life quality indicators, $x_{t-1}^1, \ldots, x_{t-1}^7$. This partitioning proceeds due to some partitioning rule denoted $A_{t-1}(u_{t-1}, i)$ which assigns some part (percentage) of u_{t-1} to u_{t-1}^i, $i = 1, \ldots, 7$. The partitioning may proceed in various ways as, e.g., by some routine or bureaucratic procedures, bargaining, compromise seeking, etc. Moreover, the partitioning rule for the investment up to the planned one and beyond it is usually different.

The partitioning problem is very difficult and somehow beyond the scope of this model as the partitioning rules are usually external, e.g. imposed by some regulation or higher level authorities. For readability, the partitioning rule will not be dealt with here, and we will only assume that some rational rule is used.

The temporal evolution of the particular life quality indicators is assumed to be governed by the state transition equation

$$x_t^i = f_{t-1}^i(x_{t-1}^i, u_{t-1}^i), \qquad i = 1, \ldots, 7; t = 1, \ldots, N \tag{8.1}$$

which describes the transition from X_{t-1} to X_t under investments $u_{t-1}^1, \ldots, u_{t-1}^7$. This state transition equation may be derived by, e.g., using experts' opinions, past experience, etc.

The transition from X_{t-1} to X_t should now be assessed, or better, evaluated. This is not a straightforward task and requires a more thorough analysis to be presented below.

8.1.2 Evaluation of development

An adequate assessment, and more so evaluation of regional development, is by no means easy. First, there are many bodies and parties involved, including mainly the inhabitants for whom the development does virtually proceed, regional authorities who are responsible for the development, some supervising higher level authorities as, e.g., provincial or governmental agencies, who consider the development of the region as part of a larger problem of, say, the development of the country. The very interests of those parties may be different, even divergent to a large extent. Second, the evaluation has to consider both tangible and intangible, and objective and subjective aspects. This does greatly complicate the evaluation, and it will be shown that a fuzzy approach here may provide adequate and simple tools.

The development is evidently a goal oriented task that is aimed at the satisfaction of some needs. The essence of its evaluation is therefore to provide some measures of how well some predetermined goals are fulfilled, i.e. to find how *effective* the development is. This should then evidently be related to the investment spent, to give an idea of how *efficient* it is. Moreover, as the practice indicates, the above evaluation of "costs" and "benefits" may often lead to outcomes being inconsistent with the social perception of the development's "goodness". For instance, the developments may be satisfactory taking the above cost–benefit analysis alone, but may be viewed as unsatisfactory because, e.g., of too high a variability of some development indicators.

These two important aspects, referred to as:

- the effectiveness of development, and
- the stability of development

will be discussed below.

8.1.2.1 Effectiveness of development

The *effectiveness* of regional development means how well the fuzzy constraints are satisfied and the fuzzy goals are attained. It evidently involves two aspects:

- the effectiveness of a particular development stage, and
- the effectiveness of the whole development (trajectory), i.e. of all the development stages over the whole planning horizon,

and these two aspects will now be discussed in detail.

Effectiveness of a development stage

Let us discuss first the former aspect, i.e. the *effectiveness of a particular development stage*. It has both an *objective* and *subjective* aspect.

Objective evaluation

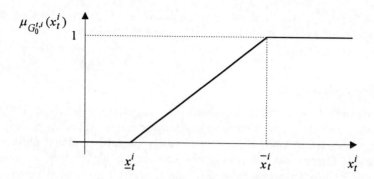

Figure 32   Objective fuzzy subgoal

The objective evaluation (of a particular development stage t) is basically the determination of how well the fuzzy constraints are fulfilled, and fuzzy goals are attained. The objective fuzzy goals concern desired values of the life quality indicators as, e.g., the mean salary, maximally admissible environmental pollution levels, etc. That is, they concern some objective entities. However, in spite of this objectivity, the goal attainment process is evidently not clear-cut, and a fuzzy goal should rather be used.

Namely, for each particular life quality indicator at development stage $t = 1, \ldots, N$, x_t^i, we define an *objective fuzzy subgoal* $G_o^{t,i}$ characterized by $\mu_{G_o^{t,i}}(x_t^i)$ as shown in Figure 32. It is to be read as follows: $G_o^{t,i}$ is fully satisfied for $x_t^i \geq \overline{x}_t^u$, where $\overline{x}_o^i$ is some *aspiration level* for the indicator x_t^i; therefore, $\mu_{G_o^{t,i}}(x_t^i) = 1$, for $x_t^i \geq \overline{x}_t^i$. Less preferable are $\underline{x}_t^i < x_t^i < \overline{x}_t^i$ for which $0 < \mu_{G_o^{t,i}}(x_t^i) < 1$, and $x_t^i \leq \underline{x}_t^i$ are assumed to be impossible, hence $\mu_{G_o^{t,i}}(x_t^i) = 0$.

Such an objective fuzzy subgoal does therefore simply and adequately formalize the "softness" of the goal attainment process, and a clear satisfaction type attitude that occurs in many real situations. Needless to say such a fuzzy (sub)goal may be relatively easily determined by experts as two values, $\underline{x}t_t^i$ and $\overline{x}_t^i$ only need to be specified.

To obtain the objective evaluation of the whole life quality index at development stage t $X_t = [x_t^1, \ldots, x_t^7]$, we need to aggregate the partial assessments of the particular life quality indicators which may be generally written as

$$\mu_{G_o^t}(X_t) = \mathrm{AGG}[\mu_{G_o^{t,1}}(x_t^1), \ldots, \mu_{G_o^{t,7}}(x_t^7)] \tag{8.2}$$

where $\mathrm{AGG} : [0,1]^7 \longrightarrow [0,1]$ is some aggregation operation. In the source works of Kacprzyk and Straszak (1980–86) this aggregation operation was equated with a connective (corresponding to "and") in fuzzy sets theory, notably with the minimum operation ["$\wedge$" - cf. (2.31), page 30], i.e.

$$\mu_{G_o^t}(X_t) = \mu_{G_o^{t,1}}(x_t^1) \wedge \ldots \wedge \mu_{G_o^{t,7}}(x_t^7) \tag{8.3}$$

and, as we may remember from Section 2.1.3, "$\wedge$" may be replaced by another suitable operation as, e.g., a t-norm [cf. (2.33), page 31], i.e.

$$\mu_{G_o^t}(X_t) = \mu_{G_o^{t,1}}(x_t^1) \, t \ldots t \, \mu_{G_o^{t,7}}(x_t^7) \tag{8.4}$$

or, say, the weighted sum, i.e.

$$\mu_{G_o^t}(X_t) = w_1 \cdot \mu_{G_o^{t,1}}(x_t^1) + \cdots + w_7 \cdot \mu_{G_o^{t,7}}(x_t^7) \tag{8.5}$$

where $w_1, \ldots, w_7 \in [0,1]$, and $w_1 + \cdots + w_7 = 1$. Evidently, other operations can also be used such as the maximum ["$\vee$" – cf. (2.32), page 30], i.e.

$$\mu_{G_o^t}(X_t) = \mu_{G_o^{t,1}}(x_t^1) \vee \ldots \vee \mu_{G_o^{t,7}}(x_t^7) \tag{8.6}$$

or an s-norm [cf. (2.37), page 32], i.e.

$$\mu_{G_o^t}(X_t) = \mu_{G_o^{t,1}}(x_t^1) \, s \, \ldots \, s \, \mu_{G_o^{t,7}}(x_t^7) \tag{8.7}$$

It may be seen that the use of the minimum (8.3) reflects a pessimistic, safety-first attitude, and a lack of substitutability (i.e. that a low value of one life quality indicator cannot be compensated by a higher value of another), the weighted sum (8.5) represents a moderate, but controllable attitude as to the pessimism/optimism and substitutability, while the maximum (8.6) – an optimistic attitude, and a high (full) substitutability. Other t-norms and s-norms correspond to intermediate cases.

An interesting aggregation is here provided by the fuzzy linguistic quantifiers presented in Section 2.3, and whose use in control under fuzziness has been presented in Section 4.1.5.

A fuzzy linguistic quantifier based aggregation may be generally written as

$$\mu_{G_o^t}(X_t) = \text{AGG}_Q[\mu_{G_o^{t,1}}(x_t^1), \ldots, \mu_{G_o^{t,7}}(x_t^7)] \tag{8.8}$$

where Q is a fuzzy linguistic quantifier exemplified by *most, almost all, much more than 50%*, ..., and in such an aggregation we take into account, say, $Q=$"*most*" objective fuzzy subgoals.

The above fuzzy linguistic quantifier based aggregation may proceed via fuzzy logic based calculi of linguistically quantified statements presented in Section 2.3 or via the ordered weighted averaging (OWA) operators presented in Section 2.3.3.

In general, the choice of a proper aggregation procedure is a crucial element of the model development. It should carefully be chosen to reflect a real perception of the problem specifics as seen by experts, or even better, by "customers" exemplified by, say, regional authorities.

Finally, note that the objective evaluation, by its very essence, concerns the authorities more than the inhabitants as it somehow "mechanically" checks the values of life quality indicators attained against some predefined desired levels. The inhabitants' perception of the above may unfortunately be different, as we will see below.

Subjective evaluation

The inhabitants' assessment of the "goodness" of development concerns in fact the (perception of) *social satisfaction* resulting from the life quality index attained. This is clearly *subjective*.

The attained value of a particular life quality indicator at development stage t, x_t^i, implies its corresponding partial *social satisfaction* s_t^i that is derived as shown in

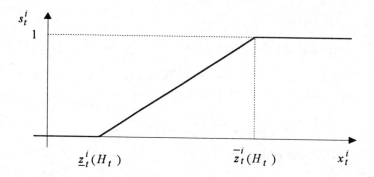

Figure 33 Partial social satisfaction

Figure 33. Its interpretation is basically the same as in the case of objective evaluation shown in Figure 32. That is, the partial social satisfaction resulting from x_t^i is complete, i.e. $s_t^i = 1$, above some aspiration level $\bar{z}_t^i$. Its is partial (intermediate), i.e. $0 < s_t^u < 1$, for s_t^i between $\underline{z}_t^i$ and $\bar{z}_y^i$, and is null (i.e. full dissatisfaction) for s_t^i below $\underline{z}_t^i$. Note that both $\underline{z}_t^i$ and $\bar{z}_t^i$ are now assumed to be some functions of the trajectory (history) of development, H_t, defined as

$$H_t = [(X_1, s_1, \mu_{G_o^1}(X_1), \mu_{G_s^1}(s_1)), \ldots, (X_t, s_t, \mu_{G_o^t}(s_t), \mu_{G_s^t}(s_t))] \qquad (8.9)$$

where $s_k = [s_k^1, \ldots, s_k^7], k = 1, \ldots, t$, is the social satisfaction resulting from X_k, i.e. the (vector of) values of the partial social satisfactions resulting from the values of the life quality indicators attained at development stage k.

The trajectory involves therefore the consecutive (values of) life quality indices attained, their resulting social satisfactions, the objective fuzzy goals, $G\mu_{G^k}(X_k)$, and the subjective fuzzy goals (to be defined a little later), $\mu_{G^k}(s_k), k = 1, \ldots, t$.

The trajectory of development H_t subsumes therefore the goodness of the past development. If the trajectory, i.e. development, is encouraging, then the inhabitants may become more demanding, and $\underline{z}_t^i(H_t)$ and $\bar{z}_t^i(H_t)$ may move up. On the other hand, if H_t is discouraging, then $\underline{z}_t^i(H_t)$ and $\bar{z}_t^i(H_t)$ may move down.

The dynamics of this shifting process is evidently very difficult to identify and express formally. A good practical solution (cf. Kacprzyk, 1983b) may be to use some known smoothing technique, and the first order exponential smoothing and logarithmic smoothing seem to be well suited since they are simple and have proven their usefulness in a wide spectrum of diverse areas.

They are:

- the exponential smoothing

$$\begin{cases} \underline{z}_{t+1}^i = \underline{z}_t^i + \frac{1}{r}(s_t^i - \underline{z}_t^i) \\ \bar{z}_{t+1}^i = \bar{z}_t^i + \frac{1}{r}(s_t^i - \bar{z}_t^i) \end{cases} \qquad (8.10)$$

- the logarithmic smoothing

$$\begin{cases} \underline{z}_{t+1}^i = \underline{z}_t^i (\frac{s_t^i}{\underline{z}_t^i})^{1/r} \\ \bar{z}_{t+1}^i = \bar{z}_t^i (\frac{s_t^i}{\bar{z}_t^i})^{1/r} \end{cases} \qquad (8.11)$$

where $r > 1$ stands for the case of a gradual adaptation, and $r < 0.5$ for the case of fast oscillation, with intermediate values corresponding to the situations in between.

For simplicity and practical reasons, we will basically assume that $\underline{z}_t^i$ and $\overline{z}_t^i$ depend an the "reduced trajectory" defined as

$$h_t = [(X_{t-1}, s_{t-1}, \mu_{G_o^{t-1}}(X_{t-1}), \mu_{G_s^{t-1}}(s_{t-1})), (X_t, s_t, \mu_{G_o^t}(s_t), \mu_{G_s^t}(s_t))] \quad (8.12)$$

i.e. which takes into account the outcomes of the two recent development stage only, t and $t-1$.

The social satisfaction at development stage t is now

$$s_t = \text{AGG}(s_t^1, \ldots, s_t^7) \quad (8.13)$$

where "AGG" is an aggregation operation, that may be defined similarly as (8.2)–(8.7), i.e.

- for the minimum operation

$$s_t = s_t^1 \wedge \ldots \wedge s_t^7 \quad (8.14)$$

- for a t-norm, in general,

$$s_t = s_t^1 \, t \, \ldots \, t \, s_t^7 \quad (8.15)$$

- for the weighted sum

$$s_t = w_1 \cdot s_t^1 + \cdots + w_7 \cdot s_t^7 \quad (8.16)$$

where $w_1, \ldots w_7 \in [0, 1]$, and $w_1 + \cdots + w_7 = 1$,
- for the maximum

$$s_t = s_t^1 \vee \ldots \vee s_t^7 \quad (8.17)$$

- for an s-norm, in general,

$$s_t = s_t^1 \, s \, \ldots \, s \, s_t^7 \quad (8.18)$$

Note that, as for the objective evaluation (8.2)–(8.7), the use of the minimum again reflects a pessimistic, safety-first attitude, and a lack of substitutability, the weighted sum represents a moderate, but controllable attitude as to the pessimism/optimism and substitutability, while the maximum an optimistic attitude, and a high (full) substitutability. Other t-norms and s-norms correspond to intermediate cases.

Moreover, as in (8.8), one can also use the fuzzy linguistic-quantifier-based aggregation generally written as

$$s_t = \text{AGG}_Q(s_t^1, \ldots, s_t^7) \quad (8.19)$$

where Q is a fuzzy linguistic quantifier. Then, analogously as for (8.8), this aggregation may proceed both by using a fuzzy-logic-based calculus of linguistically quantified statements or by employing an OWA operator.

The social satisfaction s_t is now subjected to a subjective fuzzy goal $\mu_{G_s^t}(s_t)$ shown in Figure 34 to be meant as similar previous figures.

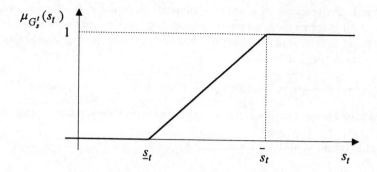

Figure 34   Subjective fuzzy goal

We have therefore developed some measures of the inhabitants' evaluation of the development that has clearly a subjective character. We will now summarize our discussion combining the objective and subjective aspects of the evaluation.

Effectiveness of development

The effectiveness of a development stage t is basically meant as a relation of what has been attained (in terms of the life quality indices and their respective social satisfactions) to what has been "paid for" (in terms of respective investments); it is therefore some benefit–cost relationship.

Formally, the (fuzzy) effectiveness of stage t is expressed as

$$\mu_{E^t}(u_{t-1}, X_t, s_t) = \text{AGG}[\mu_{C^{t-1}}(u_{t-1}), \mu_{G_o^t}(X_t), \mu_{G_s^t}(s_t)] \qquad (8.20)$$

where $\text{AGG} : [0,1]^3 \longrightarrow [0,1]$ is some aggregation operation of the partial scores, analogous to (8.2)–(8.7) and (8.14)–(8.18), exemplified by

- for the minimum operation

$$\mu_{E^t}(u_{t-1}, X_t, s_t) = \mu_{C^{t-1}}(u_{t-1}) \wedge \mu_{G_o^t}(X_t) \wedge \mu_{G_s^t}(s_t) \qquad (8.21)$$

- for a t-norm, in general,

$$\mu_{E^t}(u_{t-1}, X_t, s_t) = \mu_{C^{t-1}}(u_{t-1}) \, t \, \mu_{G_o^t}(X_t) \, t \, \mu_{G_s^t}(s_t) \qquad (8.22)$$

- for the weighted sum

$$\mu_{E^t}(u_{t-1}, X_t, s_t) = w_1 \cdot \mu_{C^{t-1}}(u_{t-1}) + w_2 \cdot \mu_{G_o^t}(X_t) + w_3 \cdot \mu_{G_s^t}(s_t) \qquad (8.23)$$

where $w_1, w_2, w_3 \in [0,1]$, and $w_1 + w_2 + w_3 = 1$,
- for the maximum operation

$$\mu_{E^t}(u_{t-1}, X_t, s_t) = \mu_{C^{t-1}}(u_{t-1}) \vee \mu_{G_o^t}(X_t) \vee \mu_{G_s^t}(s_t) \qquad (8.24)$$

- for an s-norm, in general,

$$\mu_{E^t}(u_{t-1}, X_t, s_t) = \mu_{C^{t-1}}(u_{t-1}) \, s \, \mu_{G_o^t}(X_t) \, s \, \mu_{G_s^t}(s_t) \qquad (8.25)$$

In this case the aggregation operations reflect the very nature of a compromise between the interests of the authorities (for whom the fuzzy constraints and the objective fuzzy goal matter), and those of the inhabitants (for whom the subjective fuzzy goal, and to some extent the objective fuzzy goal, matter). The use of the minimum reflects a safety-first attitude, hence a "more just" compromise, i.e. acceptable to both the parties, while the maximum stands for an opposite case, with the other aggregation operations representing intermediate cases.

Moreover, one can also use the fuzzy linguistic quantifier based aggregation generally written as

$$\mu_{E^t}(u_{t-1}, X_t, s_t) = \mathrm{AGG}_Q[\mu_{C^{t-1}}(u_{t-1}), \mu_{G_o^t}(X_t), \mu_{G_s^t}(s_t)] \qquad (8.26)$$

where Q is a fuzzy linguistic quantifier. This aggregation can again proceed by using a fuzzy logic based calculus of linguistically quantified statements or an OWA operator.

Finally, the effectiveness measures of the particular development stages $t = 1, \ldots, N$, the $\mu_{E^t}(u_{t-1}, X_t, s_t)$'s given by (8.20)–(8.26), are to be aggregated to yield the fuzzy effectiveness measure for the whole development (i.e. over the whole planning horizon N) which can be generally written as

$$\mu_E(H_N) = \mathrm{AGG}[\mu_{E^1}(u_0, X_1, s_1), \ldots, \mu_{E^N}(u_{N-1}, X_N, s_n)] \qquad (8.27)$$

where $\mathrm{AGG} : [0,1]^3 \longrightarrow [0,1]$ is some aggregation operation of the partial scores, analogous to (8.2), (8.13) or (8.20), exemplified by the following:

- for the minimum operation

$$\mu_E(H_N) = \mu_{E^1}(u_0, X_1, s_1) \wedge \ldots \wedge \mu_{E^N}(u_{N-1}, X_N, s_n) \qquad (8.28)$$

- for a t-norm, in general,

$$\mu_E(H_N) = \mu_{E^1}(u_0, X_1, s_1) \, t \, \ldots \, t \, \mu_{E^N}(u_{N-1}, X_N, s_n) \qquad (8.29)$$

- for the weighted sum

$$\mu_E(H_N) = w_1 \cdot \mu_{E^1}(u_0, X_1, s_1) + \cdots + w_n \cdot \mu_{E^N}(u_{N-1}, X_N, s_n) \qquad (8.30)$$

where $w_1, \ldots, w_n \in [0, 1]$, and $w_1 + \cdots + w_n = 1$,
- for the maximum operation

$$\mu_E(H_N) = \mu_{E^1}(u_0, X_1, s_1) \vee \ldots \vee \mu_{E^N}(u_{N-1}, X_N, s_n) \qquad (8.31)$$

- for an s-norm, in general,

$$\mu_E(H_N) = \mu_{E^1}(u_0, X_1, s_1) \, s \, \ldots \, s \, \mu_{E^N}(u_{N-1}, X_N, s_n) \qquad (8.32)$$

In this case the aggregation operations reflect the very attitude toward the relevance of what will happen in the future, from a pessimistic and cautious one (represented by "$\wedge$") to an optimistic one (represented by "$\vee$"), through all intermediate attitudes.

Moreover, one can also use the fuzzy linguistic quantifier based aggregation generally written as

$$\mu_E(H_N) = \mathrm{AGG}_Q[\mu_{E^1}(u_0, X_1, s_1), \ldots, \mu_{E^N}(u_{N-1}, X_N, s_n)] \qquad (8.33)$$

where Q is a fuzzy linguistic quantifier, and this aggregation may proceed – similarly as in all previous cases – by using a fuzzy logic based calculus of linguistically quantified propositions or an OWA operator.

We have therefore a fuzzy measure of effectiveness of the whole development path. Unfortunately, even if the development is positively evaluated against this measure, its actual perception – by both the authorities and inhabitants – may be much less favorable. One of main reasons may be too high a variability of some development indices and characteristics. This important issue, termed the stability of regional development, will now be considered.

8.1.3   Stability of development

The *stability* of regional development is here meant to concern the variability of (crucial) development indicators, characteristics, conditions, etc. Namely, it is known from experience, and this is confirmed by psychological investigations, that a limited variability usually implies a higher acceptance of development, by both the inhabitants and authorities, than a high variability.

The stability in the above sense is here meant to involve:

- the *stability of development trajectory* that concerns the variability of development outcomes, i.e. the life quality indicators attained and the resulting social satisfactions, and
- the *stability of development "policy"* that concerns the variability of some development prerequisites, i.e. the imposed fuzzy constraints, fuzzy goals, and investment partitioning rules.

It is quite clear that both the above stability types are "soft" concepts, and their measures will now be developed by using fuzzy tools.

8.1.3.1   Stability of development trajectory

This type of regional development stability is actually equivalent to the requirement that the variability of development outcomes (life quality indicators and their resulting social satisfactions) should be possibly low.

The variability of development trajectory now has the following four aspects:

- the variability of life quality indicators over time (development stages),
- the variability of social satisfactions over time,
- the variability "across" the life quality indices, and
- the variability "across" the social satisfactions.

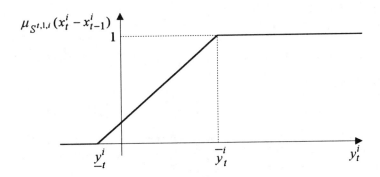

Figure 35   Fuzzy limitation on the variability of the i-th life quality indicator, x_t^i

The variability of a particular life quality indicator x_t^i – assuming for simplicity, but what is sufficient in virtually all applications, that the variability is meant only as a difference between the current and the previous value – is $x_t^i - x_{t-1}^i$, and is subjected to a *fuzzy limitation* $\mu_{S^{t,1,i}}(x_t^i - x_{t-1}^i)$ of the form shown in Figure 35 may be explained as follows. We are fully satisfied (degree of satisfaction equals 1) with the variability (change) $x_t^i - x_{t-1}^i \geq \overline{y}_t^i$, we are partially satisfied (degree of satisfaction between 0 and 1) with $\underline{y}_t^i \leq x_t^i - x_{t-1}^i < \overline{y}_t^i$, and we are fully dissatisfied (degree of satisfaction equals 0) with the values $x_t^i - x_{t-1}^i < \underline{y}_t^i$, i.e. such values are unacceptable. Note that $\underline{y}_t^i < 0$ may occur, i.e. x_t^i may fall slightly below x_{t-1}^i, but not too much – we allow therefore some interruption in growth (though with a low satisfaction) if it is sufficiently small.

So, such a measure of variability may also be viewed as a *growth requirement*. The higher the growth is required or envisaged, the higher $\underline{y}_t^i$; $\underline{y}_t^i$ can even be positive if a constant growth only is permissible. On the other hand, if not only the growth may occur, $\underline{y}_t^i$ may be assumed appropriately low. Clearly, the above stability limitation, as well as other stability limitations to be discussed, is some form of fuzzy constraint which is termed in a different way to differentiate it from other elements of the model.

The variability limitation for the life quality index $X_t = (x_t^1, \ldots, x_t^7)$ is now derived as

$$\mu_{S^{t,1}}(X_t - X_{t-1}) = \mathrm{AGG}[\mu_{S^{t,1,1}}(x_t^1 - x_{t-1}^1), \ldots, \mu_{S^{t,1,7}}(x_t^7 - x_{t-1}^7)] \qquad (8.34)$$

where $\mathrm{AGG} : [0,1]^7 \longrightarrow [0,1]$ is some aggregation operation of the partial scores, analogous to (8.2), (8.13), (8.20) or (8.27), and exemplified by the following:

- for the minimum operation

$$\mu_{S^{t,1}}(X_t - X_{t-1}) = \mu_{S^{t,1,1}}(x_t^1 - x_{t-1}^1) \wedge \ldots \wedge \mu_{S^{t,1,7}}(x_t^7 - x_{t-1}^7) \quad (8.35)$$

- for a t-norm, in general,

$$\mu_{S^{t,1}}(X_t - X_{t-1}) = \mu_{S^{t,1,1}}(x_t^1 - x_{t-1}^1)\, t \, \ldots \, t \, \mu_{S^{t,1,7}}(x_t^7 - x_{t-1}^7) \quad (8.36)$$

- for the weighted sum

$$\mu_{S^{t,1}}(X_t - X_{t-1}) = w_1 \cdot \mu_{S^{t,1,1}}(x_t^1 - x_{t-1}^1) + \cdots + w_7 \cdot \mu_{S^{t,1,7}}(x_t^7 - x_{t-1}^7)$$
$$(8.37)$$

where $w_1, w_2, w_3 \in [0, 1]$, and $w_1 + w_2 + w_3 = 1$,

- for the maximum operation

$$\mu_{S^{t,1}}(X_t - X_{t-1}) = \mu_{S^{t,1,1}}(x_t^1 - x_{t-1}^1) \vee \ldots \vee \mu_{S^{t,1,7}}(x_t^7 - x_{t-1}^7) \quad (8.38)$$

- for an s-norm, in general,

$$\mu_{S^{t,1}}(X_t - X_{t-1}) = \mu_{S^{t,1,1}}(x_t^1 - x_{t-1}^1) \, s \, \ldots \, s \, \mu_{S^{t,1,7}}(x_t^7 - x_{t-1}^7) \quad (8.39)$$

In this case the aggregation operations reflect, analogously as in (8.2)–(8.7) mainly the substitutability of life quality indicators, from "$\wedge$" standing for the lowest (lack of) substitutability, through the weighted sum and t-norms for an intermediate one, to "$\vee$" for the highest (full) one. Moreover the aggregation operation may be used to reflect some preferred growth pattern, from a balanced growth on the one extreme (for "$\wedge$"), to an imbalanced one on the other extreme ("$\vee$"), though all intermediate cases.

We can also use the fuzzy linguistic quantifier based aggregation generally written as

$$\mu_{S^{t,1}}(X_t - X_{t-1}) = \text{AGG}_Q[\mu_{S^{t,1,1}}(x_t^1 - x_{t-1}^1), \ldots, \mu_{S^{t,1,7}}(x_t^7 - x_{t-1}^7)] \quad (8.40)$$

where Q is a fuzzy linguistic quantifier, and this aggregation may proceed as in previous cases via a fuzzy logic based calculus of linguistically quantified statements or an OWA operator.

An analogous discussion remains valid for the variability of social satisfaction s_t over time which, for simplicity, is also assumed to be evaluated for the current and previous stage, i.e. t and $t-1$, only. The fuzzy limitation on this variability is in this case given by $\mu_{S^{t,1}}(s_t - s_{t-1})$ which is defined and elicited similarly as $\mu_{S^{t,1,i}}(x_t^i - x_{t-1}^i)$ shown in Figure 35.

The second type of variability considered is what might be called "across." Basically it concerns the variability of mutual proportions between the life quality indicators and their resulting partial social satisfactions. Namely, the *variability "across" the life quality index* is evaluated by the

$$\mu_{S^{t,3}}(X_t, X_{t-1}) = 1 - \text{AGG}[\mu_{Z_x^{t,1}}(x_t^1 - x_{t-1}^1), \ldots, \mu_{Z_x^{t,7}}(x_t^7 - x_{t-1}^7)] \quad (8.41)$$

and the *variability "across" the resulting social satisfaction* is evaluated by the fuzzy limitation

$$\mu_{S^{t,4}}(s_t, s_{t-1}) = 1 - \text{AGG}[\mu_{Z_s^{t,1}}(s_t^1 - s_{t-1}^1), \ldots, \mu_{Z_s^{t,7}}(s_t^7 - s_{t-1}^7)] \quad (8.42)$$

where $\text{AGG} : [0, 1]^7 \longrightarrow [0, 1]$ is some aggregation operation of the partial scores analogous to (8.2), (8.13), (8.20), (8.27) or (8.34), and exemplified by the following:

- for the minimum operation

$$\mu_{S^{t,3}}(X_t, X_{t-1}) = 1 - [\mu_{Z_x^{t,1}}(x_t^1 - x_{t-1}^1) \wedge \ldots \wedge \mu_{Z_x^1}(x_t^7 - x_{t-1}^7)] \quad (8.43)$$

and

$$\mu_{S^{t,4}}(s_t, s_{t-1}) = 1 - [\mu_{Z_s^{t,1}}(s_t^1 - s_{t-1}^1) \wedge \ldots \wedge \mu_{Z_s^1}(s_t^7 - s_{t-1}^7)] \quad (8.44)$$

- for a t-norm, in general,

$$\mu_{S^{t,3}}(X_t, X_{t-1}) = 1 - [\mu_{Z_x^{t,1}}(x_t^1 - x_{t-1}^1)\, t \,\ldots\, t\, \mu_{Z_x^1}(x_t^7 - x_{t-1}^7)] \quad (8.45)$$

and

$$\mu_{S^{t,4}}(s_t, s_{t-1}) = 1 - [\mu_{Z_s^{t,1}}(s_t^1 - s_{t-1}^1)\, t \,\ldots\, t\, \mu_{Z_s^1})(s_t^7 - s_{t-1}^7)] \quad (8.46)$$

- for the weighted sum

$$\mu_{S^{t,3}}(X_t, X_{t-1}) = $$
$$= 1 - [w_1 \cdot \mu_{Z_x^{t,1}}(x_t^1 - x_{t-1}^1) + \cdots + w_7 \cdot \mu_{Z_x^{t,7}}(x_t^7 - x_{t-1}^7)]\,(8.47)$$

and

$$\mu_{S^{t,4}}(s_t, s_{t-1}) = 1 - [w_1 \cdot \mu_{Z_s^{t,1}}(s_t^1 - s_{t-1}^1) + \cdots + w_7 \cdot \mu_{Z_s^{t,7}}(s_t^7 - s_{t-1}^7)] \quad (8.48)$$

where $w_1, \ldots, w_7 \in [0, 1]$, and $w_1 + \cdots + w_7 = 1$,

- for the maximum operation

$$\mu_{S^{t,3}}(X_t, X_{t-1}) = 1 - [\mu_{Z_x^{t,1}}(x_t^1 - x_{t-1}^1) \vee \ldots \vee \mu_{Z_x^1}(x_t^7 - x_{t-1}^7)] \quad (8.49)$$

and

$$\mu_{S^{t,4}}(s_t, s_{t-1}) = 1 - [w_1 \cdot \mu_{Z_s^{t,1}}(s_t^1 - s_{t-1}^1) + \cdots + w_7 \cdot \mu_{Z_s^{t,7}}(s_t^7 - s_{t-1}^7)] \quad (8.50)$$

- for an s-norm

$$\mu_{S^{t,3}}(X_t, X_{t-1}) = 1 - [\mu_{Z_x^{t,1}}(x_t^1 - x_{t-1}^1)\, s \,\ldots\, s\, \mu_{Z_x^1}(x_t^7 - x_{t-1}^7)] \quad (8.51)$$

and

$$\mu_{S^{t,4}}(s_t, s_{t-1}) = 1 - [\mu_{Z_s^{t,1}}(s_t^1 - s_{t-1}^1)\, s \,\ldots\, s\, \mu_{Z_s^1})(s_t^7 - s_{t-1}^7)] \quad (8.52)$$

Here the aggregation operations reflect, analogously as in, e.g., (8.34)–(8.39), mainly the substitutability of life quality indicators, from "∧" standing for the lowest (lack of) substitutability, through the weighted sum and t-norms for an intermediate one, to "∨" for the highest (full) one. Moreover the AGG operation may be used to reflect some preferred growth pattern, from a balanced growth on the one extreme (for "∧"), to a imbalanced one on the other extreme ("∨"), through all intermediate cases. On the other hand, the respective $\mu_{Z_x^{t,i}}(.)$'s and $\mu_{Z_s^{t,i}}(.)$'s, which are represented analogously as in Figure 35, reflect a "flexible" growth oriented attitude, i.e. that a growth is preferable but a small decrease may sometimes be acceptable, though not clearly preferable.

Moreover, one can also use the fuzzy linguistic quantifier based aggregation generally written as

$$\mu_{S^{t,1}}(X_t - X_{t-1}) = \text{AGG}_Q[\mu_{S^{t,1,1}}(x_t^1 - x_{t-1}^1), \ldots, \mu_{S^{t,1,7}}(x_t^7 - x_{t-1}^7)] \quad (8.53)$$

where Q is a fuzzy linguistic quantifier, and this aggregation may proceed via a fuzzy logic based calculus of linguistically quantified statements or by using an OWA operator.

The total fuzzy limitation on the stability of development stage t is therefore

$$
\begin{aligned}
\mu_{S^{t,d}}(X_t, s_t) = \\
= \ \mathrm{AGG}[\mu_{S^{t,1}}(X - t - X_{t-1}), \mu_{S^{t,2}}(s_t - s_{t-1}), \\
\mu_{S^{t,3}}(X_t, X_{t-1}), \mu_{S^{t,4}}(s_t, s_{t-1})]
\end{aligned}
\tag{8.54}
$$

where the essence and choice of the AGG $: [0,1]^4 \longrightarrow [0,1]$ is analogous to the case of the variabilities across the life quality indicators and their resulting satisfactions given just before, and one may use the minimum, t-norm, weighted sum, maximum or s-norm operation. Moreover, a fuzzy linguistic quantifier based aggregation, $\mathrm{AGG}_Q(.)$ may also be used.

And, finally, the fuzzy limitation on the stability of the whole development trajectory, i.e. for $t = 1, \dots, N$, is evaluated by the following fuzzy limitation:

$$
\mu_{S_d}(H_t) = \mathrm{AGG}[\mu_{S^{1,d}}(X_1, s_1), \dots, \mu_{S^{N,d}}(X_N, s_N)]
\tag{8.55}
$$

where the essence and choice of the AGG $: [0,1]^N \longrightarrow [0,1]$ is analogous to the case of the variabilities across the life quality index and the resulting social satisfactions given by (8.41) and (8.42), respectively.

We have therefore derived a measure for evaluating the stability of development strategy. Notice again that this stability is more relevant for the inhabitants than for the authorities. For the latter, the type of stability to be discussed below is more important.

8.1.3.2   Stability of development policy

This important aspect of development stability concerns some development prerequisites, or elements determining and conditioning the (pattern of) development. They are called here the *development policy* or *policy*, for short, and are meant to be the following sequence (from the beginning, i.e. $t = 0$, up to the development stage t, $t = 1, \dots, N$):

$$
B_t = [(C^0, A_0(u_0, i), G_o^1, G_s^1), \dots, (C^{t-1}, A_{t-1}(u_{t-1}, i), G_o^t, G_s^t)]
\tag{8.56}
$$

Therefore, the policy is here basically meant as how the successive investments are to be limited and partitioned (into parts devoted to the improvement of the particular life quality indicators), and how the successive life quality indices and their resulting partial social satisfactions should appear. It is quite evident that such a meaning of policy has nothing to do with the concept of a (control) policy employed in previous chapters of this book, i.e. a function expressing the current control as a function of the current state or of the trajectory.

As previously, for simplicity we will prefer to use the *reduced development policy*

$$
b_t = [(C^{t-2}, A_{t-2}(u_{t-2}, i), G_o^{t-1}, G_s^{t-1}), (C^{t-1}, A_{t-1}(u_{t-1}, i), G_o^t, G_s^t)]
\tag{8.57}
$$

i.e. involving the last two development stages only. This is, fortunately enough, quite sufficient in virtually all applications.

The *stability of policy* concerns the variability of the following elements:

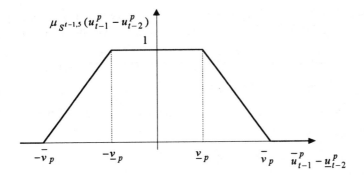

Figure 36 Fuzzy limitation on the variability of the planned investment u_{t-1}^p,

i.e. $u_{t-1}^p - u_{t-2}^p$

- the fuzzy constraints,
- the investment partitioning rules,
- the objective fuzzy goals, and
- the subjective fuzzy goals.

The first aspect is the stability of a fuzzy constraint. If we look at the form of a fuzzy constraint (cf. Figure 31), we can recognize the following two elements: u_{t-1}^p and u_{t-1}^c. The planned investment, u_{t-1}^p, is much more crucial here. Namely, it is quite clear that high fluctuations of the planned investment, which may result from a varying attitude of higher level authorities with respect to the particular region, make the intra-regional planning, and then management, more difficult, and are therefore undesirable. The *variability of the planned investment* u_{t-1}^p, i.e. $u_{t-1}^p - u_{t-2}^p$, is hence subjected to a fuzzy limitation $\mu_{S^{t-1,s}}(u_{t-1}^p - u_{t-2}^p)$ of the form shown in Figure 36.

This limitation may be explained as follows: some variability, up to $\underline{v}_p$ in both directions (i.e. between $-\underline{v}_p$ and $\underline{v}_p$), is fully allowable and acceptable (with the grade of membership equal to 1). A higher variability, i.e. between $\underline{v}_p$ and $\bar{v}_p$ (and between $\underline{v}_p$ and $\bar{v}_p$) is still possible though less desirable (with the grade of membership between 0 and 1). Finally, the variability above $\bar{v}_p$ and below $-\bar{v}_p$ is unacceptable (with the grade of membership equal to 0). Notice that we define here again the variability with respect to the current and the last development stage only, i.e. $t-1$ and $t-2$, which is fully sufficient.

The second aspect of the stability of a fuzzy constraint is the variability of its "softness" (fuzziness). First, we define a *degree of softness* of a fuzzy constraint at stage $t-1$ as

$$ds_{t-1} = \frac{u_{t-1}^c - u_{t-1}^p}{u_{t-1}^p} \qquad (8.58)$$

that is, the higher the relative difference between the maximum contingency investment and the planned investment, i.e. the "flatter" the membership function of the fuzzy constraint, the higher its softness [cf. (4.197), page 168].

The *variability of softness* of a fuzzy constraint may now be expressed as

$$vf_{t-1} = |ds_{t-1} - ds_{t-2}| \qquad (8.59)$$

which is subjected to a fuzzy limitation $\mu_{S^{t-1,6}}(vf_{t-1})$ defined analogously as in Figure 36. Some variability of the softness is therefore also allowable but not too high.

The stability limitation on the variability of a fuzzy constraint at stage $t-1$ is now

$$\mu_{S^{t-1},c}(b_t) = \text{AGG}[\mu_{S^{t-1,5}}(u^p_{t-1} - u^p_{t-2}), \mu_{S^{t-1,6}}(vf_{t-1})] \qquad (8.60)$$

where AGG $: [0,1]^2 \longrightarrow [0,1]$ is an aggregation operation exemplified by:

- the minimum operation

$$\mu_{S^{t-1},c}(b_t) = \mu_{S^{t-1,5}}(u^p_{t-1} - u^p_{t-2}) \wedge \mu_{S^{t-1,6}}(vf_{t-1}) \qquad (8.61)$$

- a t-norm, in general,

$$\mu_{S^{t-1},c}(b_t) = \mu_{S^{t-1,5}}(u^p_{t-1} - u^p_{t-2}) \, t \, \mu_{S^{t-1,6}}(vf_{t-1}) \qquad (8.62)$$

- the weighted sum

$$\mu_{S^{t-1},c}(b_t) = w_1 \cdot \mu_{S^{t-1,5}}(u^p_{t-1} - u^p_{t-2}) + w_2 \cdot \mu_{S^{t-1,6}}(vf_{t-1}) \qquad (8.63)$$

where $w_1, w_2 \in [0,1]$, and $w_1 + w_2 = 1$,
- the maximum operation

$$\mu_{S^{t-1},c}(b_t) = \mu_{S^{t-1,5}}(u^p_{t-1} - u^p_{t-2}) \vee \mu_{S^{t-1,6}}(vf_{t-1}) \qquad (8.64)$$

- an s-norm, in general,

$$\mu_{S^{t-1},c}(b_t) = \mu_{S^{t-1,5}}(u^p_{t-1} - u^p_{t-2}) \, s \, \mu_{S^{t-1,6}}(vf_{t-1}) \qquad (8.65)$$

and the aggregation operations range from the most conservative, pessimistic, and of a safety-first type, i.e. "$\wedge$," on the one extreme, to the most optimistic one, i.e. "$\vee$," on the other extreme. Evidently, the linguistic quantifier based and OWA based aggregation makes here no sense as there are only two terms to be aggregated.

The fuzzy limitation (8.60) expresses therefore a desire of the authorities to operate under possibly low varying available investments and possibly high allowable flexibility.

An analogous argument can be applied for the objective and subjective fuzzy goals. Namely:

- the stability limitation on the objective fuzzy goal at stage t is

$$\mu_{S^t,G_o}(b_t) =$$
$$= \text{AGG}\{\text{AGG}_1[\mu_{S^t,G_o,1}(\overline{x}^1_t - \overline{x}^1_{t-1}), \ldots, \mu_{S^t,G_o,7}(\overline{x}^7_t - \overline{x}^7_{t-1})],$$
$$\text{AGG}_2[\mu_{S^t,G_o,f_1}(gf^1_t), \ldots, \mu_{S^t,G_o,f_7}(gf^7_t)]\} \qquad (8.66)$$

where

$$gf^i_t = | \frac{(\overline{x}^i_t - x^i_t)}{\overline{x}^i_t} - \frac{(\overline{x}^i_{t-1} - x^i_{t-1})}{\overline{x}^i_{t-1}} |, \qquad i = 1, \ldots, 7 \qquad (8.67)$$

and both the aggregation operations AGG $: [0,1]^2 \longrightarrow [0,1]$ and $\text{AGG}_1, \text{AGG}_2 : [0,1]^7 \longrightarrow [0,1]$ are meant as (8.60), and may be represented by, e.g., the minimum, t-norm, weighted sum, maximum, s-norm etc. [cf. (8.61)–(8.65)]; moreover, for AGG_1 and AGG_2 the fuzzy linguistic quantifier based and OWA based aggregation may be employed;

- the stability limitation on the subjective fuzzy goal at stage t is

$$\mu_{S^t,G_s}(b_t) = \text{AGG}[\mu_{S^t,G_s,s}(\overline{s}_t - \overline{s}_{t-1}), \mu_{S^t,G_s,f}(sf_t)] \qquad (8.68)$$

where

$$sf_t = \left| \frac{(\overline{s}_t - \underline{s}_t)}{\overline{s}_t} - \frac{(\overline{s}_{t-1} - \underline{s}_{t-1})}{\overline{s}_{t-1}} \right| \qquad (8.69)$$

and AGG $: [0,1]^2 \longrightarrow [0,1]$ is an aggregation operations analogous to the case of (8.66) which can be exemplified by the minimum, t-norm, weighted sum, maximum, s-norm etc. [cf. (8.61)–(8.65)].

Therefore

$$\mu_{S^t,G}(b_t) = \text{AGG}[\mu_{S^t,G_o}(b_t), \mu_{S^t,G_s}(b_t)] \qquad (8.70)$$

and the aggregation operations AGG $: [0,1]^2 \longrightarrow [0,1]$ and $\text{AGG}_1, \text{AGG}_2 : [0,1,]^7 \longrightarrow [0,1]$ are meant as before, and may be represented by, e.g., the minimum, t-norm, weighted sum, maximum, s-norm, etc.

The expression (8.70) expresses a desire that the development (objective and subjective) goals be possibly low varying with respect to their aspiration levels and softness.

Finally, let us proceed to the last, very important aspect of the stability of policy, namely the *stability of the investment partitioning rules*.

First, we define a *degree of variability of an investment partitioning rule* at stage $t-1$ as

$$vp_{t-1} = \sum_{i=1}^{7} | A_{t-1}(u_{t-1}, i) - A_{t-2}(u_{t-2}, i) | \qquad (8.71)$$

where the i-th partitioning rule $A_{t-1}(u_{t-1}, i)$, $i = 1, \ldots, 7$, is meant as a part (percentage) of the total investment at stage $t-1$ allotted to the improvement of the life quality indicator x_t^i.

The stability limitation on the investment partitioning rule $A_{t-1}(.)$ is now $\mu_{S^{t-1},A}(b_{t-1}) = \mu_{S^{t-1},A}(vp_{t-1})$ whose form is analogous to that shown in Figure 36, i.e. that some variability is allowed but should be basically kept as low as possible.

Hence, the *stability of development policy* at development stage t is

$$\mu_{S^t,I}(b_t) = \text{AGG}[\mu_{S^{t-1},C}(b_{t-1}), \mu_{S^{t-1},A}(b_{t-1}), \mu_{S^t,G_o}(b_t), \mu_{S^t,G_s}(b_t)] \qquad (8.72)$$

where the aggregation operation AGG $: [0,1]^4 \longrightarrow [0,1]$ is meant as previously, and also the fuzzy linguistic quantifier based and OWA based aggregations may be employed.

The *stability of development policy* over the whole development planning horizon, i.e. for $t = 1, \ldots, N$, is therefore

$$\mu_{S_p}(B_N) = \text{AGG}[\mu_{S^1,p}(b_1), \ldots, \mu_{S^N,p}(b_N)] \qquad (8.73)$$

where the aggregation operation AGG $: [0,1]^N \longrightarrow [0,1]$ has at least a twofold character. First, it may be viewed as related to how the partial (from the particular stages) scores are aggregated as in the fuzzy decision [cf. (3.34)–(3.40), page 76]. One can also employ aggregation based on a fuzzy linguistic quantifier or an OWA operator as before. Second, the aggregation operation may also be intended to capture the discounting aspect in that the terms concerning earlier development stages should have more impact on the value of $\mu_{S_p}(.)$. This may be dealt with via the method presented in Section 3.2, and also here a fuzzy linguistic quantifier based aggregation with discounting discussed in Section 4.1.5.3 may be used.

We have therefore developed a measure for the evaluation of the stability of policy meant as some basic conditions under which the development is to proceed, and this type of stability is clearly more relevant for the authorities and other bodies responsible for the development than for the inhabitants.

The *total stability limitation of the development* is now defined as

$$\mu_S(H_N, B_N) = \text{AGG}[\mu_{S_d}(N_N), \mu_{S_p}(B_N)] \tag{8.74}$$

where the aggregation operation AGG : $[0, 1]^2 \longrightarrow [0, 1]$, as previously, is intended to properly reflect a compromise between the interests of authorities and inhabitants, ranging from a more "just" and balanced one using the minimum, to an "unjust," unbalanced one using the maximum, through a whole array of intermediate cases.

We are now in a position to formulate the multistage control problem for the regional development planning considered.

8.1.4   A multistage fuzzy control model of regional development planning

The fuzzy decision which is used for the evaluation of both the effectiveness and stability of the regional development is as follows:

$$
\begin{aligned}
\mu_D(u_0, \ldots, u_{N-1} \mid X_0, B_N) &= \\
&= \text{AGG}_1[\mu_E(H_N), \mu_S(H_N, B_N)] = \\
&= \text{AGG}_2[\mu_{C^0}(u_0), \mu_{G_o^1}(X_1), \mu_{G_s^1}(s_1), \mu_{S^{1,d}}(X_1, s_1), \mu_{S^{1,p}}(b_1), \ldots \\
&\qquad \ldots, \mu_{C^{N-1}}(u_{N-1}), \mu_{G_o^N}(X_N), \mu_{G_s^N}(s_N), \mu_{S^{N,d}}(X_N, s_N), \mu_{S^{N,p}}(b_N)] \tag{8.75}
\end{aligned}
$$

where $\text{AGG}_1 : [0, 1]^2 \longrightarrow [0, 1]$ and $\text{AGG}_2 : [0, 1]^N \longrightarrow [0, 1]$ are some aggregation operations meant analogously as their respective counterparts presented in Section 8.1.3.

The fuzzy decision expresses therefore some crucial compromises between, e.g.:

- the fuzzy constraints, fuzzy goals, and fuzzy stability limitations,
- the effectiveness and stability of development,
- the interests of the authorities and inhabitants, etc.

An adequate choice of the aggregation operation(s) is very important not only to properly reflect the intended types of the above compromises, but also to reflect many other factors as, e.g., general attitudes as to various regional development factors, preferred growth patterns, a possibility of compensating false decisions (by future decisions), etc. Rules for choosing an appropriate aggregation operation follow the lines of reasoning for choosing these operations sketched in the previous section.

The problem is now to find an optimal sequence of controls (investments) $u_0^*, \ldots, u_{N-1}^*$ such that (under a given policy B_N)

$$\mu_D(u_0^*, \ldots, u_{N-1}^* \mid X_0, B_N) =$$

$$= \max_{u_0,\dots,u_{N-1}} \mu_D(u_0,\dots,u_{N-1} \mid X_0, B_N) \qquad (8.76)$$

or, in the case when additionally the policy optimization is involved, to find an optimal sequence of controls (investments) $u_0^*,\dots,u_{N-1}^*$ and an optimal policy B_N^* such that

$$\mu_D(u_0^*,\dots,u_{N-1}^* \mid X_0, B_N^*) =$$
$$= \max_{u_0,\dots,u_{N-1},B_N} \mu_D(u_0,\dots,u_{N-1} \mid X_0, B_N) \qquad (8.77)$$

It is obvious that the policy optimization is in virtually all practical cases limited to the consideration of a very limited number of policy scenarios.

The problem (8.76) [and eventually also (8.77)] may be solved, at least for some types of the aggregation operation. For instance, in the case of employing the minimum operation the problem (8.76) becomes that of finding an optimal sequence of controls $u_0^*,\dots,u_{N-1}^*$ such that

$$\mu_D(u_0,\dots,u_{N-1} \mid X_0, B_N) = \mu_E(H_N) \wedge \mu_S(H_N, B_N) =$$
$$= \mu_{C^0}(u_0) \wedge \mu_{G_o^1}(X_1) \wedge \mu_{G_s^1}(s_1) \wedge$$
$$\wedge \mu_{S^1,d}(X_1, s_1) \wedge \mu_{S^1,p}(b_1) \wedge \dots$$
$$\dots \wedge \mu_{C^{N-1}}(u_{N-1}) \wedge \mu_{G_o^N}(X_N) \wedge \mu_{G_s^N}(s_N) \wedge$$
$$\wedge \mu_{S^N,d}(X_N, s_N) \wedge \mu_{S^N,p}(b_N) \qquad (8.78)$$

and similarly for the case with the additional policy optimization (8.77).

This problem, due to the "$\wedge$" employed, can be solved, e.g., by using the branch-and-bound technique presented in detail in Section 4.1.2. This procedure is efficient enough, and – for particularly large problems – can provide a good basis for constructing various faster heuristic procedures. Moreover, the basic model of the type (8.77)–(8.78) may be extended by introducing other elements discussed in this book as, say, the implicitly specified or fuzzy termination time, fuzzy transitions of life quality indicators, etc.

8.1.5 *Example of application to the development planning of a rural region*

To illustrate the use of the model presented in the previous section, we will apply it now (Kacprzyk, 1983b) to a "real" case of planning the development of a rural, or – better – a predominantly agricultural region. By necessity, this example will be considerably simplified to make the issues, entities, relations, etc. involved understandable to a wider audience.

Moreover, since in the regional data specific features of a region are very clearly and strongly manifested, hence very specific by their very nature, they may be unclear without a long introductory part. We will therefore use a mixture of real data, but from different regions, to make the example show better how the data are derived and represented, how the model operates, and what insight may be gained via the model as to some crucial regional development issues and aspects.

The region considered is predominantly agricultural. Its population is ca. 120,000 inhabitants, and its arable land is ca. 450,000 acres. For simplicity, the region's development will be considered over the next 3 development stages (assumed here to be years, for simplicity). We assume that during this time the population and arable land will remain the same.

Suppose that the life quality index consists of the following four life quality indicators:

- x_t^{I} – average subsidies in US$ per acre,
- x_t^{II} – sanitation expenditures (water and sewage) in US$ per capita,
- x_t^{III} – health care expenditures in US$ per capita, and
- x_t^{IV} – expenditures for paved roads (new roads and maintenance of the existing ones) in US$,

and the index t denotes the value of the particular life quality indicator at development stage $t = 0, 1, 2, 3$.

The above expenditure will be considered on yearly basis. Note that the life quality indicators assumed above do reflect some important issues occurring in rural regions, as, e.g.: x_t^{I} – a need for external subsidies that exists in virtually all countries, x_t^{II} – the crucial importance of water (for both crop and livestock production), x_t^{III} – the crucial importance of adequate health care in view of the aging of the rural population (at least in the region considered), and x_t^{IV} – the crucial importance of an adequate infrastructure.

Suppose now that the investments meant for regional development are partitioned into parts devoted to the improvement of the above-mentioned life quality indicators due to some fixed partitioning rule $A_{t-1}(u_{t-1}, i)$ which gives the percentage of the whole investment assigned at stage $t - 1$, to the i-th life quality indicator.

Let this investment partitioning rule be as follows:

- 5% for subsidies,
- 25% for sanitation,
- 45% for health care, and
- 25% for infrastructure.

Let the initial, i.e. at $t = 0$, values of the life quality indicators be:

$$x_0^{\mathrm{I}} = 0.5 \qquad x_0^{\mathrm{II}} = 15 \qquad x_0^{\mathrm{III}} = 27 \qquad x_0^{\mathrm{IV}} = 1,700,000$$

For clarity, we will not consider here the optimization of the investment policy, i.e. of the sequence of investments u_0, u_1, u_2, but will only take into account the following two options, or *scenarios*, of investment policy (in fact, this is the way such problems are solved in practice):

- Policy 1

$$u_0 = \$8,000,000 \quad u_1 = \$8,000,000 \quad u_2 = \$8,000,000$$

- Policy 2

$$u_0 = \$7,500,000 \quad u_1 = \$8,000,000 \quad u_2 = \$8,500,000$$

that is, in the first case, the investments are evenly distributed over time, while in the second case they increase; notice that in both the cases the total amount is the same.

Under the above two investment policies, Policy 1 and Policy 2, the values of the life quality indicators attained at the subsequent years are:

- Policy 1

Year(t)	u_t	x_t^{I}	x_t^{II}	x_t^{III}	x_t^{IV}
0	\$8,000,000				
1	\$8,000,000	0.88	16.7	30	\$2,000,000
2	\$8,000,000	0.88	16.7	30	\$2,000,000
3		0.88	16.7	30	\$2,000,000

- Policy 2

Year(t)	u_t	x_t^{I}	x_t^{II}	x_t^{III}	x_t^{IV}
0	\$7,500,000				
1	\$8,000,000	0.83	15.6	28.1	\$1,875,000
2	\$8,500,000	0.88	16.7	30	\$2,000,000
3		0.94	17.7	31.9	\$2,125,000

The above two development trajectories are now to be evaluated. For simplicity and readability we will only take into account the effectiveness of development. Moreover, the only objective evaluation will be considered.

Therefore, we need to specify the consecutive fuzzy constraints and objective fuzzy subgoals. Its is easy to see that since the piecewise linear form of them is assumed, then their definition requires the specification of the two values only (cf. Figure 31, page 252, and Figure 32, page 254): the aspiration level (i.e. the fully acceptable value) and the lowest (or highest) possible (still acceptable) value). Suppose that they are as follows:

t

$0 \quad C^0: \quad u_0^p = \$7,500,000$
$\qquad\qquad u_0^c = \$8,500,000$

$1 \quad C^1: \quad u_1^p = \$7,750,000$
$\qquad\qquad u_1^c = \$9,000,000 \qquad G_o^{1,\text{I}} : \underline{x}_1^{\text{I}} = 0.6 \qquad\qquad \overline{x}_1^{\text{I}} = 0.85$
$\qquad\qquad\qquad\qquad\qquad\qquad\quad G_o^{1,\text{II}} : \underline{x}_1^{\text{II}} = 14 \qquad\qquad \overline{x}_1^{\text{II}} = 16$
$\qquad\qquad\qquad\qquad\qquad\qquad\quad G_o^{1,\text{III}} : \underline{x}_1^{\text{III}} = 27 \qquad\qquad \overline{x}_1^{\text{III}} = 29$
$\qquad\qquad\qquad\qquad\qquad\qquad\quad G_o^{1,\text{IV}} : \underline{x}_1^{\text{IV}} = \$1,800,000 \quad \overline{x}_1^{\text{IV}} = \$1,900,000$

$2 \quad C^2: \quad u_2^p = \$8,000,000$
$\qquad\qquad u_1^c = \$10,000,000 \quad G_o^{2,\text{I}} : \underline{x}_2^{\text{I}} = 0.7 \qquad\qquad \overline{x}_1^{\text{I}} = 0.9$
$\qquad\qquad\qquad\qquad\qquad\qquad\quad G_o^{2,\text{II}} : \underline{x}_2^{\text{II}} = 15 \qquad\qquad \overline{x}_1^{\text{II}} = 17$
$\qquad\qquad\qquad\qquad\qquad\qquad\quad G_o^{2,\text{III}} : \underline{x}_2^{\text{III}} = 28 \qquad\qquad \overline{x}_1^{\text{III}} = 30$
$\qquad\qquad\qquad\qquad\qquad\qquad\quad G_o^{2,\text{IV}} : \underline{x}_2^{\text{IV}} = \$1,900,000 \quad \overline{x}_2^{\text{IV}} = \$2,000,000$

$3 \qquad\qquad\qquad\qquad\qquad\qquad\quad G_o^{3,\text{I}} : \underline{x}_3^{\text{I}} = 0.75 \qquad\qquad \overline{x}_3^{\text{I}} = 1$
$\qquad\qquad\qquad\qquad\qquad\qquad\quad G_o^{3,\text{II}} : \underline{x}_3^{\text{II}} = 16 \qquad\qquad \overline{x}_1^{\text{II}} = 18.5$
$\qquad\qquad\qquad\qquad\qquad\qquad\quad G_o^{3,\text{III}} : \underline{x}_3^{\text{III}} = 29 \qquad\qquad \overline{x}_1^{\text{III}} = 31$
$\qquad\qquad\qquad\qquad\qquad\qquad\quad G_o^{3,\text{IV}} : \underline{x}_3^{\text{IV}} = \$1,950,000 \quad \overline{x}_3^{\text{IV}} = \$2,100,000$

Now, if we use "$\wedge$" to reflect a safety-first attitude, which is clearly preferable in the situation considered (a rural region plagued by the aging of the society, out-migration to neighboring urban areas, economic decay, etc.), then the evaluation of the two investment policies is:

- Policy 1

$$\mu_D(\$8,000,000; \$8,000,000; \$8,000,000 \mid .) =$$
$$= \mu_{C^0}(\$8,000,000) \wedge (\mu_{G_o^{1,\mathrm{I}}}(0.88) \wedge$$
$$\wedge \mu_{G_o^{1,\mathrm{II}}}(16.7) \wedge \mu_{G_o^{1,\mathrm{III}}}(30) \wedge \mu_{G_o^{1,\mathrm{IV}}}(\$2,000,000)) \wedge$$
$$\wedge \mu_{C^1}(\$8,000,000) \wedge (\mu_{G_o^{2,\mathrm{I}}}(0.88) \wedge$$
$$\wedge \mu_{G_o^{2,\mathrm{II}}}(16.7) \wedge \mu_{G_o^{2,\mathrm{III}}}(30) \wedge \mu_{G_o^{2,\mathrm{IV}}}(\$2,000,000)) \wedge$$
$$\wedge \mu_{C^2}(\$8,000,000) \wedge (\mu_{G_o^{3,\mathrm{I}}}(0.88) \wedge$$
$$\wedge \mu_{G_o^{3,\mathrm{II}}}(16.7) \wedge \mu_{G_o^{3,\mathrm{III}}}(30) \wedge \mu_{G_o^{3,\mathrm{IV}}}(\$2,000,000)) =$$
$$= 0.5 \wedge (1 \wedge 1 \wedge 1 \wedge 1) \wedge 0.8 \wedge$$
$$\wedge (0.9 \wedge 0.85 \wedge 1 \wedge 1) \wedge 1 \wedge (0.52 \wedge 0.28 \wedge 0.5 \wedge 0.33) =$$
$$= 0.5 \wedge 0.8 \wedge 0.28 = 0.28$$

- Policy 2

$$\mu_D(\$7,500,000; \$8,000,000; \$8,500,000 \mid .) =$$
$$= \mu_{C^0}(\$7,500,000) \wedge (\mu_{G_o^{1,\mathrm{I}}}(0.83) \wedge$$
$$\wedge \mu_{G_o^{1,\mathrm{II}}}(15.6) \wedge \mu_{G_o^{1,\mathrm{III}}}(28.1) \wedge \mu_{G_o^{1,\mathrm{IV}}}(\$1,875,000)) \wedge$$
$$\wedge \mu_{C^1}(\$8,000,000) \wedge (\mu_{G_o^{2,\mathrm{I}}}(0.88) \wedge$$
$$\wedge \mu_{G_o^{2,\mathrm{II}}}(16.7) \wedge \mu_{G_o^{2,\mathrm{III}}}(30) \wedge \mu_{G_o^{2,\mathrm{IV}}}(\$2,000,000)) \wedge$$
$$\wedge \mu_{C^2}(\$8,500,000) \wedge (\mu_{G_o^{3,\mathrm{I}}}(0.94) \wedge$$
$$\wedge \mu_{G_o^{3,\mathrm{II}}}(17.7) \wedge \mu_{G_o^{3,\mathrm{III}}}(31.9) \wedge \mu_{G_o^{3,\mathrm{IV}}}(\$2,125,000)) =$$
$$= 1 \wedge (0.92 \wedge 0.8 \wedge 0.55 \wedge 0.75) \wedge 0.8 \wedge$$
$$\wedge (0.9 \wedge 0.85 \wedge 1 \wedge 1) \wedge 0.75 \wedge (0.76 \wedge 0.68 \wedge 1 \wedge 1) =$$
$$= 0.55 \wedge 0.8 \wedge 0.68 = 0.55$$

The second policy is therefore better.

In a similar manner one can deal with other elements of the regional development model presented in this section as, e.g., the development stability. Moreover, one can employ diverse aggregation operators to express crucial attitudes, compromises etc. related to the particular case considered.

This concludes our analysis of the fuzzy multistage control model for regional development planning. The model has some relevant advantages. First, it makes it possible to simply and adequately formalize and reflect an inherent softness of the problem. The crucial elements of the model – the fuzzy constraints, fuzzy goals, and fuzzy stability limitations – may be elicited quite easily, e.g., from the experts. Moreover, the need for precise data, which are practically unavailable, is clearly alleviated by the very essence of the fuzzy model.

8.2 OPTIMAL FLOOD CONTROL IN A WATER RESOURCES SYSTEM

Water resources systems are vital for virtually all countries or regions, and hence have been for a long time a subject of intensive research efforts. Unfortunately, these problems are very difficult to deal with since they concern many diverse social, economic, technological, political, etc. aspects and issues, involve both quantitative and qualitative, and tangible and intangible aspects, crucial human (often subjective) assessments, etc. They are typical "soft" problems, which are of the same type as those related to (socioeconomic) regional development extensively discussed in Section 8.1 (cf. Esogbue, 1989).

These above inherent characteristic features of water resources problems do make their analysis difficult, and call for an interdisciplinary, "soft" systems analytic approach (cf. Checkland, 1976, 1979).

And, indeed, potentials of fuzzy sets based tools and techniques for the analysis and solution of water resources related problems have been recognized for a long time. One can mention here the works of Alley, Bacinello and Hipel (1979), Esogbue and Ahipo (1982a, b), Morin, Meier and Nagaraj (1989), Ragade, Hipel and Unny (1976) or Esogbue, Theologidu and Guo (1992).

In this section we will present a fuzzy approach to flood control proposed by Esogbue, Theologidu and Guo (1992). This will provide us, first, with another "hands on" example of how a fuzzy multistage decision making (control) model is developed, and it will complement a similar exposition in Sections 8.1 and 8.4. Second, it will show an advanced application of a branch-and-bound procedure, similar to the one discussed in Section 4.1.2, to a real-world problem of universal relevance.

For facilitating the following of this model's presentation by an interested reader, and his or her possible consulting of the source paper for more detail, we will in principle follow the authors' notation and exposition; these are not always the same as those employed throughout the book but will be clear.

8.2.1 A fuzzy formulation of the flood control problem

The system under control is assumed to be a geographical unit (a region, to be more specific) which is plagued by flooding. This causes huge losses, and some measures should be undertaken to minimize these losses. In general, one distinguishes between:

- *structural measures* exemplified by constructing storage reservoirs, flood walls, levees, channel improvements, etc., and
- *non-structural measures* exemplified by floodplain zoning, land use allocation, flood proofing, flood insurance, extraordinary emergency procedures, etc.

The above measures should evidently be employed simultaneously to yield desirable effects, and hence the problem is to find an optimal mix of them. Needless to say the determination of such a mix is very difficult.

In Esogbue, Theologidu and Guo's (1992) approach the problem is broken down for analysis into two phases (complemented by a third one); it is therefore similar to the class of Esogbue's (1983) models for R&D planning to be described in Section 8.3.

The first phase, which concern a regional level, is meant for the determination of optimal sequencing and optimal timing of combinations of structural and non-structural measures to reduce the regional flood damage under some budget limitations. This is solved by a branch-and-bound procedure. The second phase, which concerns the national level, is also concerned with an optimal scheduling and sequencing of flood protection measures. Each region constitutes a stage in the fuzzy multistage control formulation developed. The third phase is basically concerned with the linking, or coordination, of the models developed in the above two phases.

First, the region is assumed to be represented as a fuzzy system under control [cf.(2.140), page 61] whose state is equated with an index describing the level of total flood damage observed (or expected to occur) as a result of a mix (combination) of structural and non-structural measures which plays the role of control.

Imprecision (fuzziness) in the above system may be caused by various reasons such as:

- we are unable to *exactly* or *probabilistically* assess damage in quantitative (mainly monetary) terms, in particular in the case of the loss of human life,
- it is practically impossible to *exactly* predict the effects (utility) of the structural and non-structural measures employed.

Now, we will develop a multistage fuzzy control model for flood control.

8.2.2 A multistage fuzzy control model for flood control

As mentioned in Section 8.2.1, the regional system considered is represented by a fuzzy system under control [cf. (2.140), page 61] whose dynamics is governed by the state transtion equation

$$X_{t+1} = F(X_t, U_t), \qquad t = 0, 1, \ldots \tag{8.79}$$

where: X_t, X_{t+1} are fuzzy states at stage t which denote the level of flood damage before the control U_t and after U_{t+1} has been applied.

First, let us list the symbols and notation to be used in the model developed. The main indices are:

n – the index of a region,

k – the index of a flood control measure,

j – the index of a flood control investment level, and

i – the index of a flood damage level.

At the national level, i.e. in phase 2 mentioned in Section 8.2.1, the following notation is employed:

$C(j)$ – the membership function of a constraint for the nation,

$G(j)$ – the membership function of a goal for the nation,

$C_n(j)$ – the membership function of a constraint for region n,

$G)_n(j)$ – the membership function of a goal for region n,

$\overline{J}$ – the upper bound of total investment for the nation, and

W_n – criticality (relevance) of region n.

Here $C(j)$ and $G(j)$ are defined on the set of all the feasible investment levels for the nation, while $C_n(j)$ and $G_n(j)$ – on the set of all feasible investment levels for region n.

At the regional level, we have the following notation:

$I_n(i)$ – the membership function of initial states in region n,

$F_n(i)$ – the membership function of final states in region n,

$G_n(i)$ – the membership function of a goal on states in region n,

$\overline{J}_n$ – the upper bound for total investment for region n, and

$C_{nk}(j)$ – the membership function of constraint for measure k for region n.

Observe that at this level $I_n(i)$, $F_n(i)$ and $G_n(i)$ are defined in the set of all possible flood damage levels for region n (i.e. states), while $C_{nk}(j)$ is defined in the set of all the feasible investment levels for measure k in region n (i.e. decisions).

Moreover, let us denote by $T_{nkj}(i,i)$ a fuzzy $I \times I$ matrix representing the state transitions (8.79) for measure k in region n with investment level j, and I is the dimension of the state space (all the possible flood damage levels for region n).

8.2.2.1   The core and expanded fuzzy model of flood control for a region – Phase 1

The roots of the first model, which constitutes a core of the approach presented, are basically the multistage decision making (control) under fuzzy constraints and fuzzy goals in Bellman and Zadeh's (1970) setting presented in Section 3, and employed throughout this book.

The main elements of this model are refined as follows:

stage – the (structural or non-structural) measure for flood control,

decision (control) – the level of investment level (in dollars) for a measure, and

state – the level of flood damage (in US$).

The data used in the model are:

$I_n(i)$ – the membership function of initial states,

$G_n(i)$ – the membership function of goals on states,

$C_{nk}(j)$ – the membership function of constraint for measure k, $k = 1, 2, \ldots, K$, and

$T_{nkj}(i, i)$ – the fuzzy matrix of state transitions for measure k with investment level j, $j = 0, 1, \ldots, J$.

The model is now postulated to find optimal investment levels for region n for the particular measures $k = 1, 2, \ldots, K$, $j_{n1}^*, \ldots, i_{nK}^*$, such that they maximize

$$\Phi_n = \max_{j_{n1}, \ldots, j_{nK}} \{[C_n(j_{n1}) \wedge \ldots \wedge C_{nK}(j_{nK})] \wedge \overline{G}(F_n)\} \tag{8.80}$$

subject to:

$$F_n = T_{nKj_{nK}} \star \cdots \star T_{n1j_{n1}} \star I_n \tag{8.81}$$

where "$\star$" is the max–min composition (2.53), page 36, and

$$\overline{G}(F_n) = 1 - \parallel G_n, F_n \parallel \tag{8.82}$$

In fact, (8.80) reflects that we wish to satisfy the particular fuzzy constraints, and fuzzy goals [cf. (3.4), page 69], (8.81) represents the state transitions (8.79), and (8.82) is necessary in the case of a fuzzy state that cannot directly enter the fuzzy goal, and is the same as proposed in Kacprzyk (1979) – cf. [(4.159) and (4.160), page 153)].

The solution of the fuzzy mathematical programming problem (8.80)–(8.82) yields the following data:

j_{nk}^* – the optimal investment level for measure k, $k = 1, 2, \ldots, K$, and

Φ_n – the optimal effect of flood control program for region n.

The above core model is then expanded since, due to budgeting constraints, it may be necessary to impose a (crisp) limit on the total investment available for region n, i.e. j_n. The model is therefore modified to include this constraint by adding to (8.80)–(8.82)

$$i_{n1} + \cdots + j_{nK} \leq j_n \tag{8.83}$$

One can also add an upper bound for the total investment in region n, $\overline{J}_n$, which cannot be exceeded (we assume that j_n, though crisp, may be fuzzified, i.e. it may be exceeded to some extent – cf. virtually all definitions of fuzzy constraints, fuzzy goals and fuzzy limitations in Section 8.1).

To be more specific, we construct the following membership function of a fuzzy goal for region n:

$$G_n(j_n) = \begin{cases} 1 & \text{for } j_n < 0 \\ \Phi_n & \text{for } 0 \leq j_n \leq \overline{J}_n \\ 0 & \text{for } \overline{J}_n < j_n \end{cases} \tag{8.84}$$

The $G_n(j_n)$ determined by (8.84) is then transferred to the phase 2 model, and the j_{nk}^*'s are stored for each j_n.

8.2.2.2 The core and expanded fuzzy model of flood control for a region – Phase 2

The core and expanded models for phase 2, i.e. at the national level, are similar to their counterparts developed in Section 8.2.2.1 for the regional level.

First, we define:

stage – the region for flood control,

decision (control) – the level of total investment for region (in US$), and

state – the effect of flood control for region,

and the necessary data are:

$C_n(j)$ –the membership function of constraint for region n, $n = 1, 2, \ldots, 10$,

$G_n(j)$ – the membership function of goal for region n, $n = 1, 2, \ldots, 10$, and

W_n – a relative importance of region n, $n = 1, 2, \ldots, 10$.

The problem to be solved is then [cf. (8.80)–(8.82)] to find an optimal investment level for the particular regions, $j_1^*, \ldots, j_{10}^*$, such that

$$\Phi = \max_{j_1, \ldots, j_{10}} [R_1(j_1) + \cdots + R_{10}(j_{10})] \qquad (8.85)$$

subject to:

$$R_n(j_n) = W_n \cdot [G_n(j_n) \wedge C_n(j_n)] \qquad (8.86)$$

and

$$W_1 + \cdots + W_{10} = 1 \qquad (8.87)$$

and the sense of (8.85)–(8.87) is similar to that of (8.80)–(8.82).

By solving (8.85)–(8.87), we obtain:

j_n^* – the optimal investment level for region n, $n = 1, 2, \ldots, 10$, and

Φ – the optimal weighted sum of effects of flood control for the nation.

The above model (8.85)–(8.87) can be extended, as in the case of (8.80)–(8.82), by assuming a (crisp) limit on the total invetment for the country, j, i.e. by adding to (8.85)–(8.87)

$$j_1 + \cdots + j_{10} \leq j \qquad (8.88)$$

And again, analogously as for (8.84), if we denote the upper bound for the investment for the nation by $\overline{J}$, then we obtain the following membership function of goal at the national level

$$G(j) = \begin{cases} 1 & \text{for } j < 0 \\ \Phi & \text{for } 0 \leq j \leq \overline{J} \\ 0 & \text{for } \overline{J} < j \end{cases} \qquad (8.89)$$

and $G(j)$ is sent to the phase 3 model, while the j_n^*'s are stored for each j.

8.2.2.3   A fuzzy model for coordination – Phase 3

Finally, we will describe a linkage model whose purpose is to coordinate the two phases described in Sections 8.2.2.1 and 8.2.2.2.

The input data are:

$C(j)$ – the membership function of a constraint at the national level, and

$G(j)$ – the membership function of a national flood control goal.

The problem is now to find an optimal investment level, j^*, such that it maximizes

$$\max_{j \in [0,J]} [G(j) \wedge C(j)] \qquad (8.90)$$

and this yields as a solution:

j^* – the optimal investment level for flood control for the country, and

$\overline{\Phi}$ – the degree to which the optimal flood control plan satisfies the national objective.

The phase 3 coordination model works now as follows:

Step 1. Run the model (8.90) and find j^*.

Step 2. Using j^* and the solution stored in phase 2, find j_n^* for region $n = 1, 2, \ldots, 10$.

Step 3. Using j_n^* and the solution stored in phase 1, find j_{nk}^* for measure k in region n; $k = 1, 2, \ldots, K$, $n = 1, 2, \ldots, 10$.

8.2.3 Solution using a branch-and-bound algorithm

Esogbue, Theologidu and Guo (1992) propose to use a branch-and-bound procedure to solve the problems formulated in Sections 8.2.2.1, 8.2.2.2 and 8.2.2.3. This idea is similar to those proposed by Kacprzyk (1979), and presented in Section 4.1.2, but since the problems considered here are more complicated than those in Section 4.1.2, then a different form of the branch-and-bound procedure is given here.

We will only outline below these two branch-and-bound procedures, for the regional and national models, and will refer the interested reader for details to the source paper (Esogbue, Theologidu and Guo, 1992).

8.2.3.1   A branch-and-bound algorithm for the regional model

First, let us slightly reformulate the regional control model (8.80)–(8.83): find optimal investment levels for region n for the particular measures $k = 1, 2, \ldots, K, j_{n1}^*, \ldots, i_{nK}^*$, such that they maximize

$$\Phi_n = \max_{j_{n1}, \ldots, j_{nK}} \{ [C_n(j_{n1}) \wedge \ldots \wedge C_{nK}(j_{nK})] \wedge \overline{G}(F_n) \} \qquad (8.91)$$

subject to:

$$F_n = T_{nK j_{nK}} \star \cdots \star T_{n1 j_{n1}} \star I_n \qquad (8.92)$$

where "$\star$" is the max–min composition (2.53), page 36, and

$$\overline{G}(F_n) = 1 - \frac{1}{l} \sum_{i=1}^{l} |G_n(i), F_n(i)| \qquad (8.93)$$

and

$$j_{n1} + \cdots + j_{nK} \leq j_n \tag{8.94}$$

Let us observe that the first part of the objective function (8.91) is

$$C = C_n(j_{n1}) \wedge \ldots \wedge C_{nk}(j_{nk}) \wedge \ldots \wedge C_{nK}(j_{nK}) \tag{8.95}$$

If we denote

$$C^{(k)} = C_n(j_{n1}) \wedge \ldots \wedge C_{nk}(j_{nk}) \tag{8.96}$$

then $C^{(k)}$ is a nondecreasing function of k, i.e.

$$C^{(k(} \geq C^{(k+1)}, \qquad k = 1, \ldots, K - 1 \tag{8.97}$$

Since the second part of the objective function (8.91), $\overline{G}(F_n)$, is nondecreasing in k too, we may employ the branch-and-bound method by using $C_{nk}(j_{nk})$ as the upper bound as in Section 4.1.2 [cf. (4.17), page 92]. The flow diagram of the branch-and-bound procedure is similar to that given in Figure 18 on page 94 – see the source paper by Esogbue, Theologidu and Guo (1992) for details.

8.2.3.2   A branch-and-bound algorithm for the national model

First, let us reformulate slightly the flood control model for the national level, (8.85)–(8.87), page 277, as follows: find optimal investment levels for the particular regions, $j_1^*, \ldots, j_{10}^*$, such that

$$\Phi = \max_{j_1, \ldots, j_{10}} [R_1(j_1) + \cdots + R_n(j_n) + \cdots + R_{10}(j_{10})] \tag{8.98}$$

subject to:

$$R_n(j_n) = W_n[G_n(j_n) \wedge C_n(j_n)] \tag{8.99}$$

$$W_1 + \cdots + W_n + \cdots + W_{10} = 1 \tag{8.100}$$

and

$$j_1 + \cdots + j_n + \cdots + j_{10} \leq j \tag{8.101}$$

A branch-and-bound procedure is again employed for solving this problem. Namely, the first part of the objective function (8.98) is

$$R_1(j_1) + \cdots + R_n(j_n) \tag{8.102}$$

while the second part of (8.98) is

$$R_{n+1}(j_{n+1}) + \cdots + R_{10}(j_{10}) \tag{8.103}$$

The upper bound of the branch in the corresponding decision tree (cf. Figure 17, page 91), when in stage n, is the sum of the elements in the first part of the objective function, (8.102), and the upper bound of the second part of the objective function, (8.103), is defined as

$$H^{(n)} = \begin{cases} \sum_{n'=n+1}^{10} \max_j[R_{n'}(j)] & \text{for } n = 1, \ldots, 9 \\ 0 & \text{for } n = 10 \end{cases} \tag{8.104}$$

When, for a particular branch in the decision tree, the sum of the first part and $H^{(n)}$ is not greater than the best solution at present, this branch is not traversed; moreover, the branch for which (8.101) does not hold is not traversed either. Finally, i.e. at stage $n = 10$, a new solution is obtained and the present best solution is updated. The flow diagram of the branch-and-bound procedure is similar to that given in Figure 18 on page 94 – see the source paper by Esogbue, Theologidu and Guo (1992) for details.

8.2.4 A numerical example

In this section we will briefly present an interesting example used by Esogbue, Theologidu and Guo (1992) to illustrate the model presented.

8.2.4.1 Example of regional allocation – Phase 1

We will show below the calculations for one region, and an analogous line of reasoning can be followed for other regions.

Suppose that we have the following data:

state space – the flood damage levels labeled by fuzzy descriptors {no, slight, moderate, severe, disastrous} which correspond to, respectively, $\{1,2,3,4,5\}$;

decision space – the investment levels labeled by fuzzy descriptors {no, low, medium, high} which correspond to, respectively, $\{0,1,2,3\}$;

limit on total investment – corresponds to 4 (in a scale in which the decision space is defined).

Now we will define the fuzzy goals, constraints, etc. Notice that we do not use all the indices as in the respective models described in previous sections. For simplicity, for the indices not accounted for, the respective elements are assumed equal.

The membership function of the initial state is

$$X_0 = 0.13/\text{no} + 0.45/\text{slight} + 0.79/\text{moderate} + 1/\text{severe} + 0.88/\text{disastrous}$$

the membership function of the fuzzy goal is

$$G = 1/\text{no} + 0.75/\text{slight} + 0.5/\text{moderate} + 0.25/\text{severe} + 0/\text{disastrous}$$

and the membership functions of fuzzy constraints for measures i, $i = 1, 2, 3, 4$, are:

$$C_1 = 1/\text{no} + 0.92/\text{low} + 0.64/\text{medium} + 0.37/\text{high}$$
$$C_2 = 1/\text{no} + 0.62/\text{low} + 0.83/\text{medium} + 0.44/\text{high}$$
$$C_3 = 1/\text{no} + 0.35/\text{low} + 0.71/\text{medium} + 0.89/\text{high}$$
$$C_4 = 1/\text{no} + 0.75/\text{low} + 0.85/\text{medium} + 0.48/\text{high}$$

Since, due to lack of space, we are unable to show all the fuzzy state transition matrices corresponding to (8.79), we will just show two of them, and these are:

$$T_1(\text{no}) = T_2(\text{no}) = T_3(\text{no}) = T_4(\text{no}) = I$$

where I is the unit matrix, and, for instance:

$$T_1(\text{low}) =$$

	no	slight	moderate	severe	disastrous
no	0.6	0.9	0.7	0.5	0.1
slight	0.1	0.6	0.9	0.7	0.5
moderate	0.0	0.1	0.6	0.9	0.7
severe	0.0	0.0	0.1	0.6	0.9
disastrous	0.0	0.0	0.0	0.1	'0.6

. . .

$$T_4(\text{high}) =$$

	no	slight	moderate	severe	disastrous
no	0.4	0.6	0.7	0.4	0.1
slight	0.3	0.4	0.6	0.7	0.4
moderate	0.2	0.3	0.4	0.6	0.7
severe	0.1	0.2	0.3	0.4	0.6
disastrous	0.0	0.1	0.2	0.3	'0.4

The optimal solution, for the particular region considered, is:

1–low   2–medium   0–no   1–low

which should be read as: the measures 1 and 4 should be invested at level 1 (i.e. low investment), measure 2 at level 2 (i.e. medium investment), and measure 3 at level 0, i.e. with no investment at all.

The value of the objective function (8.80) for this solution is 0.75.

A similar calculation can be performed for other regions but this will not be considered due to lack of space.

8.2.4.2   Example of national allocation – Phase 2

We have now the following data:

decision space – the regional investment levels labeled {no, little, low, medium, much, high} or, respectively, {0,1,2,3,4,5};

limit on total investment – corresponds to 16 in the scale introduced above for the regional investment levels;

weights – weights assigned to the particular 10 regions which are equal to: 0.08, 0.105, 0.12, 0.095, 0.13, 0.07, 0.117, 0.083, 0.112 and 0.088, respectively.

The membership functions of the fuzzy goals and fuzzy constraints for region n – G_n and C_n, respectively – are (here also, due to lack of space, we show some only):

$$G_1 = 0.6/\text{no} + 0.8/\text{little} + 1/\text{low} + 0.9/\text{medium} + 0.8/\text{much} + 0.7/\text{high}$$
$$C_1 = 1/\text{no} + 1/\text{little} + 0.8/\text{low} + 0.6/\text{medium} + 0.4/\text{much} + 0.2/\text{high}$$

$$G_2 = 0.4/\text{no} + 0.5/\text{little} + 6/\text{low} + 0.7/\text{medium} + 0.8/\text{much} + 0.9/\text{high}$$
$$C_2 = 0.8/\text{no} + 0.9/\text{little} + 0.8/\text{low} + 0.7/\text{medium} + 0.6/\text{much} + 0.5/\text{high}$$

$$\ldots$$

$$G_{10} = 0.3/\text{no} + 0.5/\text{little} + 0.7/\text{low} + 0.9/\text{medium} + 1/\text{much} + 1/\text{high}$$
$$C_{10} = 0.5/\text{no} + 0.9/\text{little} + 1/\text{low} + 0.8/\text{medium} + 0.6/\text{much} + 0.4/\text{high}$$

The optimal solution obtained is:

Region 1:	little
Region 2:	no
Region 3:	medium
Region 4:	low
Region 5:	little
Region 6:	little
Region 7:	medium
Region 8:	low
Region 9:	no
Region 10:	low

which should be read as follows: in region 1 we should invest at level 1 (i.e. little), in region 2 at level 0 (i.e. no), ..., and in region 10 at level 2 (i.e. low). The value of the objective function for such optimal investment decisions, (8.85), is 0.6436.

As reported by Esogbue, Theologidu and Guo (1992), the branch-and-bound solution procedure is numerically very efficient, and relatively easy to implement.

In this section we have presented an exposition of a two-phase (two-level) flood control model. As opposed to other models presented in virtually all sections of this chapter in which fuzzy dynamic programming was employed for the solution of problems formulated, in this section a branch-and-bound procedure was used.

8.3   RESEARCH AND DEVELOPMENT (R&D) PLANNING

In this section we will briefly present a fuzzy dynamic programming model for research and development (R&D) planning which has been proposed by Esogbue (1983).

R&D planning problems are essentially (optimal) allocation problems in which an optimal utilization of some available financial, human, technical, etc. resources is sought to meet some (corporate, regional, national, ...) needs for new products, technologies, etc.

It may easily be seen that such problems are pervaded with uncertainty and imprecision because, first, they concern the future for which, say, available resources, costs, etc. are highly uncertain. Second, the objectives are usually imprecisely stated, and there is no consensus among the experts as to what should be achieved and when, what is the most important, etc.

Traditionally, only probabilistic uncertainty has been accounted for, and a whole array of probabilistic models for the R&D planning has been proposed.

Esogbue (1983) was first to advocate the need for dealing with imprecision of fuzzy type, and he proposed some fuzzy dynamic programming models which will now be briefly discussed.

Esogbue's (1985) fuzzy models apply to situations in which all that may significantly be said is that:

- a budget allotted to a R&D project is *in the range of 1-2 billion US$*,
- goals of a R&D project can only be achieved if *much more than 100* top specialists will be employed, etc.

Basically, the model of a R&D planning problem may be stated as follows. Suppose that the R&D system consist of s divisions, $S_1, S_2, \ldots, S_s$, the total budget available is B US$, and this is assumed nonfuzzy, for simplicity but without loss of generality. Each division S_i, $i = 1, 2, \ldots, s$, consists of programs $C_{i1}, C_{i2}, \ldots, C_{ij}, \ldots, C_{in}$, all of which are to be developed. Each program, C_{ij}, $i = 1, 2, \ldots, s$, $j = 1, 2, \ldots, n$, requires the application of at least one of the strategies $T_{ij1}, T_{ij2}, \ldots, T_{ijk}, \ldots, T_{ijm}$, $i = 1, 2, \ldots, s$, $j = 1, 2, \ldots, n$. This corresponds evidently to a three-level system (i.e., division – program – strategy).

The fuzziness is assumed to occur in the values or the objectives associated with the strategies, programs and divisions. The three-level optimization problem, the maximization of achievement of the objectives, is formulated in Bellman and Zadeh's (1970) spirit, and solved using the following three fuzzy dynamic programming models concerning:

- optimal allocation of tasks,
- optimal funding of programs, and
- optimal (R&D) system's funding,

which will now be given.

8.3.1 *Optimal selection of tasks*

For the first model, concerning the optimal selection of tasks, we consider a program C_{ij} which requires at least one of the strategies $T_{ij1}, T_{ij2}, \ldots, T_{ijm}$. Suppose that x_{ij} is the budget available for program C_{ij}, and x_{ijk} is the amount allotted to strategy T_{ijk}. Let the utility associated with strategy T_{ijk} be represented by a fuzzy set $\mu_{G_{ijk}}(x_{ijk})$, $x_{ijk} \in [\underline{x}_{ijk}, \overline{x}_{ijk}]$, i.e. the amounts allotted are within a certain interval.

The constraints for program C_{ij} and strategy T_{ijk} are given as fuzzy sets $\mu_{G_{ijk}}(x_{ijk})$, and the budget constraint is

$$\sum_{k=1}^{m} x_{ijk} \leq x_{ij} \tag{8.105}$$

The fuzzy objective function for program C_{ij} is given as

$$g_{ij}(x_{ij1}, \ldots, x_{ijm}) = \mu_{G_{ij1}}(x_{ij1}) \vee \ldots \vee \mu_{G_{ijm}}(x_{ijm}) \tag{8.106}$$

The problem is clearly to maximize the fuzzy objective function (8.106) subject to the budgeting constraint (8.105), i.e.

$$
\begin{cases}
\max_{x_{ij1},\ldots,x_{ijm}} g_{ij}(x_{ij1},\ldots,x_{ijm}) \\[2mm]
\text{subject to:} \\[2mm]
\sum_{k=1}^{m} x_{ijk} \leq x_{ij} \\[1mm]
x_{ijk} \in [\underline{x}_{ijk}, \overline{x}_{ijk}]
\end{cases}
\tag{8.107}
$$

If we denote by $f_{ijk}(x_{ijk}$ the optimum value of (8.106) for program C_{ij} when x_{ij} US\$ are allotted to strategies $T_{ijk},\ldots,T_{ijm}$, and an optimal policy is employed throughout, then we obtain the following set of dynamic programming recurrence equations:

- for $k = m$, $j = 1, 2, \ldots, n$; $i = 1, 2, \ldots, s$:

$$
f_{ijm}(x_{ij}) = \max_{x_{ijm} \in [\underline{x}_{ijm}, \overline{x}_{ijm}]; x_{ijm} \leq x_{ij}} \mu_{G_{ijm}}(x_{ijm})
\tag{8.108}
$$

- for $k = 1, 2, \ldots, m-1$; $j = 1, 2, \ldots, n$; $i = 1, 2, \ldots, s$:

$$
f_{ijk}(x_{ij}) =
$$
$$
= \max_{x_{ijk} in [\underline{x}_{ijk}, \overline{x}_{ijk}]; x_{ijk} \leq x_{ij}} \{\mu_{G_{ijk}}(x_{ijk}) \vee \vee f_{ij(k-1)}(x_{ij} - x_{ijk})\}
\tag{8.109}
$$

8.3.2 Optimal funding of technologies

This model concerns the funding of various programs which contribute to objectives of the divisions. It is assumed that the subsets of programs employed to accomplish these objectives are disjoint.

Formally, suppose that in a division S_i there exist programs $C_{i1}, C_{i2}, \ldots, C_{ij}, \ldots, C_{in}$, and all the programs must succeed in attaining the S_i's objective, $j = 1, 2, \ldots, n$, $i = 1, 2, \ldots, s$.

We denote by $h_{ij}(x_{ij})$ the maximum membership value of program C_{ij} provided x_{ij} US\$ are alotted, and if an optimal policy is employed.

From the model developed in the preceeding section, Section 8.3.1, it may be noticed that [cf. (8.108) and (8.109)], for $k = 1, 2, \ldots, m$, $j = 1, 2, \ldots, n$, $i = 1, 2, \ldots, s$:

$$
h_{ij}(x_{ij}) = \max_{x_{ij} \in [0,B]} [f_{ijk}(x_{ij})]
\tag{8.110}
$$

and this may be seen to play the role of a constraint for division S_i and its associated programs C_{ij}.

The objective function for division S_i is clearly

$$
G_i(x_{i1}, x_{i2}, \ldots, x_{in_i}) =
$$
$$
= h_{i1}(x_{i1}) \wedge h_{i2}(x_{i2}) \wedge \ldots \wedge h_{ik}(x_{ik}) \wedge \ldots \wedge h_{in_i}(x_{in_i})
\tag{8.111}
$$

which is implied by the requirement that all the programs are to be fulfilled to attain the divisional objective.

If the budget available for division S_i is $x_i \in [0, B]$, then the optimization problem is, for $i = 1, 2, \ldots, s$:

$$
\begin{cases}
\max_{x_{i1}, x_{i2}, \ldots, x_{in_i}} G_i(x_{i1}, x_{i2}, \ldots, x_{in_i}) \\[2mm]
\text{subject to:} \quad \sum_{j=1}^{n_i} x_{ij} \leq x_i \in [0, B] \\[2mm]
x_{ij} \in [0, B]; j = 1, 2, \ldots, n_i
\end{cases}
\tag{8.112}
$$

If we define $f_{ij}(x_i)$ as the optimal membership value associated with division S_i when x_i US\$ are allotted to the programs (technologies) $C_{i1}, \ldots, C_{in_i}$, then – similarly as problem (8.107)– we can devise the following dynamic programming recurrence equations for solving the problem considered (8.112):

- for $j = n_i$, $i = 1, 2, \ldots, s$:

$$f_{in_i}(x_i) = \max_{x_{in_i} \in [0,B], x_{in_i} \leq x_i} [h_{in_i}(x_{in_i})] \qquad (8.113)$$

- for $j = 1, 2, \ldots, m$, $i = 1, 2, \ldots, s$:

$$f_{ij}(x_i) = \max_{x_{ij} \in [0,B], x_{ij} \leq x_i} [h_{ij}(x_{ij}) \wedge f_{i(j+1)}(x_i - x_{ij})] \qquad (8.114)$$

8.3.3   Optimal system funding

This model concerns the allocation of available resources among the divisions $S_1, S_2, \ldots, S_s$. We define $h_i(x_i)$ to be the maximal membership values of division S_i when x_i US\$ are allotted to it. From the model derived in Section 8.3.2, we have, for $i = 1, 2, \ldots, s$:

$$h_i(x_i) = \max_{x_i \in [0,B]} f_{ij}(x_{ij}) \qquad (8.115)$$

where the $f_{ij}(x_{ij})$'s are derived by solving (8.113) and (8.114).

The objective function for the entire R&D system, if we assume that the objectives of all the divisions, $S_1, S_2, \ldots, S_s$, must be fulfilled to achieve the objective of the whole R&D system, is

$$g(x_1, x_2, \ldots, x_s) = h_1(x_1) \wedge h_2(x_2) \wedge \ldots \wedge h_s(x_s) \qquad (8.116)$$

and the optimization problem is

$$\left\{ \begin{array}{l} \text{subject to:} \quad \begin{array}{l} \max_{x_1, x_2, \ldots, x_s \in [0,B]} g(x_1, x_2, \ldots, x_s) \\ \sum_{i=1}^{s} x_i \leq B \\ x_i \in [0, B]; i = 1, 2, \ldots, s \end{array} \end{array} \right. \qquad (8.117)$$

If we define $f_i(B)$ to be the maximal membership value associated with the entire R&D system with the total budget of B US\$, then the optimization problem (8.117) can be solved by the following recurrence equations:

- for $i = 3$:

$$f_s(B) = \max_{x_s \in [0,B]} h_s(x_s) \qquad (8.118)$$

- for $i = 1, 2, \ldots, s - 1$:

$$f_i(B) = \max_{x_i \in [0,B]} [h_i(x_i) - f_{i+1}(B - x_i)] \qquad (8.119)$$

These three models represent a hierarchical fund allocation procedure employed in R&D planning. Esogbue (1983) described then fuzzy dynamic programming procedured for their solution, with flow diagrams. He also gives details of a practical problem in the realm of health care planning: fund allocation in cancer research, including a cancer research appropriation process, which has been successfully solved by using the fuzzy dynamic programming models developed in Sections 8.3.1–8.3.3.

8.4 SCHEDULING GENERATION UNIT COMMITMENT IN A POWER SYSTEM

An interesting application of fuzzy dynamic programming to the determination of the so-called (generation) unit commitment in a power system was proposed by Su and Hsu (1991). We will present it below. For convenience of the reader, who may wish to consult the source paper for more detail (Su and Hsu, 1991), we will basically use the original notation, with slight changes only to make it more in line with that used in this book.

8.4.1 A general problem formulation

The problem is basically as follows. In a (national, regional, ...) power system – which is meant to cover the needs of population, industry, etc. for electrical energy – there are diverse electricity generation units: nuclear, coal-fired, oil-fired, gas-fired, water-powered, etc. Each of them clearly has its specific characteristics with respect to the power output, operational costs, efficiency, start-up time, shut-down time, etc. The demand for power varies over time, and is clearly only imprecisely known. Evidently, not all the generating units need to work all the time, and the problem is to find the most economically justified schedule of their switching on and off, the so-called *unit commitment*.

The problem sketched above is an optimization problem, which has so far been solved mainly by using dynamic programming and branch-and-bound. However, since the unit commitment has to be determined in advance, there is an inherent uncertainty and imprecision in the knowledge of basic entities, mainly the power system's (future, of course) load demand. This makes the solution of the problem difficult.

Su and Hsu's (1991) fuzzy dynamic programming model can best be presented starting from its nonfuzzy counterpart, i.e. the (nonfuzzy) dynamic programming formulation. We will give below its basic form only.

Suppose that the planning horizon is one day, divided into 24 hours, i.e. $N = 24$. The objective function to be minimized is the *total cost* which is

$$\text{COST} = \sum_{t=1}^{N} \sum_{i=1}^{I} [\text{FCOST}_i(G_i(t)) + \text{SCOST}_i] \qquad (8.120)$$

where:

- COST – the total cost over the planning horizon,
- I – the number of thermal units,
- $\text{FCOST}_i(G_i(t))$ – the cost of producing G_i (kWh) by thermal unit i at stage t, and
- SCOST_i – the start-up cost for thermal unit i.

The first constraint is the balance between the power to be generated and that to be used, called the *generation–load balance*, which is expressed as

$$\sum_{i=1}^{I} G_i(t) + \text{GHYDRO}(t) \geq L(t), \qquad t = 1, 2, \dots, N \qquad (8.121)$$

where:

- GHYDRO(t) – the total power generated by hydro units at stage t, and
- $L(t)$ – the load demand at stage t;

and, clearly, the transmission losses are neglected in (8.121).

The second constraint is the so-called *spinning reserve requirement* given as

$$\sum_{i=1}^{I} \text{RES}_i(t) + \text{RES}(t) \geq \text{REQ}(t), \qquad t = 1, 2, \ldots, N \qquad (8.122)$$

where:

- $\text{RES}_i(t)$ – the spinning reserve of thermal unit i at stage t,
- $\text{RESH}(t)$ – the total spinning reserve from hydro units, and
- $\text{REQ}(t)$ – the required spinning reserve at stage t.

And, similarly, one can introduce into the objective function and constraints elements related to the minimum start-up and shut-down times required, and to the crew size available.

It may be shown (cf. Su and Hsu, 1991) that the minimization of the objective functions (8.120) subject to the constraints (8.121) and (8.122) may proceed by employing dynamic programming.

In the fuzzification of the problem of unit commitment scheduling outlined above which has been proposed by Su and Hsu (1991) the fuzzy objective function related to the total cost is characterized by a fuzzy set C. The constraints are divided into two groups. First, the generation–load balance and spinning reserve requirements are considered fuzzy since they depend on the imprecisely (fuzzily) known predictions of hourly loads (demand). The other constraints, i.e. those related to the minimum start-up time, minimum shut-down time and the crew size are still considered crisp, i.e. nonfuzzy, and this is justified in the situation considered.

Moreover, as we may remember from our discussion in Chapter 4, the type of system under control is a key ingredient in the (fuzzy) dynamic programming formulations. Here we assume a deterministic system under control, i.e. whose states at the particular stages t are crisp (nonfuzzy).

The fuzzy dynamic programming model developed has therefore nonfuzzy states, and some nonfuzzy constraints imposed on the states. Moreover, it has two fuzzy constraints: L which is related to the load–generation balance, and S which is related to the spinning reserve, and a fuzzy objective function C.

The fuzzy decision [cf. (3.7), page 70] D, which serves the role of a (fuzzy) objective function, is the intersection [cf. (2.31), page 30, or (2.33)–(2.36), page 31] of C, L and S, i.e.

$$D = C \cap L \cap S \qquad (8.123)$$

and, evidently, to be able to use the intersection in (8.123), C, L and S must be defined in the same universe of discourse. Normally, as in virtually all applications, a standard universe of discourse is assumed as, e.g., the intervals $[0, 1$, $[-1, 1]$. Then, such standard universes of discourse are rescaled to obtain real values of entities in question. This is also the case here.

Note that the very essence of the problem considered, whose fuzzy objective function is expressed by (8.123), may be linguistically stated as

$$
\left\{
\begin{array}{ll}
\text{IF} & \text{the total cost is } \textit{very low} \\
 & \text{and} \\
 & \text{the hourly loads are } \textit{generally} \text{ met} \\
 & \text{and} \\
 & \text{the amount of spinning reserve is } \textit{very large} \\
 & \\
\text{THEN} & \text{the resultant commitment schedule is } \textit{desirable}
\end{array}
\right. \tag{8.124}
$$

and, clearly, the first issue to be solved is how to define the membership functions of the linguistic terms shown above in italics.

8.4.2 Determination of the membership function of the fuzzy objective function

We seek a commitment schedule with the total cost being as low as possible, and hence the total cost of a high value should have a low degree of membership in the fuzzy decision, while that with a low value a high degree of membership. The membership function for the fuzzy total cost proposed by Su and Hsu (1991) is

$$
\mu_{C(t,j)} = \exp\{-w \cdot \Delta C(t,j)\} \tag{8.125}
$$

where:

- (t,j) – the state j at stage t,
- $\Delta C(t,j) = \frac{\mathrm{COST}(t,j) - \mathrm{COST}_{\min}(t)}{\mathrm{COST}_{\min}(t)}$,
- $\mathrm{COST}(t,j)$ – the least total cost attained at state (t,j),
- $\mathrm{COST}_{\min} = \min_j \mathrm{COST}(t,j)$ – the minimum cost attained at stage t,
- w – a weighting factor.

8.4.3 Determination of the membership function of the fuzzy load demand

Since the hourly loads are to be forecast, then there is evidently an error between the actual load, L_a, and the forecast one, L_f, i.e.

$$
L_a = L_f + \Delta L \tag{8.126}
$$

with ΔL being the resulting forecast error.

It is now assumed that the forecast load L_f is crisp (nonfuzzy) while both the actual load L_a and the forecast error ΔL are fuzzy. Thus, if we know ΔL, the fuzzy set L_a can be determined.

By analyzing short-term load forecasting results over the last couple of years, the mean absolute error for each hour of the day has been calculated, and the results are shown in linguistic terms in Table 2 in which: VL – *very large*, L – *large*, M – *medium*, S – *small*, and VS – *very small*.

The load forecast error can obviously be both positive and negative (and also even 0 if we are lucky), and the values of the forecast errors can be different in these two cases.

Table 2 Linguistic description of the load forecast error

hour	error	hour	error	hour	error
1	VL	9	S	17	S
2	L	10	S	18	S
3	L	11	S	19	S
4	L	12	VS	20	S
5	L	13	M	21	S
6	M	14	S	22	S
7	M	15	S	23	S
8	M	16	S	24	M

Suppose therefore that we consider results at some hour. We divide all samples into two groups: one with the positive forecast error, and one with the negative forecast error. Then, for each particular linguistic value VL, L, M, S, and VS, we obtain – using the mean absolute percentage error of the two above groups – the values: $M_+(VL)$ and $M_-(VL)$ for the samples with very large (VL) errors, $M_+(L)$ and $M_-(L)$ for those with large (L) errors, etc.

In the model it is assumed that the membership function of the load forecast error is given as

$$\mu_L(\Delta l) = \begin{cases} \frac{1}{1+2.333(\frac{\Delta l}{M_+})^2} & \text{for } \Delta l \geq 0 \\ \frac{1}{1+2.333(\frac{\Delta l}{M_-})^2} & \text{for } \Delta l < 0 \end{cases} \qquad (8.127)$$

where:

$$\Delta l = \frac{\Delta L}{L_f} \times 100[\%] = \frac{L_a - L_f}{L_f} \times 100[\%] \qquad (8.128)$$

is the percentage error.

8.4.4   Determination of the membership function for the spinning reserve requirement

It is common practice for the utilities to specify the required spinning reserve for ensuring a safe and smooth operation during outage events. Basically this is an additional power that may become available within 30 minutes, if needed.

Such a requirement is clearly imprecisely specified, and its membership function is defined as

$$\mu_S(t) = \begin{cases} 1 & \text{if } \text{RES}(t) \geq \text{REQ}(t) \\ \exp\{R\frac{\text{RES}(t)-\text{REQ}(t)}{\text{REQ}(t)}\} & \text{otherwise} \end{cases} \qquad (8.129)$$

where:

- $\text{RES}(t)$ – the total spinning reserve at stage (hour) t,
- $\text{REQ}(t)$ – the total required spinning reserve at stage t, and
- R – a weighting factor,

and the value of R is chosen by the operator in that if he or she feels that a small amount of the spinning reserve is acceptable, then a small R may be assumed; otherwise, a larger R may be chosen.

8.4.5 A fuzzy dynamic programming algorithm

Now, having the three membership functions: $\mu_C(.)$ of the total cost defined by (8.125), $\mu_L(.)$ of the load demand defined by (8.127), and $\mu_S(.)$ of the spinning reserve defined by (8.129), the membership functions of the fuzzy commitment decision is determined as the intersection (8.123).

Then, the recurrent equation of fuzzy dynamic programming [cf. (4.7), page 87] is devised as (Su and Hsu, 1991)

$$\mu_D(t,j) = \max_{k \in K}[\mu_C(t,j) \wedge \mu_L(t,j) \wedge \mu_D(t-1,k)] \tag{8.130}$$

where:

- $\mu_D(t,j)$ – the maximum membership value of D attained at state (t,j),
- $\mu_C(t,j)$ – the membership value for the total cost of state (t,j),
- $\mu_L(t,j)$ – the membership value for the load demand of state (t,j),
- $\mu_S(t,j)$ – the membership value for the spinning reserve of state (t,j),
- $\{k\}$ – the set of feasible states at stage $t-1$.

This fuzzy dynamic programming formulation is similar to (4.7), page 87; however Su and Hsu (1991) simplify it by taking into account at each stage t only a predefined number of states leading to the highest membership value. This clearly reduces the computational requirement of fuzzy dynamic programming but, on the other hand, may lead to suboptimal solutions only.

8.4.6 A numerical example

To illustrate the method proposed above, Su and Hsu (1991) consider a power generation system in Taiwan consisting of 6 nuclear units, 48 thermal units and 44 hydro units; in fact, in that paper cited they discuss the commitment scheduling of the thermal units only, assuming that the hydro units have been optimized separately.

For lack of space we will not provide a detailed description of technical data of all the generating units (cf. Su and Hsu, 1991). Basically, they are divided into 8 distinct types:

- nuclear,
- coal-fired,
- oil-fired,
- run-of-river hydro,
- non-run-of-river hydro,
- gas turbines,
- pumped-storage (generating), and
- pumped-storage (pumping),

and their maximum outputs vary from 6 to 985 MW, while their pickup rates range from 0 to 55 MW/min.

Table 3   Forecast of hourly loads

Hour (t)	1	2	3	4	5	6	7	8
Load (MW)	5807	5777	5650	5604	5616	5732	5900	6546
Hour (t)	9	10	11	12	13	14	15	16
Load (MW)	7375	8034	8322	8377	7085	8167	8390	8386
Hour (t)	17	18	19	20	21	22	23	24
Load (MW)	8404	7909	8053	8083	7691	7041	6900	6771

Table 4   Average errors $M_+(.)$ and $M_-(.)$

Linguistic description	$M_+(.)$	$M_-(.)$
VL	0.05418	-0.03271
L	0.04668	-0.02786
M	0.01979	-0.02115
S	0.01488	-0.01678
VS	0.01267	-0.01425

The forecast hourly loads are shown in Table 3, while the values of average errors $M_+(.)$ and $M_-(.)$ [cf. (8.127)] are given in Table 4.

Then, the membership functions $\mu_C(.)$, $\mu_L(.)$ and $\mu_S(.)$ have been assumed as (8.125), (8.127) and (8.129), respectively, with the weighting factors $W = 14.048$ and $R = 200$. Next, the fuzzy dynamic programming algorithm expressed by (8.130) has been employed.

The results obtained are given, first, in Table 5 which lists the maximum grades of membership of obtained for $\mu_C(.)$, $\mu_L(.)$ and $\mu_S(.)$ [and hence for the fuzzy decision $\mu_D(.)$ too – cf. (8.123)] for the optimal commitment schedule determined for the particular 24 stages (hours).

On the other hand, Table 6, page 293, shows the hourly load demands, spinning reserves and production costs for the optimal commitment schedule determined by using fuzzy dynamic programming. Moreover, for illustration, the optimal results yielded by the traditional (nonfuzzy) dynamic programming approach are listed.

In general, the results obtained do clearly indicate that the fuzzy dynamic programming approach is very effective for dealing with unit commitment problems under imprecise knowledge. These results compare very well with those yielded by the conventional (nonfuzzy) dynamic programming technique.

8.5   INTRA-OPERATIVE ANESTHESIA ADMINISTRATION

This is an example of medical applications of fuzzy dynamic programming, and was proposed by Esogbue (1985). The rationale for that work was a need for automation of some repetitive tasks of anesthesiologists during surgery to better utilize their

Table 5 The membership values of $\mu_L(.)$, $\mu_S(.)$, $\mu_C(.)$ and $\mu_D(.)$ for the optimal commitment schedule determined

Hour(t)	$\mu_L(.)$	$\mu_S(.)$	$\mu_C(.)$	$\mu_D(t-1)$	$\mu_D(t)$
1	0.91	1	0.907	1	0.907
2	0.91	1	0.942	0.907	0.907
3	0.91	1	0.948	0.907	0.907
4	0.91	1	0.951	0.907	0.907
5	0.91	1	0.975	0.907	0.907
6	0.91	1	0.978	0.907	0.907
7	0.91	1	0.966	0.907	0.907
8	0.91	1	0.937	0.907	0.907
9	0.91	1	0.954	0.907	0.907
10	0.91	1	0.952	0.907	0.907
11	0.91	1	0.951	0.907	0.907
12	0.91	1	0.966	0.907	0.907
13	0.91	1	0.951	0.907	0.907
14	0.88	1	0.980	0.907	0.880
15	0.88	1	0.963	0.880	0.880
16	0.88	1	0.981	0.880	0.880
17	0.88	1	0.983	0.880	0.880
18	0.88	1	0.978	0.880	0.880
19	0.88	1	0.984	0.880	0.880
20	0.88	1	0.985	0.880	0.880
21	0.88	1	0.996	0.880	0.880
22	0.88	1	0.983	0.880	0.880
23	0.88	1	0.988	0.880	0.880
24	0.88	1	0.989	0.880	0.880

Table 6 The load demands, spinning reserves, and production costs for the optimal commitment schedule obtained by using fuzzy dynamic programming

Hour (t)	Load demand (MW)	Spinning reserve (MW)	Production cost (NT$)
1	5768	3816	2138765
2	5744	3845	1958122
3	5617	3985	1933568
4	5572	4017	1944686
5	5583	4008	1947506
6	5707	3909	1985061
7	5874	3702	2123212
8	6517	3576	2585599
9	7349	2624	2826448
10	9006	2416	3169934
11	8293	2203	3167070
12	8352	2149	3230344
13	7054	3319	2770938
14	8134	2391	3278328
15	8356	2633	3814072
16	8352	2617	3238464
17	8370	2591	3331276
18	7877	3077	3215652
19	8020	2927	3262328
20	8050	2913	3313260
21	7659	2820	3003252
22	7012	2975	2673548
23	6872	3153	2718076
24	6736	3611	2620544

time, and hence help alleviate the shortage in the profession. It was quite clear that such a "human-centered" problem – in which non-quantitative knowledge and experience does play a considerable role, and human reactions, limitations, etc. are very imprecisely known – is a good candidate for the use of fuzzy tools, namely fuzzy dynamic programming.

We will now briefly present Esogbue's (1985) model, following the author's notation and exposition.

Let $k = 0, 1, \ldots, K$ be the stages (time moments) considered. According to the standards of the American Medical Association, the observation interval is (at most) $\delta k = 5$ min, so that $k = 0, 5, 10, \ldots, K$ [min]. The planning horizon is assumed finite, $K < \infty$, and fixed and specified (cf. Chapter 4), though Esogbue (1985) indicates that the use fuzzy dynamic programming model with Kacprzyk's (1978b, c) fuzzy termination time (cf. Chapter 6) may be expedient because the termination time of a surgical operation is usually only imprecisely known.

A relevant function of anesthesia administration is to monitor, at regular time moments k mentioned above, the status of a patient's health. This proceeds by monitoring some critical indicators. For instance, these can be:

- systolic blood pressure,
- diastolic blood pressure,
- pulse rate,
- body temperature,
- electrocardiograph measurements,
- respiration rate,
- response to external stimuli,
- dilation of the pupil, etc.

Among these indicators, the first four may be relatively precisely measured, while the last three cannot be precisely measured.

Suppose therefore that indicators of the above type, which constitute a *state variable vector* at time k, are denoted by

$$\overline{p}_k = [p_{k1}, p_{k2}, \ldots, p_{kr}, p_{kr+1}, p_{kr+2}, \ldots, p_{kn}]$$

where the first r variables, i.e. $p_{k1}, p_{k2}, \ldots, p_{kr}$, are nonfuzzy, while the rest of them, i.e. $p_{kr+1}, p_{kr+2}, \ldots, p_{kn}$, are fuzzy.

First, in the model described the nonfuzzy variables are "fuzzified." Basically, fuzziness is introduced by considering a *severity index* which is a measure of the degree of normality of these (nonfuzzy) variables.

This degree of severity is derived as follows. Let the i-th state variable take on its values in the interval $[\underline{p}_i, \overline{p}_i]$, $i = 1, \ldots, r$. Then its desired mean value can be generally written as

$$\hat{p}_i = \frac{\underline{p}_i + \overline{p}_i}{2} \qquad (8.131)$$

If now p_{ki}, $k = 0, 1, \ldots, K$, $i = 1, \ldots, r$, is a desired mean value of the i-th state variable at time k, then one can represent the severity of each of the otherwise nonfuzzy state variables by the fuzzy set whose membership function is

$$\mu_{G_{ki}}(p_{ki}) = \frac{p_{ki} - \hat{p}_{ki}}{\delta}, \qquad \text{for each } p_{ki} \in [\underline{p}_i, \overline{p}_i] \qquad (8.132)$$

where

$$\delta = \begin{cases} \frac{1}{2}(\overline{p_i} - \underline{p_i}) & \text{if } \underline{p_i} \le p_{ki} \le \overline{p_i} \\ \overline{p_i} - \hat{p_i} & \text{if } \hat{p_i} \le p_{ki} \le \overline{p_i} \\ \hat{p_i} - \underline{p_i} & \text{if } \underline{p_i} \le p_{ki} \le \hat{p_i} \end{cases} \qquad (8.133)$$

Notice that the membership function (8.132) is derived by some normalization which is similar to that in (3.3), page 69.

The second aspect of the model is the representation of intrinsically fuzzy variables as, e.g., respiration rate, dilation of the pupil, response to external stimuli, etc. The severities of these variables, measured by their degrees are represented by the membership functions of fuzzy sets which are similar to those defined by (8.132), but assuming the interval $[\underline{p}_{ki}, \overline{p}_{ki}] = [0, 1]$ such that the usual level corresponds to 0, and the most severe level corresponds to 1.

The next step is to define the control variable q. Namely, depending on the patient's status measured by the state variables discussed above, the anesthesiologist is to decide upon an appropriate control action to maintain or restore homeostasis. These actions may be both quantitative as, e.g., administration of exact dosages of drugs or anesthetic gas, blood transfusion rate, etc., while other are inherently fuzzy as, e.g., the change of patient's position, insertion of an endotracheal tube, etc.

Thus the vector of control variables at stage $k = 0, 1, \ldots, K$ is

$$q_k = [q_{k1}, q_{k2}, \ldots, q_{ks}, q_{ks+1}, q_{ks+2}, \ldots, q_{km}] \qquad (8.134)$$

where $q_{k1}, q_{k2} \ldots, q_{ks}$ are nonfuzzy, and $q_{ks+1}, q_{ks+2} \ldots, q_{km}$ are fuzzy.

The transition function, which corresponds to the deterministic system under control of type (2.130), page 59, is given in general as a function

$$p_{k+1} = f(p_k, q_k), \qquad k = 0, 1, \ldots, K \qquad (8.135)$$

which can be determined by using, for example, regression analysis.

Moreover, Esogbue (1985) advocates the use of a fuzzy transition function of type (2.140), page 61, in case when the transitions are inherently imprecise as, for example, when a novel surgical procedure is applied or a new drug is administered.

The objective (return, performance, ...) function reflects the main goal of the attending anesthesiologist which is to maintain or restore each patient's state variable to be within a range of desirable values, discouraging greater deviations.

Therefore, if we define, for stage $k = 0, 1, \ldots, K$:

- g_k – a weighted sum of the squared deviations of all variables from a desired base level (homeostasis),
- $\hat{p_i}$ – a desired value of the i-th state variable,
- p_{ki} – an actual (measured) value of this variable, and
- w_i – a consensory weight attached by a panel of experts to the relevance of deviation of the i-th variable,

then the objective function is

$$g_k = \sum_{i=1}^{N} w_i \cdot (p_{ki} - \hat{p_i}), \qquad k = 1, 2, \ldots, K \qquad (8.136)$$

which, in turn, if we include the severities of both nonfuzzy and fuzzy variables of type (8.132), is equivalent to

$$g_k = \sum_{i=1}^{N} w_i \cdot [\mu_{G_{ki}}(p_{ki})]^2 = \sum_{i=1}^{N} \left(w_i \frac{\mid p_{ki} - \hat{p}_i \mid}{\delta} \right)^2, \qquad k = 1, 2, \ldots, K \quad (8.137)$$

where δ is given by (8.133).

All the elements for a fuzzy dynamic programming formulation of the problem considered are now available. The recurrence equation is as follows:

- for $k = 0$

$$v_1(p_1) = \min_{q_0} \left(\sum_{i=1}^{N} w_i \frac{\mid p_{1i} - \hat{p}_i \mid}{\delta} \right) \qquad (8.138)$$

- for $k = 1, 2, \ldots, K$

$$v_k(p_k) = \min_{q_k} (\sum_{i=1}^{N} w_i \left(\frac{\mid p_{ki} - \hat{p}_i \mid}{\delta} \right)^2 + f_{k-1}[f(p_k, q_k)]) \qquad (8.139)$$

where the particular p_{ki}'s and q_k's, $k = 0, 1, \ldots, K$, are to be chosen from their respective sets of admissible values.

For more detail on this problem, its fuzzy dynamic programming formulation and solution, we refer the reader to the source paper (Esogbue, 1985).

8.6   RESOURCE ALLOCATION

In its most general form, *resource allocation* is a large class of diverse problems whose essence is to find how to allocate (partition, divide, etc.) some resource (e.g., money, manpower, natural resources, etc.) so that to obtain the highest possible effect (return, output, ...). It is easy to see that such a description may characterize an extremely wide array of problems; in fact, virtually all problems of operations research or systems analysis are in their very essence of this type.

Problems of resource allocation have attracted, due to their generality, much research interest, and all these efforts have resulted in a wide array of effective tools. Moreover, resource allocation was also one of the earliest applications of fuzzy dynamic programming, with the pioneering Esogbue and Ramesh's (1970) work.

In this book we will present one of the recent attempts at using fuzzy dynamic programming to formulate and solve a multicriterion resource allocation problem which is due to Hussein and Abo-Sinna (1995). This will show the reader both how to deal with a resource allocation problem under fuzziness, and – even more so – how to deal with its multicriterion version.

The point of departure in Hussein and Abo-Sinna's (1995) work is the following nonfuzzy multicriterion resource allocation problem. Suppose that S is an available quantity of some resource. We wish to find an optimal allocation of this resource over n activities (stages), $u_1^*, \ldots, u_n^*$ such that some performance functions Z_j,

$j = 1, \ldots, m \geq 2$, are maximized, and some constraints are satisfied. This may be written, in widely used notation of mathematical programming problems, as

$$
\begin{cases}
\overline{max}_{u_1,\ldots,u_n} Z_j(u_1^*, \ldots, u_n^*) = \sum_{i=1}^n f_i^j(u_i) \\
\text{subject to:} \quad 0 \leq \sum_{i=1}^n g_i(u_i) \leq S \\
u_1, \ldots, u_n \geq 0
\end{cases}
\tag{8.140}
$$

where $f_i^j(u_i)$'s and $g_i(u_i)$'s are some constituents of the performance functions and constraints, respectively.

Evidently, due to the existence of multiple performance functions the maximization in (8.140) cannot be meant literally, but in the sense of multicriterion optimization, i.e. we are looking first for the so-called *efficient solutions*, i.e. those which cannot improve all the criteria, and among such solutions there should be the one(s) which will be "really" optimal; however, the determination of such optimal solution(s) is another story, outside the scope of our analysis.

It can be shown – and this can be found in virtually all books on optimization or operations research – that an efficient solution to problem (8.140) can be obtained by solving the following parametric optimization problem:

$$
\begin{cases}
\max_{u_1,\ldots,u_n} \sum_{j=1}^m w_j Z_j(u_1, \ldots, u_n) \\
\text{subject to:} \quad 0 \leq \sum_{i=1}^n g_i(u_i) \leq S; u_1, \ldots, u_n \geq 0 \\
\sum_{j=1}^m w_j = 1; w_1, \ldots, w_m \geq 0
\end{cases}
\tag{8.141}
$$

If we introduce now the following real-valued function, for each $i = 1, \ldots, n$, $\sum_{j=1}^m w_j = 1$, and for each $0 \leq s_i \leq S$:

$$
R_i^*(w_1, \ldots, w_m, s_n) =
$$
$$
= \max_{u_1,\ldots,u_n} \left\{ \sum_{i=1}^n \sum_{j=1}^m w_j f_i^j(u_i); \text{ subject to: } \sum_{i=1}^n g_i(u_i) \leq s_n \right\}
\tag{8.142}
$$

then, using the same line of reasoning as (4.4)–(4.7) in Section 4.1.1, because in $R_i^*(.)$ given by (8.142) the constituent terms depend just on the successive u_i's, we obtain the following set of dynamic programming recurrence equations:

$$
\begin{cases}
R_i^*(w_1, \ldots, w_m, x_i) = \\
\quad = \max_{u_i} \{ \sum_{j=1}^m w_j f_i^j(u_i) + R_{i+1}^*(w_1, \ldots, w_m, x_{i+1}) \} \\
x_{i+1} = x_i - u_i, \qquad i = 1, \ldots, N-1
\end{cases}
\tag{8.143}
$$

which yields the solution to the problem (8.141).

Similar to the case of many mathematical programming problems, including the multistage ones, it is clear that the problem formulation (8.141) may often be too rigid, and its fuzzified version may be more adequate [cf. Kacprzyk and Orlovski's (1987a) or Delgado, Kacprzyk, Verdegay and Vila's (1994) volumes].

Hussein and Abo-Sinna (1995) perform the fuzzification of (8.141) [or, in fact, of its corresponding set of dynamic programming recurrence equations (8.143)] by using the following model.

Let:

- $x_i \in S$, $i = 0, 1, \ldots, n$, be (crisp, nonfuzzy) state variables at stage i where $S = \{s_1, \ldots, s_N\}$ is the set of their state of possible values,
- $u_{i-1} \in U$, $i = 1, \ldots, n$, be (crisp, nonfuzzy) controls (decision variables) at stage $i - 1$ where $U = \{c_1, \ldots, c_M\}$ is the set of their possible values,
- $x_{i+1} = x_i - u_i$, $i = 0, 1, \ldots, n - 1$ is a (crisp, nonfuzzy) state transition function [which corresponds to the deterministic system under control of type (2.130), page 59].

For each stage i, $i = 0, 1, \ldots, N - 1$, a fuzzy goal $\tilde{F}_i(w_1, \ldots, w_m, u_i, x_i)$ is defined by the following membership function:

$$\mu_{\tilde{F}_i}(w_1, \ldots, w_m, u_i, x_i) =$$

$$= \begin{cases} 1 & \text{if } F_i(w_1, \ldots, w_m, u_i, x_i) \geq \\ & \geq \overline{F}_i(w_1, \ldots, w_m, x_i) + d_i(x_i) \\ \dfrac{F_i(w_1, \ldots, w_m, u_i, x_i) - \underline{F}_i(w_1, \ldots, x_i)}{\overline{F}_i(w_1, \ldots, w_m, x_i) + d_i(x_i) - \underline{F}_i(w_1, \ldots, w_m, x_i)} & \text{if } \underline{F}_i(w_1, \ldots, w_m, x_i) < \\ & < F_i(w_1, \ldots, w_m, u_i, x_i) < \quad (8.144) \\ & < \overline{F}_i(w_1, \ldots, w_m, x_i) + d_i(x_i) \\ 0 & \text{if } F_i(w_1, \ldots, w_m, u_i, x_i) \leq \\ & \leq \underline{F}_i(w_1, \ldots, w_m, x_i) + d_i(x_i) \end{cases}$$

where:

$$\overline{F}_i(w_1, \ldots, w_m, x_i) = \max_{u_i \in U} F_i(w_1, \ldots, w_m, u_i, x_i)$$

$$\underline{F}_i(w_1, \ldots, w_m, x_i) = \min_{u_i \in U} F_i(w_1, \ldots, w_m, u_i, x_i)$$

and:

- $d_i(x_i)$ is a tolerance interval, and
- $F(w_1, \ldots, w_m, u_i, x_i)$ is the (value of) performance function in problem (8.141), i.e. of $w_j Z_j(.)$.

Observe that this is so far a typical fuzzification of a mathematical programming problem in the sense of Zimmermann (1976).

Now, if we define, for $i = N$,

$$\mu_{\tilde{R}_N^*}(w_1, \ldots, w_m, S) = \max_{u_N \in U} \mu_{\tilde{F}_N}(w_1, \ldots, w_m, u_N, S) \qquad (8.145)$$

then

$$\mu_{\tilde{R}_i^*}(w_1, \ldots, w_m, u_i, x_i) =$$
$$= \mu_{\tilde{F}_i}(w_1, \ldots, w_m, u_i, x_i) + \mu_{\tilde{R}_{i+1}^*}(w_1, \ldots, w_m, u x_{i+1}) +$$
$$- \mu_{\tilde{F}_i}(w_1, \ldots, w_m, u_i, x_i) \cdot \mu_{\tilde{R}_{i+1}^*}(w_1, \ldots, w_m, x_{i+1}) \qquad (8.146)$$

and $\mu_{\tilde{R}_i^*}(w_1, \ldots, w_m, S) = 0$.

Notice that in (8.146) the probabilistic product (algebraic sum) [cf. (2.39), page 32] is used, and for a reasoned explanation the reader is referred to the source paper of Hussein and Abo-Sinna (1995).

Table 7   Expected efficiency of allocating i workers to the particular jobs j

		Jobs			
		$j=1$	2	3	4
Number	$i=0$	0	0	0	0
of	1	25	20	33	13
workers	2	42	38	43	24
	3	55	54	47	32
	4	63	65	50	39
	5	69	73	52	45
	6	74	80	54	50

Table 8   Expected efficiency of allocating i workers to the particular jobs j

		Jobs			
		$j=1$	2	3	4
Number	$i=0$	70	90	85	130
of	1	60	60	60	115
workers	2	50	50	50	100
	3	40	40	55	100
	4	40	30	40	90
	5	45	20	30	80
	6	50	25	25	80

Therefore, the set of fuzzy dynamic programming recurrent equations is

$$
\begin{cases}
\mu_{\tilde{R}_i^*}(w_1,\ldots,w_m,x_i) = \max_{u_i \in U}[\mu_{F_i}(w_1,\ldots,w_m,u_i,x_i)+ \\
\quad + \mu_{\tilde{R}_{i+1}^*}(w_1,\ldots,w_m,x_{i+1}) - \mu_{F_i}(w_1,\ldots,w_m,u_i,x_i)\times \\
\quad \times \mu_{\tilde{R}_{i+1}^*}(w_1,\ldots,w_m,x_{i+1})] \\
x_{i+1} = x_i - u_i, \qquad i = 0,1,\ldots,N-1
\end{cases} \tag{8.147}
$$

As shown by Hussein and Abo-Sinna (1995), under some very mild assumptions (mainly the monotonicity of the membership functions), the optimal solution obtained by solving (8.147) is an optimal solution of problem (8.141).

Example 8.1 Suppose that we have two performance functions $Z_1(u_i)$ and $Z_2(u_i)$, $i = 1,\ldots,N$ which represent the expected efficiency and cost for allocating $S = 6$ workers to $N = 4$ jobs; evidently, the first function is to be maximized while the second one is to be minimized.

The expected efficiency of allocating $i = 0,1,\ldots,6$ workers to the particular jobs $j = 1,2,3,4$ – in some currency units equivalent to what they would "produce" – is given in Table 7.

On the other hand, the expected cost of allocating $i = 0,1,\ldots,6$ workers to the particular jobs $j = 0,1,2,3,4$ is given in Table 8.

We have therefore a bicriterion resource allocation problem, and we will use the parametric optimization problem (8.141) to find the efficient solutions.

Table 9 Values of the $F_n(w_1, \ldots, w_m, u_j)$'s in the performance function of the
parametric problem (8.141)

		Jobs			
		$j=1$	2	3	4
Number	$i=0$	$70(w\text{-}1)$	$90(w\text{-}1)$	$85(w\text{-}1)$	$130(w\text{-}1)$
of	1	$85(w\text{-}60)$	$80(w\text{-}60)$	$93(w\text{-}60)$	$128(w\text{-}115)$
workers	2	$92(w\text{-}50)$	$88(w\text{-}50)$	$93(w\text{-}50)$	$124(w\text{-}100)$
	3	$95(w\text{-}40)$	$94(w\text{-}40)$	$102(w\text{-}55)$	$132(w\text{-}100)$
	4	$103(w\text{-}40)$	$95(w\text{-}30)$	$90(w\text{-}1)$	$120(w\text{-}90)$
	5	$114(w\text{-}1)$	$93(w\text{-}20)$	$82(w\text{-}30)$	$125(w\text{-}80)$
	6	$124(w\text{-}50)$	$105(w\text{-}25)$	$78(w\text{-}25)$	$130(w\text{-}80)$

Table 10 Tolerance intervals $d_n(x_n)$ in the definition of the fuzzy goal

		Jobs			
		$j=1$	2	3	4
Number	$i=0$	$d_1(0) = 10$	$d_2(0) = 10$	$d_3(0) = 15$	$d_4(0) = 10$
of	1	$d_1(1) = 15$	$d_2(1) = 20$	$d_3(1) = 7$	$d_4(1) = 12$
workers	2	$d_1(2) = 8$	$d_2(2) = 12$	$d_3(2) = 7$	$d_4(2) = 16$
	3	$d_1(3) = 5$	$d_2(3) = 6$	$d_3(3) = 8$	$d_4(3) = 18$
	4	$d_1(4) = 7$	$d_2(4) = 5$	$d_3(4) = 10$	$d_4(4) = 21$
	5	$d_1(5) = 6$	$d_2(5) = 7$	$d_3(5) = 18$	$d_4(5) = 15$
	6	$d_1(6) = 10$	$d_2(6) = 15$	$d_3(6) = 22$	$d_4(6) = 20$

It is easy to see that since we wish to maximize the efficiency and to minimize the cost, then the parametric return values $F_n(w_1, \ldots, w_m, u_n)$, which correspond to the sum of the $w_j Z_j(u)$'s in the performance function of (8.141), are as shown in Table 9, for the particular number of workers allocated to some set of jobs.

Suppose now that the tolerance intervals $d_n(s_n)$ in the definition of the membership function of the fuzzy goal (8.144) are as given in Table 10.

Therefore, using the parametric return values given in Table 9 and the tolerance intervals given in Table 10, we arrive at the parametric membership function of the fuzzy goal, $\mu_{\tilde{F}_i}(w_1, \ldots, w_m, u_i, x_i)$ defined by (8.144), as shown in Table 8.1.

Now, we employ the fuzzy dynamic programming recurrence equations (8.147). Because of lack of space, we cannot show here the intermediate results which may be found in the source Hussein and Abo-Sinna's (1995) work. These results are summarized in Table 12 in which the consecutive efficient solutions $u_4^*, u_3^*, u_2^*, u_1^*$, and their corresponding ranges of w are given.   □

The next step is evidently to choose such an efficient solution from among those obtained ones which is the "best." We will not however deal here with this problem as it is beyond the scope of this work. We may refer the interested reader to, say, many articles in Fedrizzi, Kacprzyk and Roubens (1991), Kacprzyk and Orlovski (1987a) or Delgado, Kacprzyk, Verdegay and Vila (1994).

Table 11   The parametric return values $\mu_{\bar{F}_i}(w_1, \ldots, w_n, u_i, x_i)$

		Jobs			
		$j{=}1$	2	3	4
Number	$i{=}0$	$0.875w$	$0.9w$	$0.85w$	$0.93w$
of	1	$0.85w$	$0.8w$	$0.93w$	$0.9w$
workers	2	$0.92w$	$0.88w$	$0.93w$	$0.89w$
	3	$0.95w$	$0.94w$	$0.93w$	$0.88w$
	4	$0.936w$	$0.95w$	$0.9w$	$0.86w$
	5	$0.95w$	$0.93w$	$0.82w$	$0.89w$
	6	$0.925w$	$0.875w$	$0.78w$	$0.87w$

Table 12   The efficient solutions obtained

		Efficient solution (Number of workers allocated to job j)			
		$j{=}1$	2	3	4
Range	$0 \le w < 0.16$	2	3	1	0
of	$0.16 \le w < 0.85$	3	0	3	0
w	$0.16 \le w < 0.85$	5	0	1	0
	$0.85 \le w \le 1$	2	4	0	0

This concludes our brief analysis of a fuzzy dynamic programming model for multicriteria resource allocation. The model does extend the classical analysis by providing more flexibility and expressive power (in the sense of being able to more adequately express real human decision maker's intentions, wishes, attitudes, etc.). It is, on the other hand, relatively simple conceptually, and can be efficiently implemented.

8.7   INVENTORY CONTROL

The second application to be considered in this chapter concerns some class of inventory control. To be more specific, it concerns the determination of an *optimal inventory (replenishment*, in fact) *policy* as seen from the viewpoint of top management. Or, in other words, it concerns some "global" corporate inventory policy.

The problem is clearly of a "soft" character since one faces an obvious necessity of an aggregated evaluation of an inventory level, replenishment, demand, etc. as well as a lack of detailed knowledge of costs, losses, etc. Needless to say that all this does complicate to a large extent the analysis of the problem, and makes the application of conventional, "hard" techniques questionable.

To overcome this inherent difficulty, a fuzzy multistage control model was proposed by Kacprzyk and Staniewski (1982). The two variants of the model to be discussed consider similar situations. The inventory is dealt with from the "global" viewpoint (i.e. from the whole company's, or the top management's level). Though the real inventory may consist of a great number of different items, from such a global point of view it may be regarded as an aggregate, i.e. one good (or commodity), expressed in monetary terms as the (total) value of inventory. The same concerns evidently the (value of) replenishment and demand. We have therefore a one-item inventory, with characteristic features that are only approximately known.

The next important issue is the determination of costs which are the main ingredients of all inventory control models, including this one, which are basically formulated as the minimization of some appropriately defined costs. However, it is easy to see that the above-mentioned approximate and aggregate characterization of inventory level, replenishment and demand in monetary terms, as well as dealing with the problem from such a high level perspective, make the determination of relevant costs, and relations between them, extremely difficult, if at all possible.

The cost-related elements of the model are therefore proposed to be expressed by some fuzzy constraints and fuzzy goals. The fuzzy constraint at control stage t, $\mu_{C'}(r_t \mid z_t)$, specifies for each replenishment r_t its grade of membership (of a preference nature) as a function of the inventory level z_t. For instance, if the inventory level is high, we may be less willing to allow a higher replenishment. On the other hand, the fuzzy goal $\mu_{G'}(z_t)$ specifies for each inventory level z_t its grade of membership as, e.g., one can prefer lower inventory levels, inventory levels more or less between some limits, etc. The particular forms of the fuzzy constraints and fuzzy goal are similar to those shown in Figure 14, page 68, i.e. some values of replenishment or inventory level are fully satisfactory, some are satisfactory to some extent, and some are unsatisfactory (impossible); in practice, the piecewise linear form shown in Figure 14 is fully satisfactory (cf. Section 8.1).

In the case of many practical problems from the class of inventory control considered, the process is quite long, low varying and stable, it may be expedient to assume an infinite planning horizon and employ a control model with an infinite termination time considered in Section 7.1. The fuzzy decision is then [cf. (7.9), page 218]

$$\mu_D(r_0, r_1, \ldots \mid z_0) = \lim_{N \longrightarrow \infty} \bigwedge_{t=0}^{N} [\mu_{C'}(r_t \mid z_t) \wedge \mu_{G'}(z_{t+1})] \qquad (8.148)$$

Notice that the min-type fuzzy decision is here assumed. Such a clearly safety first attitude expressed by this type of a fuzzy decision (cf. Section 3.1) is mainly justified by high possible losses that may result from a false decision made at such a high management level.

The fuzzy decision (8.148) represents a compromise between which inventory level should preferably be kept (this is represented by the fuzzy goal) and what replenishment should preferably be made (this is represented by the fuzzy constraint). Knowledge of these fuzzy constraints and fuzzy goals can be elicited from experienced specialists.

The fuzzy constraints and fuzzy goals reflect in fact implicitly cost structure and relations in the problem considered. Note, however, that the costs are not explicitly involved which is clearly of extraordinary practical relevance.

The next important step of model building is the determination of how the inventory behaves. In general, this is equivalent to the determination of a dynamic system governing the state transitions (dynamics of inventory level changes). Such a system may be stochastic or fuzzy; a deterministic system here makes little sense of course.

First, as discussed by Kacprzyk and Staniewski (1982), we can assume that the inventory level and replenishment are deterministic but the demand is stochastic. This implies the following temporal evolution equation describing the behavior of the inventory system:

$$z_{t+1} = z_t + r_t - d_t, \qquad t = 0, 1, \dots \tag{8.149}$$

where z_t and z_{t+1} are inventory levels at stages t and $t + 1$, respectively, r_t is the replenishment at stage t, and d_t is a random demand at stage t that is assumed to be Poisson distributed.

The above state transition equation is equivalent therefore to a Markov chain described by the conditional probability $p(z_{t+1} \mid z_t, r_t)$ governing the transitions of inventory levels (cf. Section 2.4.2).

The problem considered may be therefore stated as to find an optimal sequence of replenishments $r_0^*, r_1^*, \dots$ such that

$$\mu_D(r_0^*, r_1^*, \dots \mid z_0) = \max_{r_0, r_1, \dots} E \mu_D(r_0, r_1, \dots \mid z_0) =$$

$$= \max_{r_0, r_1, \dots} E \left[\lim_{N \longrightarrow \infty} \bigwedge_{t=0}^{N} (\mu_{C^t}(r_t \mid z_t) \wedge \mu_{G^t}(z_{t+1})) \right] \tag{8.150}$$

This problem is clearly equivalent to the control of a stochastic system in a fuzzy environment with an infinite termination time discussed in Section 7.2 [problem (7.26), page 224]. By solving the problem (8.150) we evidently obtain an optimal stationary replenishment strategy relating an optimal replenishment to the current inventory level and a summary of the past trajectory.

Note that in the above problem formulation there was a strong assumption made that the inventory level, replenishment and demand are given as crisp (nonfuzzy) values, and that the probability distribution of demand is known. There was therefore an uncertainty (randomness) in the behavior of the inventory system, but the imprecision (fuzziness) was only in the environment in which the inventory operates, i.e. in the constraints and goals (costs, implicitly).

The fuzziness was additionally introduced into the behavior of the inventory system by Kacprzyk and Staniewski (1982). Namely, the entities characterizing the inventory were assumed to be given as fuzzy sets (numbers), denoted Z_t and Z_{t+1} for the fuzzy inventory levels at stages t and $t + 1$, respectively, and R_t for the fuzzy replenishment at stage t, and D for the fuzzy demand.

The introduction of these fuzzy elements of the model might be therefore viewed as an attempt at diminishing the information requirements as only some fuzzy valuations were needed. Moreover, no knowledge of the probability distribution of the demand was assumed.

The temporal evolution of the (fuzzy) inventory system is now governed by

$$Z_{t+1} = Z_t + R_t - D, \qquad t = 0, 1, \dots \tag{8.151}$$

where "+" and "−" denote the addition and subtraction of the respective fuzzy numbers as defined in Section 2.1.7 [e.g., by (2.71), page 42, for the addition, and (2.72), page 42, for the subtraction].

The inventory system is now equivalent to a fuzzy system under control with the fuzzy state transitions governed by (2.140), page 61, discussed in Section 2.4.3.

Since the numbers of possible fuzzy inventory levels, fuzzy replenishments and fuzzy demands may be very high (theoretically infinite), this can make the control problem practically unsolvable as we may remember from, e.g., Section 7.3. Then we introduce, as shown in Section 7.3.1, a relatively small number of *reference fuzzy inventory levels*, say $\overline{S}_1, \ldots, \overline{S}_r$, and *reference fuzzy replenishments*, say $\overline{C}_1, \ldots, \overline{C}_q$.

The temporal evolution equation (8.151) is then reformulated in terms of the reference fuzzy inventory levels and reference fuzzy replenishments as

$$\overline{Z}_{t+1} = A(\overline{Z}_t + \overline{R}_t - D), \qquad t = 0, 1, \ldots \tag{8.152}$$

where "+" and "−" are the addition and subtraction of fuzzy numbers [e.g., given by (2.71) for the addition, and (2.72) for the subtraction], and $A(.)$ means the approximation by reference fuzzy sets as defined, for instance, in Section 7.3.1, page 237.

The replenishment and inventory level are now fuzzy so that the fuzzy constraint should be modified due to, e.g., (7.77) or (7.78), page 239, by using, respectively, the normalized distance between two fuzzy sets, i.e.

$$\mu_C(\overline{R}_t \mid \overline{Z}_t) = 1 - d(\overline{R}_t, C'(\overline{Z}_t)) \tag{8.153}$$

or a degree of equality between two fuzzy sets, i.e.

$$\mu_C(\overline{R}_t \mid \overline{Z}_t) = e(\overline{R}_t, C'(\overline{Z}_t)) \tag{8.154}$$

And analogously, for the fuzzy goal

$$\mu_G(\overline{Z}_t) = 1 - d(G', \overline{Z}_t) \tag{8.155}$$

or

$$\mu_G(\overline{Z}_t) = e(G', \overline{Z}_t) \tag{8.156}$$

The problem considered is now to find an optimal sequence of reference fuzzy replenishments $\overline{R}_0^*, \overline{R}_1^*, \ldots$ such that [cf. (7.83), page 240]

$$\mu_D(\overline{R}_0^*, \overline{R}_1^*, \ldots \mid \overline{Z}_0) =$$
$$= \max_{\overline{R}_0, \overline{R}_1, \ldots} \lim_{N \longrightarrow \infty} \bigwedge_{t=0}^{N} b_t [\mu_C(\overline{R}_t \mid \overline{Z}_t) \wedge \mu_G(\overline{Z}_{t+1})] \tag{8.157}$$

As shown in Section 7.3.2, this problem is equivalent to the control of an auxiliary deterministic system in a fuzzy environment over an infinite horizon. It may be solved by a policy iteration algorithm which yields an optimal stationary (replenishment) strategy $a_\infty^* = (a^*, a^*, \ldots)$ such that

$$\overline{R}_t^* = a^*(\overline{Z}_t), \qquad t = 0, 1, \ldots \tag{8.158}$$

Then the optimal stationary (replenishment) policy determined, which is equivalent to

$$a^*(\overline{S}_1) = \overline{C}_a \quad a^*(\overline{S}_2) = \overline{C}_b \quad \ldots \quad a^*(\overline{S}_r) = \overline{C}_d \tag{8.159}$$

where $\overline{S}_1, \ldots, \overline{S}_r \in \mathcal{X}$ and $\overline{C}_a, \ldots, \overline{C}_d \in \mathcal{U}$, is represented by the following fuzzy conditional statement [cf. (2.62), page 38, or (7.89), page 241]:

$$
\begin{aligned}
&\text{IF } \overline{Z}_t = \overline{S}_1 \text{ THEN } \overline{R}_t = \overline{C}_1 \\
&\text{ELSE} \\
&\text{IF } \overline{Z}_t = \overline{S}_2 \text{ THEN } \overline{R}_t = \overline{C}_b \\
&\text{ELSE } \ldots \text{ ELSE} \\
&\text{IF } \overline{Z}_t = \overline{S}_r \text{ THEN } \overline{R}_t = \overline{C}_d =
\end{aligned} \tag{8.160}
$$

$$= \overline{S}_1 \times \overline{C}_a + \overline{S}_2 \times \overline{C}_b + \cdots + \overline{S}_r \times \overline{C}_d$$

where "$\times$" is the Cartesian product (2.56), page 37, and "$+$" is the union (2.32), page 30, of the respective fuzzy relations (cf. Section 2.1.4).

The fuzzy conditional statement (8.160) is equivalent to a fuzzy relation representing an optimal fuzzy replenishment rule (cf. Section 7.3.2). This fuzzy replenishment rule make it possible to find an optimal fuzzy replenishment (not necessarily reference) for an arbitrary fuzzy inventory level (not necessarily reference) due to (7.92), page 242.

Though the use of the above procedure is possible in principle, it is usually inefficient, and an interpolative reasoning presented in Section 4.3.3 may be used.

This concludes our analysis of an inventory control application. And here again, the use of a fuzzy model is very advantageous. The main reason is that the softness (fuzziness) in various aspects of the problem formulation may be dealt with in an adequate and relatively easy way. The data needed may be elicited from experts in quite a natural way, with considerably diminished information needs. The solution procedure is, on the other hand, quite efficient, in particular by using some "trickery."

8.8   A BRIEF REVIEW OF OTHER APPLICATIONS

In this section we will briefly review other more relevant applications of some of the multistage fuzzy control models discussed in this book, mainly the applications of fuzzy dynamic programming.

It should be noted that this brief account is by no means exhaustive, and is only meant to give the reader a glimpse of what has been done using fuzzy dynamic programming in diverse areas. The interested reader will then be able to read the source literature, in particular after having followed our exposition on how a fuzzy multistage model is developed in the previous sections in this chapter, in particular in Sections 8.1 or 8.4.

We will now consecutively sketch the applications in:

- scheduling of power generator maintenance,
- design of a fuzzy controller,
- designing a distillation column in chemical engineering,
- determination of shortest paths for the transportation of hazardous waste,

- scheduling of autonomous guided vehicles in flexible manufacturing systems, and
- optimizing spare parts inventories in a power station.

Huang, Lin and Huang (1992) presented a fuzzy dynamic programming model for the determination of an optimal schedule for the outage starting times of power generators for maintenance over the one-year planning horizon. They assumed five fuzzy constraints and fuzzy goals related to: manpower required, possible interval of necessary maintenance time, location of a power plant, reserve margin goal, and increased production cost goal. All these fuzzy constraints and goals are required to be satisfied.

As to the main elements of the model, the state variable is represented by the pair "generator i, maintenance starting time t, the control (decision) variable is a starting time of maintenance, the stage is the number of the generator scheduled for maintenance at the time t in the state variable.

The authors derive then a (deterministic) state transition function, a return (performance) function as the intersection of the above-mentioned fuzzy constraints and fuzzy goals, and finally a set of fuzzy dynamic programming recurrence equations.

As an example, the maintenance scheduling at the Taiwan Power Company who has 6 nuclear units, 24 thermal units, and a few other smaller units is presented. The problem has 36 possible states, 30 stages and ca. 36 possible decisions. The results obtained are very encouraging.

Hoyo, Terano and Masui (1993) considered the use of fuzzy dynamic programming for quasi-optimal design of a fuzzy controller. The main idea was to employ fuzzy dynamic programming to derive controls when a control plant is far from the goal, and then to switch to an ordinary fuzzy controller when the plant is near the goal.

Basically, the fuzzy constraint is applied at each control stage to keep the output from being too large, and prevent the bang-bang control which a crisp constraint would yield in this case. The fuzzy goal is a compromise between the termination time and the distance from a predefined target (termination state).

It seems that Hoyo, Terano and Masui's (1993) idea should be very inspiring as it tries to combine the strength of the prescriptive, model-based approach to fuzzy control which is particularly well suited when a required status should be attained, and that of a conventional fuzzy (logic) control when the purpose is to stabilize the control.

Krasławski, Górak and Vogenpohl (1989) proposed an application of fuzzy dynamic programming to the determination of distillation sequence (sequence of distillation columns) for a mixture of some number of components. At the consecutive control stages the particular components are separated, and the cost of each separation is fuzzy which is implied by the fact that (inherently imprecise) expert opinions need to be used for the estimation of these costs. Each separation sequence ("trajectory") results therefore in a fuzzy cost, and these costs are then ranked to find an optimal sequence.

Klein (1991a) considered first the general problem of how to find fuzzy shortest paths in networks. The network considered in this paper is directed, acyclic and layered. The arcs are assigned fuzzy lengths, and the fuzzy shortest path is sought. The fuzzy shortest path is determined by employing a fuzzy-dynamic-programming-like procedure which is based on the principle of optimality.

Klein (1992b) then presented an application of the above model for finding the fuzzy shortest paths to the determination of an itinerary for the transportation of hazardous waste load in a given (existing) road structure.

In the first model, the route from a predetermined point of departure to a predetermined point of destination is sought. Each road segment is described by a set of membership grades for some attributes related to, e.g, safety of the road segment, safety of road intersection, distance, driving difficulty, dangers related to population density, etc.

In the second model, the route from a predetermined point of departure to a best potential location (to be determined by the algorithm!) is sought, under the same fuzzy constraints assigned to the particular road segments.

Yuan and Wu (1991) proposed an application of fuzzy dynamic programming to real-time control of a transportation system in a flexible manufacturing system. Their work is concerned with the so-called autonomous guided vehicles which are driverless, programmable vehicles that can move around the factory. A serious difficulty in their control is that the travel time is not only a function of the distance but also of the width of the lane, how slippery the road is, which obstacles are present, etc. For each path in the network of possible lanes membership grades are assigned to the particular possible travel times (taking into account the above aspects). Then, a utility of each path is estimated taking into account the average moving time, the probability of collision, the probability of successful reaching of a destination, etc. The authors claim that the results are very encouraging, and the model can be used for real-time control.

Sugianto and Mielczarski (1995a, b, 1996) consider a spare parts inventory in a power generation plant. Basically, the availability of the power plant is crucial since its breakdown, or too long an unavailability (due to the lack of a proper spare part), may imply considerable losses. However, an obvious, conservative strategy to keep in stock as many spare parts as possible (just in case . . .) is clearly far from optimum as it leads to excessive inventory costs, and hence makes the cost of electricity generation high. On the other hand, a limited stock of spare parts, though reducing the costs, implies a risk of the power plant's unavailability, and hence interruptions in power delivery.

It is therefore clear from the above that there is some optimal spare parts inventory, which leads to an optimal comprimise. And, indeed, many optimization type approaches, aimed at the minimization of total costs, have been proposed for solving this problem.

Unfortunately, the problem considered is plagued by a high degree of uncertainty and vagueness which concern such factors as failure frequency and predictability, the repairability and repair times of certain items, and the delivery times of spare parts. Moreover, the availability of a power plant does not depend on a particular item but on other components, so that the optimization problem should be defined for the whole system instead of for a single item.

Sugianto and Mielczarski (1995a, b, 1996) have proposed a new approach for optimizing the spare parts inventory which takes into account the whole system, with the particular elements (e.g., spare parts) arranged into serial or parallel configurations. The problem is then viewed as a resource allocation process with the objective function being the availability of the power plant. The purchasing costs of spare parts are subject to a budget constraint. Both the objective function and constraints involve imprecise (fuzzy) parameters. Fuzzy dynamic programming is used as a solution technique, and a decision support system in proposed to handle this and other related problems.

This fuzzy approach deals with the utilization of surrogate data from external databases and expert knowledge expressed in terms of numbers and their associated confidence levels. A method is proposed for the use of fuzzy aggregation operations to propagate confidence levels throughout the optimization process.

This concludes our brief exposition of some more relevant application of fuzzy dynamic programming. We tried to cover as large an array of fields as possible, hoping that the readers representing diverse fields of science and technology would find those models interesting, valuable and applicable in their domains of interest and expertise.

We are deeply convinced that fuzzy dynamic programming, and more generally the prescriptive fuzzy control models developed in the preceding chapters, will find more and more relevant applications for solving real world problems.

9

Concluding Remarks

In this book we have provided an extensive exposition of multistage fuzzy control which may be viewed as an alternative paradigm to the traditional fuzzy (logic) control. We are advocating this approach.

Basically, in the traditional, descriptive approach to fuzzy logic control it is, explicitly or implicitly, assumed that the model of the process under control is unknown, for whatever reason exemplified by being really unknown, too costly to derive, etc. It is then assumed, which is usually true, that an experienced process operator knows how to control the process.

However, a performance function is not explicitly involved, and when we wish to change the mode of operation as, for example, when driving a car to switch from an economic speed to that of fast driving, which implies a change of the performance function, we should employ another set of fuzzy control rules. Since there may be numerous possible, or desirable, modes of operation which the operator may be willing to follow, the number of appropriate sets of fuzzy control rules may clearly be prohibitive.

This traditional approach may be therefore termed *descriptive* and *non-model-based*.

In the approach presented and advocated in this book we adopt another paradigm which is more in line with control theory. First, we assume that a model of the system under control is known. However, in traditional approaches we normally assume the availability of some deterministic model in the form of, say, differential or difference equations, or at most of a stochastic model, and in both cases information requirements may be prohibitive. In our approach we allow for an imprecise, or fuzzy, specification of a model of the system under control in the form of, say, linguistic IF–THEN rules dealt with in terms of fuzzy sets and fuzzy logic.

Therefore, such IF–THEN rules may be used in our approach for the specification of dynamics of the system under control, not for the specification of "good" control. We assume next that an explicit performance function is given, and changeable. An efficient Bellman and Zadeh's (1970) approach is employed in which such a performance function is represented in a distributed form as a sequence of fuzzy constraints on the controls and fuzzy goals on the states at the consecutive control stages over a predefined or otherwise planning horizon.

The controls is derived in the approach described by an algorithm, for the particular performance function assumed. These controls may be not only "good" as in the traditional fuzzy control, but even "optimal" as these algorithms are of an optimization type.

The approach presented and advocated in this book may be therefore described as *prescriptive* and *model-based*.

We hope that the alternative approach to fuzzy control will become a relevant part of a new generation of fuzzy control which should find its way to the theory and practice of fuzzy control in the near future. This hope and expectation is certainly amplified by the fact that, first, the alternative approach presented has a firm formal basis and a wide array of modelling and solution tools, and second, that relevant applications – part of which is shown in Chapter 8 – provide convincing arguments of its usefulness.

Bibliography

Adamo J.M. (1980) Fuzzy decision trees. *Fuzzy Sets and Systems* 4: 207–219.

Aizerman M.A. (1977) Some unsolved problems in the theory of automatic control and fuzzy proofs. *IEEE Transactions on Automatic Control* AC-22: 116–118.

Alley H., C.P. Bacinello and K.W. Hipel (1979) Fuzzy set approaches to planning in the Grand River Basin. *Advances in Water Resources* 2: 3–12.

d'Ambrosio B. (1989) *Qualitative Process Theory Using Linguistic Variable*. Springer-Verlag, Berlin.

Asai K., M. Sugeno and T. Terano (1994) *Applied Fuzzy Systems*. Academic Press, New York.

Asai K., H. Tanaka and T. Okuda (1977) On discrimination of fuzzy states in probability space. *Kybernetes* 6: 185–192.

Backhouse R.C. and B.A. Carré (1975) Regular algebra applied to path-finding problems. *Journal of Industrial Mathematics and Applications* 15: 109–118.

Baldwin J.F. and B.W. Pilsworth (1982) Dynamic programming for fuzzy systems with fuzzy environment. *Journal of Mathematical Analysis and Applications* 85: 1–23.

Bandemer H. (1987) From fuzzy data to functional relationships. *Mathematical Modelling* 9: 419–426.

Bandemer H. and S. Gottwald (1995) *Fuzzy Sets, Fuzzy Logic, Fuzzy Methods, with Applications*. Wiley, Chichester.

Bandemer H. and W. Näther (1992) *Fuzzy Data Analysis*. Kluwer, Dordrecht.

Bandler W. and L.J. Kohout (1980) Fuzzy power sets for fuzzy implication operators. *Fuzzy Sets and Systems* 4: 13–30.

Bellman R.E. (1975) Communication, ambiguity and understanding. *Mathematical Biosciences* 26: 347–357.

Bellman R.E. and M. Giertz (1973) On the analytic formalism of the theory of fuzzy sets. *Information Sciences* 5: 149–157.

Bellman R.E. and L.A. Zadeh (1970) Decision making in a fuzzy environment. *Management Science* 17: 141–164.

Berenji H.R. (1992) A reinforcement learning based architecture for fuzzy logic control. *International Journal of Approximate Reasoning* 6: 267–272.

Bertsekas D.P. and S.E. Shreve (1978) *Stochastic Optimal Control: The Discrete Time Case*. Academic Press, New York.

Bezdek J.C. (1981) *Pattern Recognition with Fuzzy Objective Function Algorithms*. Plenum Press, New York.

Bezdek J.C., Ed. (1987a) *Analysis of Fuzzy Information – Vol. 1: Mathematics and Logic*. CRC Press, Boca Raton, FL.

Bezdek J.C., Ed. (1987b) *Analysis of Fuzzy Information – Vol. 2: Artificial Intelligence and Decision Systems*. CRC Press, Boca Raton, FL.

Bezdek J.C., Ed. (1987c) *Analysis of Fuzzy Information – Vol. 3: Applications in Engineering and Science*. CRC Press, Boca Raton, FL.

Black M. (1937) Vagueness: an exercise in logical analysis. *Philosophy of Science* 4: 427–455.

Black M. (1963) Reasoning with loose concepts. *Dialogue* 2: 1–12.

Black M. (1970) *Margins of Precision*. Cornell University Press, Ithaca, NY.

Bolc L. and P. Borowik (1992) *Many-Valued Logics*. Springer-Verlag, Berlin.

Bouchon-Meunier B. (1992) Fuzzy logic and knowledge representation using linguistic modifiers. In L.A. Zadeh and J. Kacprzyk (Eds.): *Fuzzy Logic for the Management of Uncertainty*, Wiley, New York, pp. 399–414.

Britov G.S. and L.K. Reznik (1981) Optimal control of linear fuzzy systems (in Russian). *Automation and Remote Control* 42: 462–465.

Buckley J.J. (1992) Theory of the fuzzy controller: An introduction. *Fuzzy Sets and Systems* 51: 249–258.

Buckley J.J. (1993) Controllable processes and the fuzzy controller. *Fuzzy Sets and Systems* 53: 27–32.

Butnariu D. and E.P. Klement (1993) *Triangular Norm-based Measures and Games with Fuzzy Coalitions*. Kluwer, Dordrecht.

Cao S.G. and N.W. Rees (1995) Identification of fuzzy models. *Fuzzy Sets and Systems* 74: 307–320.

Carlsson C. (1984) Fuzzy systems: a basis for a modelling methodology. *Cybernetics and Systems* 15: 361–378.

Carlsson C. (1989) *Fuzzy Sets for Management Decision*. Verlag TÜV Rheinland, Cologne.

Celmiņš A. (1987) Least squares model fitting to fuzzy vector data. *Fuzzy Sets and Systems* 22: 245–269.

Chang R.L.P. and T. Pavlidis (1977) Fuzzy decision tree algorithms. *IEEE Transactions on Systems, Man and Cybernetics* SMC-7: 28–35.

Chang S.S.L. (1969a) Fuzzy dynamic programming and the decision making process. *Proceedings of the Third Princeton Conference on Information Sciences*, Princeton, NJ.

Chang S.S.L. (1969b) Fuzzy dynamic programming and approximate optimization of partially known systems. *Proceedings of the Second Hawaii International Conference on Systems Science*, Honolulu, HI.

Chang S.S.L. (1974) Control and estimation of fuzzy systems. *Proceedings of the IEEE Conference on Decision and Control*.

Chang S.S.L. (1977a) On fuzzy algorithm and mapping. In M.M. Gupta, G.N. Saridis and B.R. Gaines (Eds.): *Fuzzy Automata and Decision Processes*, North-Holland, Amsterdam.

Chang S.S.L. (1977b) Application of fuzzy set theory to economics. *Kybernetes* 6: 203–207.

Chang S.S.L. and L.A. Zadeh (1972) On fuzzy mapping and control. *IEEE Transactions on Systems, Man and Cybernetics* SMC-2: 30–34.

Checkland P.B. (1976) Science and the systems paradigm. *International Journal of General Systems* 3: 127–134.

Checkland P.B. (1979) Techniques in "soft" systems practice. Part I: Systems diagrams – some tentative guidelines, Part II: Building conceptual models. *Journal of Applied Systems Analysis* 6: 33-49.

Chen Y.Y. and T.C. Tsao (1989) A description of the dynamical behavior of fuzzy systems. *IEEE Transactions on Systems, Man and Cybernetics* 19: 745–755.

Cox E. (1994) *The Fuzzy System Handbook. A Practitioner's Guide to Building, Using, and Maintaining Fuzzy Systems*. Academic Press, New York.

Cumani A. (1981) On a possibilistic approach to the analysis of fuzzy feedback systems. *IEEE Transactions on Systems, Man and Cybernetics* SMC-12: 417–422.

Czogała E. (1984) *Probabilistic Sets in Decision-Making and Control*. Verlag TÜV Rheinland, Cologne.

Czogała E. and K. Hirota (1986) *Probabilistic Sets: Fuzzy and Stochastic Approach to Decision, Control and Recognition Processes*. Verlag TÜV Rheinland, Cologne.

Czogała E. and W. Pedrycz (1981) On identification of fuzzy systems and its application in control problems. *Fuzzy Sets and Systems* 6: 73–83.

Czogała E. and W. Pedrycz (1982) Control problems in fuzzy systems. *Fuzzy Sets and Systems* 7: 257–274.

Czogała E. and W. Pedrycz (1984) Identification and control problems in fuzzy systems. *TIMS Studies in the Management Sciences* 20: 447–466.

Davis L. (1991) *Handbook of Genetic Algorithms*. Van Nostrand Reinhold, New York.

Delgado M., J. Kacprzyk, J.L. Verdegay and M.A. Vila, Eds. (1994) *Fuzzy Optimization:*

Recent Advances, Physica-Verlag, Heidelberg (A Springer-Verlag Company).

Dompere K.K. (1995) The theory of social costs and costing for cost-benefit analysis in a fuzzy-decision space. *Fuzzy Sets and Systems* 76: 1–24.

Driankov D., H. Hellendoorn and M. Reinfrank (1993) *An Introduction to Fuzzy Control.* Springer-Verlag, Berlin.

Dubois D., J. Lang and H. Prade (1991) Fuzzy sets in approximate reasoning, Part 2: Logical approaches. *Fuzzy Sets and Systems* 40: 203–244.

Dubois D. and H. Prade (1978) Operations on fuzzy numbers. *International Journal of Systems Science* 9: 613–626.

Dubois D. and H. Prade (1980) *Fuzzy Sets and Systems: Theory and Applications.* Academic Press, New York.

Dubois D. and H. Prade (1982) A class of fuzzy measures based on triangular norms. *International Journal of General Systems* 8: 43–61.

Dubois D. and H. Prade (1983) Ranking of fuzzy numbers in the setting of possibility theory. *Information Sciences* 30: 183–224.

Dubois D. and H. Prade (1984) Fuzzy logic and the generalized modus ponens revisited. *Cybernetics and Systems* 15: 292–331.

Dubois D. and H. Prade (1988) *Possibility Theory – An Approach to Computerized Processing of Uncertainty.* Plenum Press, New York.

Dubois D. and H. Prade (1989) Fuzzy sets, probability and measurement. *European Journal of Operational Research* 40: 135–154.

Dubois D. and H. Prade (1991) Fuzzy sets in approximate reasoning, Part 1: Inference with possibility distributions. *Fuzzy Sets and Systems* 40: 143–202.

Dubois D. and H. Prade (1993) Fuzzy sets and probability: Misunderstandings, bridges and gaps. *Proceedings of Second IEEE International Conference on Fuzzy Systems* (San Francisco, CA). IEEE Press, New York, pp. 1059–1068.

Dubois D., H. Prade and J.M. Toucas (1990) Inference with imprecise numerical quantifiers. In Z. Raś and M. Zemankova (Eds.): *Intelligent Systems: State of the Art and Future Directions.* Ellis Horwood, Chichester, pp. 52–72.

Dubois D., H. Prade and R.R. Yager, Eds. (1993) Readings in Fuzzy Sets for Intelligent Systems. Morgan Kaufmann, San Mateo, CA.

Esogbue A.O. (1983) Dynamic programming, fuzzy sets and the modelling of R&D management control systems. *IEEE Transactions on Systems, Man, and Cybernetics* SMC-13: 18–30.

Esogbue A.O. (1985) A fuzzy dynamic programming model of intra-operative anesthesia administration. In J. Kacprzyk and R.R. Yager (Eds.): Management Decision Support Systems using Fuzzy Sets and Possibility Theory. Verlag TÜV Rheinland, Cologne, pp. 155–161.

Esogbue A.O. (1986) Optimal clustering of fuzzy data via fuzzy dynamic programming. *Fuzzy Sets and Systems* 18: 283–298.

Esogbue A.O. (1989) *Dynamic Programming for Optimal Water Resources Systems Analysis.* Prentice-Hall, Englewood Cliffs, NJ.

Esogbue A.O. (1991b) Computational aspects and applications of a branch and bound algorithm for fuzzy multistage decision processes. *Computers and Mathematics with Applications* 21: 117–127.

Esogbue A.O. and Z.M. Ahipo (1982a) Fuzzy sets in water resources planning. In R.R. Yager (Eds.): *Recent Developments in Fuzzy Sets and Possibility Theory*, Pergamon Press, New York, pp. 450–465.

Esogbue A.O. and Z.M. Ahipo (1982b) A fuzzy sets model for measuring the effectiveness of public participation in water resources planning. *Water Resources Bulletin* 18: 451–456.

Esogbue A.O. and R.E. Bellman (1981) A fuzzy dynamic programming algorithm for clustering non-quantitative data arising in water pollution control planning. *Proceedings of Third International Conference on Mathematical Modelling* (Los Angeles, CA).

Esogbue A.O. and R.E. Bellman (1984) Fuzzy dynamic programming and its extensions. *TIMS/Studies in the Management Sciences* 20: 147–167.

Esogbue A.O., M. Fedrizzi and J. Kacprzyk (1988) Fuzzy dynamic programming with stochastic systems. In J. Kacprzyk and M. Fedrizzi (Eds.): *Combining Fuzzy Imprecision with Probabilistic Uncertainty in Decision Making.* Springer-Verlag, Berlin/New York, pp. 266–285.

Esogbue A.O. and V. Ramesh (1970) Dynamic programming and fuzzy allocation processes. *Technical Memo No. 202*, Dept. of Operations Research, Case Western University, Cleveland, OH.

Esogbue A.O., M. Theologidu and K. Guo (1992) On the application of fuzzy sets theory to the optimal flood control problem arising in water resources systems. *Fuzzy Sets and Systems* 48: 155–172.

Evans G.W., W. Karwowski and M.R. Wilhelm, Eds. (1989) *Applications of Fuzzy Set Methodologies in Industrial Engineering.* North-Holland, Amsterdam.

Fedrizzi M., J. Kacprzyk and F. Molinari (1984) Determination of corporate investment policies in a "soft" setting using fuzzy decision making models. In A. Straszak (Ed.): *Large Scale Systems Theory and Applications – Proceedings of Third IFAC/IFORS Symposium* (Warsaw, Poland, 1983), Pergamon Press, Oxford, pp. 14–46.

Fedrizzi M., J. Kacprzyk and M. Roubens, Eds. (1991) *Interactive Fuzzy Optimization.* Springer-Verlag, Berlin/New York.

Filev D. (1992) Systems approach to dynamic fuzzy models. *International Journal of General Systems* 21: 311–337.

Filev D. and P. Angelov (1992) Fuzzy optimal control. *Fuzzy Sets and Systems* 47: 151–156.

Francelin R.A. and F.A.C. Gomide (1992) Neural network to solve fuzzy discrete programming problems. Tech. Rep. 018/92, DCA/FEE UNICAMP, Campinas, Brazil.

Francelin R.A. and F.A.C. Gomide (1993) A neural network for fuzzy decision making problems. *Proceedings of Second IEEE International Conference on Fuzzy Systems* (San Francisco, CA, USA), Vol. 1, pp. 655–660.

Francelin R.A., F.A.C. Gomide and J. Kacprzyk (1995) A class of neural networks for dynamic programming. *Proceedings of Sixth International Fuzzy Systems Association World Congress (Saõ Paolo, Brazil), Vol. II, pp. 221–224.*

French S. (1988) *Decision Theory: An Introduction to the Mathematics of Rationality.* Ellis Horwood, Chichester.

Fung L.W. and K.S. Fu (1973) The k-th optimal policy algorithm for decision-making in fuzzy environment. *Proceedings of the Third IFAC Symposium on Identification and System Parameter Estimation*, The Hague.

Fung L.W. and K.S. Fu (1975) An axiomatic approach to rational decision-making in a fuzzy environment. In L.A. Zadeh, K.S. Fu, K. Tanaka and M. Shimura (Eds.): *Fuzzy Sets and their Applications to Cognitive and Decision Processes*, Academic Press, New York.

Fung L.W. and K.S. Fu (1977) Characterization of a class of fuzzy optimal control problems. In M.M. Gupta, G.N. Saridis and B.R. Gaines (Eds.): *Fuzzy Automata and Decision Processes*, New York: North-Holland, pp. 209–219.

Gaines B.R. (1977) Foundations of fuzzy reasoning. *International Journal of Man-Machine Studies* 8: 623–668.

van Gigch J.P., and L.L. Pipino (1980) From absolute to probable and fuzzy in decision-making. *Kybernetes* 9: 47–55.

de Glas M. (1983) Theory of fuzzy systems. *Fuzzy Sets and Systems* 10: 65–78.

Gluss B. (1972) Fuzzy multistage decision-making, fuzzy state and terminal regulators and their relationship to non-fuzzy quadratic state and terminal regulators. *International Journal of Control* 17: 177–192.

Goguen J.A. (1967) L-fuzzy sets. *Journal of Mathematical Analysis and Applications* 18: 145–174.

Goguen J.A. (1969) The logic of inexact concepts. *Synthese* 19: 325–373.

Goodman I.R. and H.T. Nguyen (1985) *Uncertainty Models for Knowledge-Based Systems.* North-Holland, Amsterdam.

Gottwald S. (1993) *Fuzzy Sets and Fuzzy Logic.* Vieweg, Wiesbaden.

Grabisch M., T. Murofushi and M. Sugeno (1992) Fuzzy measure of fuzzy events defined by fuzzy integrals. *Fuzzy Sets and Systems* 50: 293–313.

Grabisch M., H.T. Nguyen and E.A. Walker (1995) *Fundamentals of Uncertainty Calculi with*

Applications to Fuzzy Inference. Kluwer, Dordrecht.

Gupta M.M., A. Kandel, W. Bandler and J.B. Kiszka, Eds. (1985) *Approximate Reasoning in Expert Systems*. North-Holland, Amsterdam.

Gupta M.M., R,K, Ragade and R.R. Yager, Eds. (1979) *Advances in Fuzzy Sets Theory and Applications*. North-Holland, Amsterdam.

Gupta M.M. and E. Sanchez, Eds. (1982a) *Approximate Reasoning in Decision Analysis*. North-Holland, Amsterdam.

Gupta M.M. and E. Sanchez, Eds. (1982b) *Fuzzy Information and Decision Processes*. North-Holland, Amsterdam.

Gupta M.M., G.N. Saridis and B.R. Gaines, Eds. (1977) *Fuzzy Automata and Decision Processes*. North-Holland. Amsterdam.

Gupta M.M. and T. Yamakawa, Eds. (1988a) *Fuzzy Computing: Theory, Hardware and Applications*. North-Holland, Amsterdam.

Gupta M.M. and T. Yamakawa, Eds. (1988b) *Fuzzy Logic in Knowledge-Based Systems, Decision and Control*. North-Holland, Amsterdam.

Hamacher H. (1978) Über logische Verknüpfungen unscharfer Aussagen und deren zugehörige Bewertungsfunktionen. In R. Trappl, G.J. Klir and L. Ricciardi (Eds.): *Progress in Cybernetics and Systems Research 3*, Hemisphere, Washington, DC, pp. 276–288.

Han L. (1993) The mathematical model of general multistage decision making in an uncertain environment. *Proceedings of Fifth International Fuzzy Systems Association World Congress '93*, Vol. 1, pp. 631–633,

Higashi M. and G.J. Klir (1983) Measures of uncertainty and information based on possibility distributions. *International Journal of General Systems* 9: 43–58.

Hirota K. (1980) Concepts of probabilistic sets. *Fuzzy Sets and Systems* 5: 31–46.

Hirota K. (1985) *Fuzzy Control of Intelligent Robots* (in Japanese). McGraw-Hill, Tokyo.

Hirota K., Ed. (1993) *Industrial Applications of Fuzzy Technology*. Springer-Verlag, Tokyo.

Hirota K. and W. Pedrycz (1995) A fuzzy modelling environment for designing fuzzy controllers. *Fuzzy Sets and Systems* 73: 287–302.

Holmblad L.P. and J.J. Østergaard (1982) Control of a cement kiln by fuzzy logic. In M.M. Gupta and E. Sanchez (Eds.): *Fuzzy Information and Decision Processes*. North-Holland, Amsterdam, pp. 389–400.

Howard R.A. (1960) *Dynamic Programming and Markov Processes*. MIT Press, Cambridge, MA.

Howard R.A. (1971) *Dynamic Probabilistic Systems*. Volumes 1 and 2, Wiley, New York.

Hoyo T., T. Terano and S. Masui (1991) Stability analysis of fuzzy control systems. In R. Lowen and M. Roubens (Eds.): *Proceedings of Fourth International Fuzzy Systems Association Congress* (Brussels, Belgium), Vol. AI, pp. 45–48.

Hoyo T., T. Terano and S. Masui (1993) Design of quasi-optimal fuzzy controller by fuzzy dynamic programming. *Proceedings of Second IEEE International Conference on Fuzzy Systems* (San Francisco, CA, USA), Vol. 2, pp. 1253–1258.

Huang C.J., C.E. Lin and C.L. Huang (1992) Fuzzy approach for generator maintenance scheduling. *Electric Power Systems Research* 24: 31–38.

Hussein M.L. and M.A. Abo-Sinna (1995) Decomposition of multiobjective programming problems by hybrid fuzzy dynamic programming. *Fuzzy Sets and Systems* 60: 25–32.

Işik C. and S. Ammar (1992) Fuzzy optimal search methods. *Fuzzy Sets and Systems* 46: 331–337.

Jacobson D.H. (1976) On fuzzy goals and maximizing decisions in stochastic optimal control. *Journal of Mathematical Analysis and Applications* 55: 434–440.

Jamshidi M., N. Vadiee and T.J. Ross, Eds. (1993) *Fuzzy Logic and Control*. Prentice-Hall, Englewood Cliffs, NJ.

Kacprzyk J. (1975) Application of fuzzy sets theory for the optimal assignment of work places (in Polish). *Archiwum Automatyki i Telemechaniki* XX: 247–260.

Kacprzyk J. (1977) Control of a nonfuzzy system in a fuzzy environment with a fuzzy termination time. *Systems Science* 3: 320–334.

Kacprzyk J. (1978a) A branch-and-bound algorithm for the multistage control of a nonfuzzy system in a fuzzy environment. *Control and Cybernetics* 7: 51–64.

Kacprzyk J. (1978b) Control of a stochastic system in a fuzzy environment with a fuzzy termination time. *Systems Science* 4: 291–300.

Kacprzyk J. (1978c) Decision-making in a fuzzy environment with fuzzy termination time. *Fuzzy Sets and Systems* 1: 169–179.

Kacprzyk J. (1979) A branch-and-bound algorithm for the multistage control of a fuzzy system in a fuzzy environment. *Kybernetes* 8: 139–147.

Kacprzyk J. (1982a) Multistage decision processes in a fuzzy environment: a survey. In M.M. Gupta and E. Sanchez (Eds.): *Fuzzy Information and Decision Processes*, North-Holland, Amsterdam, pp. 251–253.

Kacprzyk J. (1982b) Control of a stochastic system in a fuzzy environment with Yager's probability of a fuzzy event. *Busefal* 12: 77–88.

Kacprzyk J. (1983a) *Multistage Decision Making in Fuzzy Conditions* (in Polish). PWN Scientific Publishers, Warsaw.

Kacprzyk J. (1983b) *Multistage Decision Making under Fuzziness*, Verlag TÜV Rheinland, Cologne.

Kacprzyk J. (1983c) A generalization of fuzzy multistage decision making and control via linguistic quantifiers. *International Journal of Control* 38: 1249–1270.

Kacprzyk J., Ed. (1984a) Special Issue on Fuzzy Sets and Possibility Theory in Optimization Models. *Control and Cybernetics*, Vol. 13, No. 3.

Kacprzyk J. (1984c) Yager's probability of a fuzzy event in stochastic control under fuzziness. In M.M. Gupta and M. Sanchez (Eds.): *Fuzzy Information, Knowledge Representation and Decision Analysis*, Pergamon Press, Oxford, pp. 379–384.

Kacprzyk J. (1984d) A generalized formulation of multistage decision making and control under fuzziness. In M.M. Gupta and E. Sanchez (Eds.): *Fuzzy Information, Knowledge Representation and Decision Analysis*, Pergamon Press, Oxford, 1984, pp. 73–78.

Kacprzyk J. (1985) Zadeh's commonsense knowledge and its use in multicriteria, multistage and multiperson decision making. In M.M. Gupta *et al.* (Eds.): *Approximate Reasoning in Expert Systems*, North-Holland, Amsterdam, pp. 105–121.

Kacprzyk J. (1986a) Towards 'human-consistent' multistage decision making and control models via fuzzy sets and fuzzy logic. Bellman Memorial Issue (A.O. Esogbue, Ed.), *Fuzzy Sets and Systems* 18: 299–314.

Kacprzyk J. (1986b) A down-to-earth managerial decision making via a fuzzy-logic-based representation of commonsense knowledge. In L.F. Pau (Ed.): *Artificial Intelligence in Economics and Management*, Elsevier (North-Holland), Amsterdam, pp. 57–64.

Kacprzyk J. (1986c) Fuzzy dispositions as a tool for more human-perception-consistent models in systems analysis. In S. Bocklisch, S. Orlovski, M. Peschel and Y. Nishiwaki (Eds.): *Fuzzy Sets: Applications, Methodological Approaches and Results*, Band 30, Mathematical Research, Akademie-Verlag, Berlin, pp. 64–73.

Kacprzyk J. (1987a) *Fuzzy Sets in Systems Analysis* (in Polish). PWN Scientific Publishers, Warsaw.

Kacprzyk J. (1987b) Towards "human-consistent" decision support systems through commonsense knowledge based decision making and control models: a fuzzy logic approach. *Computers and Artificial Intelligence* 6: 97–122.

Kacprzyk J. (1987c) Stochastic systems in fuzzy environments: control. In M.G. Singh (Ed.): *Systems and Control Encyclopedia*, Pergamon Press, Oxford, pp. 4657–4661.

Kacprzyk J. (1991a) Compatibility relations for the representation of associations between variables in knowledge-based systems, and their use in approximate reasoning. *Fuzzy Sets and Systems* 42: 273–291.

Kacprzyk J. (1991b) Fuzzy linguistic quantifiers in decision making and control. *Proceedings of International Fuzzy Engineering Symposium – IFES '91* (Yokohama, Japan), Vol. 2, pp. 800–811.

Kacprzyk J. (1992a) Fuzzy optimal control revisited: toward a new generation of fuzzy control? *Proceedings of Second International Conference on Fuzzy Logic and Neural Networks – IIZUKA '92*, Vol. 1, pp. 429–432.

Kacprzyk J. (1992b) Fuzzy logic with linguistic quantifiers in decision making and control. *Archives of Control Sciences* 1 (XXXVII): 127–141.

Kacprzyk J. (1992c) Fuzzy sets and fuzzy logic. In S.C. Shapiro (Ed.): *Encyclopedia of Artificial Intelligence*, Vol. 1, Wiley, New York, pp. 537–542.

Kacprzyk J. (1993a) Interpolative reasoning in optimal fuzzy control. *Proceedings of Second IEEE International Conference on Fuzzy Systems – FUZZ–IEEE'93*, Vol. II, pp. 1259–1263.

Kacprzyk J. (1993b) Fuzzy control with an explicit performance function using dynamic programming and interpolative reasoning. *Proceedings of First European Congress on Fuzzy and Intelligent Technologies – EUFIT'93* (Aachen, Germany), Vol. 3, pp. 1459–1463.

Kacprzyk J. (1993c) A prescriptive approach to fuzzy control: a step toward a 'more mature' fuzzy control? *Proceedings of First Asian Fuzzy Systems Symposium* (Singapore), pp. 360–365.

Kacprzyk J. (1993d) In search for a new generation of fuzzy control: can a prescriptive approach based on interpolative reasoning and neural networks help? *Proceedings of ANZIIS '93 – Australian and New Zealand Conference on Intelligent Information Systems* (Perth, Australia), pp. 402–406.

Kacprzyk J. (1993e) Interpolative reasoning for computationally efficient optimal fuzzy control. *Proceedings of Fifth International Fuzzy Systems Association World Congress '93* (Seoul, Korea), Vol. II, pp. 1270–1273.

Kacprzyk J. (1994a) Fuzzy dynamic programming – basic issues. In M. Delgado, J. Kacprzyk, J.-L. Verdegay and M.A. Vila (Eds.): *Fuzzy Optimization: Recent Advances*, Physica-Verlag, Heidelberg (A Springer-Verlag Company), pp. 321–331.

Kacprzyk J. (1994b) On measuring the specificity of IF–THEN rules. *International Journal of Approximate Reasoning* 11: 29–53.

Kacprzyk J. (1994c) Fuzzy dynamic programming – battling against the curse of dimensionality via interpolative reasoning. *Procedings of Third International Conference on Fuzzy Logic, Neural Nets and Soft Computing – IIZUKA'94*, pp. 245–246.

Kacprzyk J. (1994d) Prescriptive approaches to fuzzy control: yet-to-be-rediscovered old jewels? Proceedings of Brazil–Japan Joint Symposium on Fuzzy Systems, Vol. Tutorials, pp. 22–35.

Kacprzyk J. (1994e) On prescriptive approaches to fuzzy control. *Proceedings of the First Workshop on Fuzzy Based Expert Systems – FUBEST '94* (Sofia, Bulgaria), pp. 1–3.

Kacprzyk J. (1994f) Interpolative reasoning in multistage optimal control. In R. Kulikowski, K. Szkatuła and J. Kacprzyk (Eds.): *Proceedings of 9th Polish–Italian and 5th Polish–Finnish Symposium on Systems Analysis and Decision Support in Economics and Technology*, Omnitech Press, Warsaw, pp. 144–150.

Kacprzyk J. (1995a) A genetic algorithm for the multistage control of a fuzzy system in a fuzzy environment. *Proceedings of Joint Third International IEEE Conference on Fuzzy Systems and Second International Symposium on Fuzzy Engineering – FUZZ-IEEE'95/IFES'95* (Yokohama, Japan), Vol. III, pp. 1083–1088

Kacprzyk J. (1995b) Multistage fuzzy control using a genetic algorithm. *Proceedings of Sixth World International Fuzzy Systems Association Congress* (Saõ Paolo, Brazil), Vol. II, pp. 225–228

Kacprzyk J. (1995c) A modified genetic algorithm for multistage control of a fuzzy system. *Proceedings of Third European Congress on Intelligent Techniques and Soft Computing – EUFIT'95* (Aachen, Germany), Vol. 1, pp. 463–466.

Kacprzyk J. and A.O. Esogbue (1996) Fuzzy dynamic programming: main developments and applications. *Fuzzy Sets and Systems* 81: 31–46.

Kacprzyk J. and M. Fedrizzi, Eds. (1988a) *Combining Fuzzy Imprecision with Probabilistic Uncertainty in Decision Making*, Springer-Verlag, Berlin.

Kacprzyk J., R.A. Francelin and F.A.C. Gomide (1995) Multistage fuzzy control: problem classes and their solution via dynamic programming, branch-and-bound, neural networks and genetic algorithms. *Proceedings of Sixth International Fuzzy Systems Association World Congress (Saõ Paolo, Brazil), Vol. II, pp. 213–216.*

Kacprzyk J. and C. Iwański (1987) A generalization of discounted multistage decision making and control through fuzzy linguistic quantifiers: an attempt to introduce commonsense knowledge. *International Journal of Control* 45: 1909–1930.

Kacprzyk J. and S.A. Orlovski, Eds. (1987a) *Optimization Models Using Fuzzy Sets and*

Possibility Theory, Reidel, Dordrecht.

Kacprzyk J., K. Safteruk and P. Staniewski (1981a) Control of a stochastic system in a fuzzy environment over an infinite horizon in a non-Bellman-Zadeh setting. In A. Straszak and A. Ziółkowski (Eds.): *Proceedings of First Polish-Finnish Symposium on Systems Analysis and Its Applications*, IBS PAN, Warsaw, Part I, pp. 79–91.

Kacprzyk J., K. Safteruk and P. Staniewski (1981b) On the control of stochastic systems in a fuzzy environment over infinite horizon. *Systems Science* 7: 121–131.

Kacprzyk J. and P. Staniewski (1980a) A new approach to the control of stochastic systems in a fuzzy environment. *Archiwum Automatyki i Telemechaniki* XXV: 433–443.

Kacprzyk J. and P. Staniewski (1980b) Determination of long-range inventory policies via fuzzy decision-making models. *Proceedings of Fourth Polish–Italian Conference on Applications of Systems Theory to Economy, Management and Technology*, IBS PAN, Warsaw, pp. 327–335.

Kacprzyk J. and P. Staniewski (1982) Long-term inventory policy-making through fuzzy decision-making models. *Fuzzy Sets and Systems* 8: 117–132.

Kacprzyk J. and P. Staniewski (1983a) Control of a deterministic system in a fuzzy environment over an infinite planning horizon. *Fuzzy Sets and Systems* 10: 291–298.

Kacprzyk J. and P. Staniewski (1983b) Control of a fuzzy system over infinite horizon via approximation by reference fuzzy sets. *Proceedings of Third Polish–German (GDR) Seminar on Non-Conventional Optimization Problems*, ZFR Info.- 83.01, Berlin, pp. 93–106.

Kacprzyk J. and A. Straszak (1980) Determination of 'stable' integrated regional development trajectories via fuzzy decision-making models. *Proceedings of 4th Polish–Italian Conference on Applications of Systems Theory to Economy, Management and Technology*, IBS PAN, Warsaw, pp. 337–347.

Kacprzyk J. and A. Straszak (1981) Application of fuzzy decision-making models for determining optimal policies in 'stable' integrated regional development. In P.P. Wang and S.K. Chang (Eds.): *Fuzzy Sets Theory and Applications to Policy Analysis and Information Systems*, Plenum Press, New York, pp. 321–328.

Kacprzyk J. and A. Straszak (1982a) A fuzzy approach to the stability of integrated regional development. In G.E. Lasker (Ed.): *Applied Systems and Cybernetics*, Vol. 6, Pergamon Press, New York, pp. 2997–3004.

Kacprzyk J. and A. Straszak (1982b) Determination of 'stable' regional development trajectories via a fuzzy decision-making model. In R.R. Yager (Ed.): *Recent Developments in Fuzzy Sets and Possibility Theory*, Pergamon Press, New York, pp. 531–541

Kacprzyk J. and A. Straszak (1984) Determination of stable trajectories for integrated regional development using fuzzy decision models. *IEEE Transactions on Systems, Man and Cybernetics* SMC-14: 310–313.

Kacprzyk J. and A. Straszak (1986) Strategic regional development policy making using a fuzzy model. In A. Straszak and J.W. Owsiński (Eds.): *Strategic Regional Policy: Paradigms, Methods, Issues, and Case Studies*, Part I, Warsaw, pp. 48–63.

Kacprzyk J. and R.R. Yager (1984a) Linguistic quantifiers and belief qualification in fuzzy multicriteria and multistage decision making. *Control and Cybernetics* 13: 155–173.

Kacprzyk J. and R.R Yager (1984b) "Softer" optimization and control models via fuzzy linguistic quantifiers. *Information Sciences* 34: 157–178.

Kacprzyk J. and R.R. Yager, Eds. (1985) *Management Decision Support Systems Using Fuzzy Sets and Possibility Theory*. Verlag TÜV Rheinland, Cologne.

Kacprzyk J. and R.R. Yager (1990) Using fuzzy logic with linguistic quantifiers in multiobjective decision making and optimization: a step towards more human-consistent models. In R. Słowiński and J. Teghem (Eds.): *Stochastic versus Fuzzy Approaches to Multiobjective Mathematical Programming under Uncertainty*, Kluwer, Dordrecht, pp. 331–350. Kandel A. (1986) *Fuzzy Mathematical Techniques with Applications*. Addison-Wesley, Reading, MA.

Kandel A. and G. Langholz, Eds. (1991) *Fuzzy Control Systems*. CRC Press, Boca Raton, FL.

Kandel A. and S.C. Lee (1979) *Fuzzy Switching and Automata: Theory and Applications*.

Crane Russak, New York.

Kaufmann A. (1975) *Introduction to the Theory of Fuzzy Subsets. Vol. 1: Fundamental Theoretical Elements.* Academic Press, New York.

Kaufmann A. and M.M. Gupta (1985) *Introduction to Fuzzy Mathematics – Theory and Applications.* Van Nostrand Reinhold, New York.

Kaufmann A. and M.M. Gupta (1988) *Fuzzy Mathematical Models in Engineering and Management Science.* North-Holland, Amsterdam.

Keller J.M., R.R. Yager and H. Tahani (1992) Neural network implementation of fuzzy logic. *Fuzzy Sets and Systems* 45: 1–12.

Kickert W.J.M. (1978) *Fuzzy Theories on Decision-Making.* Nijhoff, Leiden.

Klein C.M. (1991a) Fuzzy shortest paths. *Fuzzy Sets and Systems* 39: 27–41.

Klein C.M. (1991b) A model for the transportation of hazardous waste. *Decision Sciences* 22: 1091–1108.

Klir G.J. (1987) Where do we stand on measures of uncertainty, ambiguity, fuzziness, and the like? *Fuzzy Sets and Systems* 24: 141–160.

Klir G.J. and T.A. Folger (1988) *Fuzzy Sets, Uncertainty and Information.* Prentice-Hall, Englewood Cliffs, NJ.

Klir G.J. and B. Yuan (1995) *Fuzzy Sets and Fuzzy Logic: Theory and Application.* Prentice-Hall, Englewood Cliffs, NJ.

Kohler W.H. and K. Steiglitz (1974) Characterization and theoretical comparison of branch and bound algorithms for permutation problems. *Journal of the Association for Computing Machinery* 21: 165–181.

Kloeden P.E. (1982) Fuzzy dynamical systems. *Fuzzy Sets and Systems* 7: 275–298.

Kóczy L. and K. Hirota (1992) Analogical fuzzy reasoning and gradual inference rules. *Proceedings of the Second Intternational Conference on Fuzzy Logic and Neural Networks – IIZUKA'92,* Vol. 1, pp. 329–332.

Komolov S.V., S.P. Makeev, G.P. Serov and I.F. Shakhnov (1979) On the problem of optimal control of a finite automaton with fuzzy constraints and fuzzy goal (in Russian). *Kybernetika* (Kiev) 6: 30–34.

Kosko B. (1992a) Fuzzy systems as universal approximators. *Proceedings of First IEEE Conference on Fuzzy Systems,* San Diego, CA, pp. 1153–1162.

Kosko B. (1992b) *Neural Networks and Fuzzy Systems.* Prentice-Hall, Englewood Cliffs, NJ.

Krasławski A., A. Górak and A. Vogelpohl (1989) Fuzzy dynamic programming in the synthesis of distillation column systems. *Computers and Chemical Engineering* 13: 611–618.

Kreps D.M. (1977a) Decision problems with expected utility criteria, I: upper and lower convergent utility. *Mathematics of Operations Research* 2: 45–53.

Kreps D.M. (1977b) Decision problems with expected utility criteria, II: stationarity. *Mathematics of Operations Research* 2: 266–274.

Kruse R. (1984) Statistical estimation with linguistic data. *Information Sciences* 33: 197–207.

Kruse R., J. Gebhard and F. Klawonn (1994) *Foundations of Fuzzy Systems.* Wiley, Chichester, UK.

Kruse R. and K.D. Meyer (1987) *Statistics with Vague Data.* Reidel, Dordrecht.

Kruse R., E. Schwecke and J. Heinsohn (1991) *Uncertainty and Vagueness in Knowledge-Based Systems: Numerical Methods.* Springer-Verlag, Berlin.

Kurano M., M. Yasuda, J. Nakagami and Y. Yoshida (1992) A limit theorem in some dynamic fuzzy systems. *Fuzzy Sets and Systems* 51: 83–88.

Lakov D. (1985) Adaptive robot under control. *Fuzzy Sets and Systems* 17: 1–8.

Lee C.C. (1990a) Fuzzy logic in control systems: Fuzzy logic controller, Part I. *IEEE Transaction on Systems, Man and Cybernetics* SMC-20: 404–418.

Lee C.C. (1990b) Fuzzy logic in control systems: Fuzzy logic controller, Part II. *IEEE Transaction on Systems, Man and Cybernetics* SMC-20: 419–443.

Lee Y.C., C. Huang and Y.P. Shin (1994) A combined approach to fuzzy model identification. *IEEE Transactions on Systems, Man and Cybernetics* SMC-24: 736–744.

Liu B.D. and A.O. Esogbue (1996) Fuzzy criterion set and fuzzy criterion dynamic programming. *Journal of Mathematical Analysis and Applications* 199: 293–311.

Lowen R. (1996) *Fuzzy Set Theory – Basic Concepts, Techniques and Bibliography*. Kluwer, Dordrecht.

Mamdani E.H. (1974) Application of fuzzy algorithms for the control of a simple dynamic plant. *Proceedings of IEE* 121: 1585–1588.

Mamdani E.H. (1976) Advances in the linguistic synthesis of fuzzy controllers. *International Journal of Man-Machine Studies* 8: 669–678.

Mamdani E.H. and S. Assilian (1975) An experiment in linguistic synthesis with a fuzzy logic controller. *International Journal of Man-Machine Studies* 7: 1–13.

Mamdani E.H. and B.R. Gaines, Eds. (1981) *Fuzzy Reasoning and its Applications*. Acdemic Press, New York.

Mareš M. (1994) *Computation over Fuzzy Quantities*. CRC Press, Boca Raton, FL.

Michalewicz Z. (1994) *Genetic Algorithms + Data Structures = Genetic Programming*. Springer-Verlag, Heidelberg.

Mizumoto M. and K. Tanaka (1976) Some properties of fuzzy sets of type 2. *Information and Control* 31: 312–340.

Mizumoto M. and H.-J. Zimmermann (1982) Comparison of fuzzy reasoning methods. *Fuzzy Sets and Systems* 8: 253–283.

Morin T.L. and A.O. Esogbue (1971) Some efficient dynamic programming algorithms for the optimal sequencing and scheduling in water supply projects. *Water Resources* 7: 479–484.

Morin T.L. and R.E. Marsten (1976) Branch-and-bound strategies for dynamic programming. *Operations Research* 24: 611–627.

Morin T.L., W.L. Meier and K.K. Nagaraj (1989) Dynamic programming for flood control planning: the optimal mix of adjustments to floods. In A.O. Esogbue (Ed.): *Dynamic Programming for Optimal Water Resources Systems Analysis*, Prentice-Hall, Englewood Cliffs, NJ, pp. 286–306.

Munakata T. and Y. Jani (1994) Fuzzy systems: An overview. *Communications of the ACM* 37: 69–76.

Narshima Sastry V., R.N. Tiwari and K.S. Sastry (1993) Dynamic programming approach to multiple objective control problems having deterministic or fuzzy goals. *Fuzzy Sets and Systems* 57: 195–202.

Negoita C.V. (1981b) *Fuzzy Systems*. Abacus Press, Tunbridge Wells.

Negoita C.V. and D.A. Ralescu (1975) *Application of Fuzzy Sets to System Analysis*. Birkhäuser, Basel and Halstead Press, New York.

Negoita C.V. and A.C. Stefanescu (1975) On the state equation of fuzzy systems. *Kybernetes* 4: 231–234.

Nemhauser G.L. (1966) *Introduction to Dynamic Programming*. Wiley, New York.

di Nola A.D. and A.G.S. Ventre, Eds. (1986) *The Mathematics of Fuzzy Systems*. Verlag TÜV Rheinland, Cologne.

di Nola A.D., S. Sessa, W. Pedrycz and E. Sanchez (1989) *Fuzzy Relation Equations and their Application to Knowledge Engineering*. Kluwer, Dordrecht.

Norwich A.M. and I.B. Turksen (1984) A model for the measurement of membership and the consequences of its empirical implementation. *Fuzzy Sets and Systems* 12: 1–25.

Novák V. (1989) *Fuzzy Sets and Their Applications*. Hilger, Bristol and Boston.

Odanaka T. (1987) On some stochastic control processes in a fuzzy environment. *Proceedings of Second International Fuzzy Systems Assiciation World Congress – IFSA'87* (Tokyo, Japan), pp. 217–220.

Okuda T., H. Tanaka and K. Asai (1978) A formulation of fuzzy decision problems with fuzzy information using probability measures of fuzzy events. *Information and Control* 38: 135–147.

Ong S.K. and A.Y.C. Nee (1996) Fuzzy set based approach for concurrent constraint set-up planning. (*Journal of Intelligent Manufacturing* 7: 107–120.

Orlovski S.A. (1981) *Problems of Decision-Making with Fuzzy Information* (in Russian). Nauka, Moscow.

Pedrycz W. (1981) An approach to the analysis of fuzzy systems. *International Journal of Control* 34: 403–421.

Pedrycz W. (1984) Identification of fuzzy systems. *IEEE Transactions on Systems, Man and*

Cybernetics SMC-14: 361–366.

Pedrycz W. (1993) *Fuzzy Control and Fuzzy Systems.* Research Studies Press/Wiley, Taunton/New York (Second edition).

Pedrycz W. (1995a) *Fuzzy Sets Engineering,* CRC Press, Boca Raton, FL.

Pedrycz W. (1995b) Fuzzy relational control, In H.T. Nguen, M. Sugeno, R. Tong and R.R. Yager (Eds.): *Theoretical Design of Fuzzy Control*, Wiley, New York, pp. 235–260.

Pedrycz W., Eds. (1996) *Fuzzy Modelling: Paradigms and Practice.* Kluwer, Boston.

Piskunov A.I. (1988a) A structural approach to the analysis of fuzzily formalized systems – Part I: Equivalent structure transformation of fuzzy systems (in Russian). *Automation and Remote Control* 49: 500–507.

Piskunov A.I. (1988b) A structural approach to the analysis of fuzzily formalized systems – Part II: Decision making procedures in structured fuzzy systems (in Russian). *Automation and Remote Control* 49: 643–654.

Piskunov A.I. (1992a) Multistage fuzzily formalized systems I. Fuzzy goal and constraints in multistage decision-making processes under fuzzy conditions (in Russian). *Automation and Remote Control* 52: 1259–1264.

Piskunov A.I. (1992b) Multistage fuzzily formalized systems II. Methods of solution of control problems for isolated and multistage fuzzily formalized systems. *Automation and Remote Control* 52: 1404–1413.

Procyk T.J. and E.H. Mamdani (1979) A linguistic self organizing process controller. *Automatica* 15: 15–30.

Puri M.L. and D.A. Ralescu (1986) Fuzzy random variables. *Journal of Mathematical Analysis and Applications* 114: 409–422.

Puterman M.L. (1994) *Markov Decision Processes. Discrete Stochastic Dynamic Programming.* Wiley, New York.

Ragade R.R., K,W, Hipel and T.E. Unny (1976) Nonquantitative methods in water resources management. *Journal of Water Resources Planning and Management Div. ASCE* 102: 297–309.

Ralescu D.A. (1995) Cardinality, quantifiers, and the aggregation of fuzzy criteria. *Fuzzy Sets and Systems* 69: 355-365.

Rocha A.F. (1993) *Neural Nets: A Theory for Brain and Machines.* Springer-Verlag, Heidelberg.

Sanchez E. (1976) Resolution of composite fuzzy relation equations. *Information and Control* 30: 38–48.

Sanchez E. and L.A. Zadeh, Eds. (1987) *Approximate Reasoning in Intelligent Systems, Decision and Control.* Pergamon Press, Oxford.

Simpson P.K. (1989) *Artificial Neural Systems.* Pergamon Press, New York.

Skala H.-J. (1978) On many valued logics. *Fuzzy Sets and Systems* 1: 129–149.

Skala H.-J., S. Termini and E. Trillas, Eds. (1984) *Aspects of Vagueness.* Reidel, Dordrecht.

Sládky K., J. Kacprzyk and P. Staniewski (1982) On optimal control od stochastic systems in a fuzzy environment. *Proceedings of the Fourth FORMATOR Symposium*, Prague.

Smets P. (1982) Probability of a fuzzy event: an axiomatic approach. *Fuzzy Sets and Systems* 7: 153–164.

Smets P., E.H. Mamdani, D. Dubois and H. Prade, Eds. (1988) *Non-Standard Logics for Automated Reasoning.* Academic Press, London.

Smithson M. (1989) *Ignorance and Uncertainty.* Springer-Verlag, Berlin.

Stanford R.E. (1981) The set of limiting distributions for a Markov chain with fuzzy transition probabilities. *Fuzzy Sets and Systems* 7: 71–78.

Stein W.E. (1980) Optimal stopping in a fuzzy environment. *Fuzzy Sets and Systems* 3: 253–259.

Su C.C and Y.Y. Hsu (1991) Fuzzy dynamic programming: An application to unit commitment. *IEEE Transactions on Power Systems* PS-6: 1231–1237.

Sugeno M., Ed. (1985a) *Industrial Applications of Fuzzy Control.* North-Holland, Amsterdam.

Sugeno M. (1985b) An introductory survey of fuzzy control. *Information Sciences* 36: 59–83.

Sugeno M. (1987) *Fuzzy Control* (in Japanese). Nikkan Kogyo, Tokyo.

Sugeno M. and M. Nishida (1985) Fuzzy control of model car. *Fuzzy Sets and Systems* 16:

103–113.

Sugeno M. and T. Yasukawa (1993) A fuzzy-logic-based approach to qualitative modeling. *IEEE Transactions on Fuzzy Systems* FS-1: 7–31.

Sugianto L.F. and W. Mielczarski (1995a) A fuzzy logic approach to optimize inventory. In G.F. Forsyth and M. Ali (Eds.): *Proceedings of Eighth International Conference on Industrial and Engineering Applications of Artificial Intelligence and Expert Systems*, Gordon and Breach (Australia), pp. 419–424.

Sugianto L.F. and W. Mielczarski (1995b) Control of power station inventory by optimal allocation process with fuzzy logic approach. *Proceedings of Control '95 Conference*(Australia), Vol. 2, pp. 529–533.

Sugianto L.F. and W. Mielczarski (1996) Dynamic programming application to optimize spare parts inventory. *Control and Cybernetics* (in press).

H. Takagi and M. Sugeno (1985) Fuzzy identification of systems and its application to modelling and control. *IEEE Transactions on Systems, Man and Cybernetics* SMC-15: 116–132.

Tanaka H., T. Okuda and K. Asai (1974) On fuzzy mathematical programming. *Journal of Cybernetics* 3: 37–46.

Tanaka H., T. Okuda and K. Asai (1976) A formulation of fuzzy decision problem and its application to an investment problem. *Kybernetes* 5: 25–30.

Tanaka K. and M. Sugeno (1992) Stability analysis and design of fuzzy control systems. *Fuzzy Sets and Systems* 45: 135–156.

Terano T., K. Asai and M. Sugeno (1991) *Fuzzy Systems Theory and its Applications*. Academic Press, New York.

Thole U., H.-J. Zimmermann and P. Zysno (1979) On the suitability of minimum and product operator for the intersection of fuzzy sets. *Fuzzy Sets and Systems* 2: 167–180.

Tong R.M. (1980) Some properties of fuzzy feedback systems. *IEEE transactions of Systems, Man and Cybernetics* SMC-10: 716–723.

Tong R.M. (1984) A retrospective view of fuzzy control systems. *Fuzzy Sets and Systems* 14: 199–210.

Trillas E. and L. Valverde (1992) On the implication and indistinguishability in the setting of fuzzy logic. In L.A. Zadeh and J. Kacprzyk (Eds.): *Fuzzy Logic for the Management of Uncertainty*, Wiley, New York, pp. 198–212.

Turksen I.B. (1991) Measurement of membership functions and their acquisition. *Fuzzy Sets and Systems* 40: 5–38.

Vira J. (1981) Fuzzy expectation values in multistage optimization problems. *Fuzzy Sets and Systems* 6: 161–167.

Wang P.P., Ed. (1983) *Advances in Fuzzy Sets, Possibility Theory, and Applications*. Plenum Press, New York.

Wang P.P. and S.K. Chang, Eds. (1980) *Fuzzy Sets Theory and Applications to Policy Analysis and Information Systems*. Plenum, New York.

Wang Z. and G.J. Klir (1992) *Fuzzy Measure Theory*. Plenum Press, New York.

Weber S. (1983) A general concept of fuzzy connectives, negation and implication based on *t*-norms and *t*-conorms. *Fuzzy Sets and Systems* 11: 115–134.

Wenstøp F. (1979) Deductive verbal models of organizations. *International Journal of Man-Machine Studies* 8: 293–311.

Whalen T. (1984) Decision-making under uncertainty with various assumptions about available information. *IEEE Transactions on Systems, Man and Cybernetics* SMC-14: 888–900.

Willmott R. (1980) Two fuzzier implication operators in the theory of fuzzy power sets. *Fuzzy Sets and Systems* 4: 31–36.

Wygralak M. (1996) *Vaguely Defined Objects – Representations, Fuzzy Sets and Nonclassical Cardinality Theory*. Kluwer, Dordrecht.

Yager R.R. (1977) Multiple objective decision-making using fuzzy sets. *International Journal of Man-Machine Studies* 9: 375–382.

Yager R.R. (1978) Fuzzy decision-making including unequal objectives. *Fuzzy Sets and Systems* 1: 87–95.

Yager R.R. (1979) A note on probabilities of fuzzy events. *Information Sciences* 18: 113–129.

Yager R.R. (1980) On a general class of fuzzy connectives. *Fuzzy Sets and Systems* 4: 235–242.

Yager R.R., Eds. (1982) *Fuzzy Set and Possibility Theory: Recent Developments.* Pergamon Press, New York.

Yager R.R. (1981) A procedure for ordering fuzzy subsets of the unit interval. *Information Sciences* 34: 143–161.

Yager R.R. (1983a) Robot planning with fuzzy sets. *Robotica* 1: 41–50.

Yager R.R. (1983b) Quantifiers in the formulation of multiple objective decision functions. *Information Sciences* 31: 107–139.

Yager R.R. (1984b) A representation of the probability of a fuzzy subset. *Fuzzy Sets and Systems* 13: 273–283.

Yager R.R. (1988) On ordered weighted averaging aggregation operators in multi-criteria decision making. *IEEE Transactions on Systems, Man and Cybernetics* SMC-18: 183–190.

Yager R. R. (1991) Connectives and quantifiers in fuzzy sets. *Fuzzy Sets and Systems* 40: 39–76.

Yager R.R. (1992a) On the specificity of a possibility distribution. *Fuzzy Sets and Systems* 50: 279–292.

Yager R. R. (1992b) On the inclusion of importances in multi-criteria decision making in the fuzzy set framework. *International Journal of Expert Systems: Research and Applications* 5: 211–228.

Yager R.R. (1996) Quantifier guided aggregation using OWA operators. *International Journal of Intelligent Systems* 11: 49–73.

Yager R.R. and D.P. Filev (1993) On the issue of defuzzification and selection based on a fuzzy set. *Fuzzy Sets and Systems* 55: 255–271.

Yager R.R. and D.P. Filev (1994) *Foundations of Fuzzy Control.* Wiley, New York.

Yager R.R. and J. Kacprzyk, Eds. (1997) *The Ordered Weighted Averaging Operators: Theory, Methodology and Applications.* Kluwer, Boston, MA.

Yager R.R., J. Kacprzyk and M. Fedrizzi, Eds. (1994) *Advances in the Dempster–Shafer Theory of Evidence.* Wiley, New York.

Yager R.R. and L.A. Zadeh, Eds. (1992) *An Introduction to Fuzzy Logic Applications in Intelligent Systems.* Kluwer, Boston.

Yasonubu S. and S. Miyamoto (1985) Automatic train operation by predictive fuzzy control. In M. Sugeno (Ed.): *Industrial Applications of Fuzzy Control,* North-Holland, Amsterdam, pp. 1–18.

Yoshida Y. (1994) Markov chains with a transition possibility measure and fuzzy dynamic programming. *Fuzzy Sets and Systems* 66: 39–57.

Yoshinari Y., W. Pedrycz and K. Hirota (1993) Construction of fuzzy models through clustering techniques. *Fuzzy Sets and Systems* 54: 157–165.

Yuan Y. and Z. Wu (1991) Algorithm of fuzzy dynamic programming in AGV scheduling. *Proceedings of the International Conference on Computer Integrated Manufacturing – ICCIM '91,* pp. 405–408.

Zadeh L.A. (1965a) Fuzzy sets. *Information and Control* 8: 338–353.

Zadeh L.A. (1965b) Fuzzy sets and systems. In J. Fox (Ed.): *System Theory.* Polytechnic Press, Brooklyn, NY, pp. 29–37.

Zadeh L.A. (1968a) Fuzzy algorithms. *Information and Control* 12: 94–102.

Zadeh L.A. (1968b) Probability measures of fuzzy events. *Journal of Mathematical Analysis and Applications* 23: 421–427.

Zadeh L.A. (1971) Toward a theory of fuzzy systems. In R.E. Kalman and N. DeClaris (Eds.): *Aspects of Network and System Theory,* Holt, Rinehart and Winston, New York, pp. 469–490.

Zadeh L.A. (1972) A rationale for fuzzy control. *Measurement and Control* 34: 3–4.

Zadeh L.A. (1973) Outline of a new approach to the analysis of complex systems and decision processes. *IEEE Transactions on Systems, Man and Cybernetics* SMC-2: 28–44.

Zadeh L.A. (1975a) Fuzzy logic and approximate reasoning. *Synthese* 30: 407–428.

Zadeh L.A. (1975b) The concept of a linguistic variable and its application to approximate reasoning. *Information Sciences*: (Part I) 8: 199–249, (Part II) 8: 301–357, (Part III) 9:

43–80.

Zadeh L.A. (1976a) The linguistic approach and its application to decision analysis. In Y.C. Ho and S.K. Mitter (Eds.): *Directions in Large-Scale Systems*. Plenum Press, New Yoprk, pp. 335–361.

Zadeh L.A. (1976b) A fuzzy-algorithmic approach to the definition of complex or imprecise concepts. *International Journal of Man-Machine Studies* 8: 249–291.

Zadeh L.A. (1978a) PRUF – A meaning representation language for natural languages. *International Journal of Man-Machine Studies* 10: 395–460.

Zadeh L.A. (1978b) Fuzzy sets as a basis for a theory of possibility. *Fuzzy Sets and Systems* 1: 3–28.

Zadeh L.A. (1979) A theory of approximate reasoning. In J.E. Hayes, D. Michie and L.I. Mikulich (Eds.): *Machine Intelligence* 3, Wiley, New York, pp. 149–194.

Zadeh L.A. (1983a) A computational approach to fuzzy quantifiers in natural languages. *Computers and Mathematics with Applications* 9: 149–184.

Zadeh L.A. (1983b) The role of fuzzy logic in the management of uncertainty in expert systems. *Fuzzy Sets and Systems* 11: 199–228.

Zadeh L.A. (1985) Syllogistic reasoning in fuzzy logic and its application to usuality and reasoning with dispositions. *IEEE Transaction on Systems, Man and Cybernetsics* SMC-15: 754–763.

Zadeh L.A. (1986) Fuzzy probabilities. *Information Processing and Management* 20: 363–372.

Zadeh L.A. (1994) Fuzzy logic, neural networks, and soft computing. *Communications of the ACM* 37: 77–84.

Zadeh L.A. and J. Kacprzyk, Eds. (1992) *Fuzzy Logic for the Management of Uncertainty*, Wiley, New York.

Zimmermann H.-J. (1976) Description and optimization of fuzzy systems. *International Journal of General Systems* 2: 209–215.

Zimmermann H.-J. (1978) Fuzzy programming and linear programming with several objective functions. *Fuzzy Sets and Systems* 1: 45–55.

Zimmermann H.-J. (1987) *Fuzzy Sets, Decision Making, and Expert Systems*. Kluwer, Dordrecht.

Zimmermann H.-J. (1996) *Fuzzy Set Theory and its Applications*. Kluwer, Boston (Third edition).

Zimmermann H.-J, and P. Zysno (1980) Latent connectives in human decision making. *Fuzzy Sets and Systems* 4: 37–51.

Zimmermann H.-J., L.A. Zadeh and B.R. Gaines, Eds. (1987) *Fuzzy Sets and Decision Analysis*. North-Holland, Amsterdam.

Index